Disaster Resilience and Green Growth

Over the years, the relationship between environment and disasters has received significant attention. This is largely due to the emerging recognition that environmental changes - climate change, land-use and natural resource degradation make communities more vulnerable to disaster impacts. There is a need to break this nexus through environment based and sustainability inclusive interventions. Science – technology and economic measures for disaster risk management, hence, need to adapt more integrated approaches for infrastructure and social resilience. Environmental and anthropogenic factors are key contributors to hazard, risk, and vulnerability and, therefore, should be an important part of determining risk-management solutions.

Green growth approaches have been developed by emphasizing sustainability inclusion and utilizing the benefits of science-technology interventions along policy-practice linkages with circular economy and resource efficiency. Such approaches recognize the perils of traditional material-oriented economy growth models that tend to exploit natural resources, contribute to climate change, and exacerbate disaster vulnerabilities, Green growth integrated approaches are rapidly becoming as preferred investment avenue for mitigating climate change and disaster risks and for enhancing resilience. This includes ecosystem-based and nature-based solutions with potential to contribute to the resilience of infrastructure, urban, rural and peri-urban systems, livelihoods, water, and health. They can lead to food security and can further promote people-centric approaches.

Some of the synergistic outcomes of green growth approaches include disaster risk reduction, climate change mitigation and adaptation, resilient livelihoods, cities, businesses and industry. The disaster risk reduction and resilience outcome of green growth approaches deserve special attention, both for the academic and policy communities. Scholars and professionals across the domains of DRR, CCA, and green growth are in need of publications that fulfill their knowledge needs concerning the disaster resilience outcomes of green growth approaches. Keeping the above background in view, the book series offers comprehensive coverage combining the domains of environment, natural resources, engineering, management and policy studies for addressing disaster risk and resilience in the green growth context in an integrated and holistic manner. The book series covers a range of themes that highlight the synergistic outcomes of green growth approaches.

The book series aims to bring out the latest research, approaches, and perspectives for disaster risk reduction along with highlighting the outcomes of green growth approaches and including Science-technology-research-policy-practice interface, from both developed and developing parts of the world under one umbrella. The series aims to involve renowned experts and academicians as volume-editors and authors from all the regions of the world. It is curated and developed by authoritative institutions and experts to serve global readership on this theme.

Bhawana Pathak • Rama Shanker Dubey
Editors

Climate Change and Urban Environment Sustainability

Springer

Editors
Bhawana Pathak
School of Environment and Sustainable Development
Central University of Gujarat
Gandhinagar, India

Rama Shanker Dubey
Central University of Gujarat
Gandhinagar, India

ISSN 2662-4885 ISSN 2662-4893 (electronic)
Disaster Resilience and Green Growth
ISBN 978-981-19-7617-9 ISBN 978-981-19-7618-6 (eBook)
https://doi.org/10.1007/978-981-19-7618-6

This Springer imprint is published by the registered company Springer Nature Singapore Pte Ltd.
The registered company address is: 152 Beach Road, #21-01/04 Gateway East, Singapore 189721, Singapore

Preface

Climate change is a critical issue of our times and remains a serious concern. The consequences of global warming, such as an increase in temperature and uncertain weather conditions, greatly impact urban centres. Understanding the mechanism of climate change and climate variability is an important aspect that requires monitoring from regional perspectives. The fast trend in urbanization during the past couple of decades has substantially transformed the overall scenario of land (cover, use and growth pattern), resulting in many environmental, social and economic challenges. Smart and resilient urbanization is essential to tackle the growing threat of climate instability. Different analytical and practical approaches need to be adopted to foster resilience and environmental sustainability in urban areas. Effective and constructive mechanisms need to be devised to protect the life and property of human population in urban areas from anticipated deteriorating quality of environment and expected hazards and disasters resulting due to climate change. With effective and innovative urban planning, coupled with optimum use of available resources, with honest commitment of all stakeholders, sustainable development can be achieved. Sustainable urban environment is a key contributor to the growth of a successful nation.

This book focuses on our present status of understanding the relationship between climate change, urban development, and environment sustainability with an emphasis on core issues and challenges associated with urban environment sustainability. It covers various concepts related to climate-resilient urban development, effective implementation of climate change adaptation and mitigation measures to promote urbanization in a much better social, economic and environmental perspective.

Gandhinagar, India

Bhawana Pathak
Rama Shanker Dubey

Contents

Editors and Contributors

About the Editors

Bhawana Pathak works as a Professor in the School of Environment and Sustainable Development, Central University of Gujarat, Gandhinagar. She holds an M. Sc. in Botany with a Gold Medal from Kumaun University Nainital and a Ph.D. in Plant Ecology from G. B. Pant Institute of Himalayan Environment and Development, Kosi–Katarmal, Almora. She has more than 18 years of research and teaching experience in the specialized area of environmental ecology, biodiversity conservation, environmental biotechnology and environmental nanotechnology. She has contributed to the development of curriculum for different courses and published more than sixty research papers/scientific research articles, book chapters and also contributed as editor/author to four books. She involves in innovative research work in the specialized area for implementation and future policies. She has also presented scientific papers in national and international conferences and seminars. She is a recipient of the Outstanding Faculty award in Bio Science and Technology.

Rama Shanker Dubey (born December 23, 1955) serves as Vice Chancellor of the Central University of Gujarat at Gandhinagar, India. He holds M.Sc. and Ph. D. degrees in Biochemistry from Banaras Hindu University. For over 45 years he has conducted pioneering research on environmental issues focusing on soil salinity, water stress and metal toxicity effects on key metabolic alterations in plants. He has over 173 scientific publications with 18,400+ citations, **h-Index 47 and i10-Index 103.** He has guided 32 doctoral students and has worked in many reputed laboratories of Sweden, Germany and Japan. He is the recipient of several awards in environmental biochemistry and biotechnology like the BHU Medal in 1976, Young Scientist Award in 1982, Swedish Institute Fellowship in 1989, Japan Society for the Promotion of Science Fellowships in 1998, 2005, and INSA-German Science Academy Nominations in 1997, 2003, 2007. He is Fellow of Indian Society of Agricultural Biochemists and Biotech Research Society of India.

Contributors

Surbhi Agrawal Indian Institute of Management (IIM), Bengaluru, Bengaluru, Karnataka, India

P. Anantharaman Centre of Advanced Studies in Marine Biology (CASMB), Annamalai University, Parangipettai (Portnovo), India

Chandani Appadoo Department of Biosciences and Ocean Studies, Faculty of Science, University of Mauritius, Reduit, Mauritius

Arun Arya Department of Environmental Studies, Faculty of Science, The Maharaja Sayajirao University of Baroda, Vadodara, India

V. K. Bansal Department of Civil Engineering, National Institute of Technology (NIT) Hamirpur, Hamirpur, Himachal Pradesh, India

N. Bhatt Department of Biogas Research and Microbiology, Gujarat Vidyapith, Gandhinagar, Gujarat, India

Neha Chaudhary School of Environment and Sustainable Development, Central University of Gujarat, Gandhinagar, Gujarat, India

V. Chorsiya School of Physiotherapy, Delhi Pharmaceutical Sciences and Research University, New Delhi, India

Saurabh Choubey School of Environment and Sustainable Development, Central University of Gujarat, Gandhinagar, India

Deeksha Dave School of Inter-disciplinary and Trans-Disciplinary Studies, New Delhi, India

Vinars Dawane ANCHROM, HPTLC Labs, Mumbai, Maharashtra, India

Chitrangada Debsarma School of Environment and Sustainable Development, Central University of Gujarat, Gandhinagar, Gujarat, India

Rama Shanker Dubey School of Environment and Sustainable Development, Central University of Gujarat, Gandhinagar, India

P. Dutta Indian Institute of Public Health, Gandhinagar, India

Shaha Hashim Blue Marine Foundation, Malé, Maldives

Asha Humbal School of Environment and Sustainable Development, Central University of Gujarat, Gandhinagar, Gujarat, India

Akbar John Institute of Oceanography and Maritime Studies (INOCEM), International Islamic University Malaysia (IIUM), Kuantan, Pahang, Malaysia

Vaishali Joshi Department of Chemistry, Govt. P G College, Manawar, Dhar, Madhya Pradesh, India

Sandeep Kalyan School of Environment and Sustainable Development, Central University of Gujarat, Gandhinagar, India

Abu Hena Mustafa Kamal Faculty of Fisheries and Food Science, University Malaysia Terengganu, Kuala Nerus, Terengganu, Malaysia

N. Kapoor Department of Architecture, National Institute of Technology (NIT) Hamirpur, Hamirpur, Himachal Pradesh, India

Ashish Kumar Kaushal Jindal Global Business School, O P Jindal Global University, Sonipat, Haryana, India
Department of Management Studies, Indian Institute of Technology Delhi, New Delhi, India

Priyanka Kumari School of Environment and Sustainable Development, Central University of Gujarat, Gandhinagar, Gujarat, India

Pankaj Kumar Integrated Regional Office, Ministry of Environment, Forest and Climate Change (MoEFCC), Govt. of India Saifabad, Hyderabad, Telangana, India

Avantika Latwal International Institute of Information Technology, Hyderabad, Hyderabad, Telangana, India

Ashwani Luthra Guru Ramdas School of Planning, Guru Nanak Dev University, Amritsar, India

Amit Kumar Mishra School of Environment and Sustainable Development, Gandhinagar, India

Adil Muhammad Department of Chemical Engineering, Government Engineering College Kozhikode, Kozhikode, Kerala, India

E. P. Nobi Ministry of Environment, Forest and Climate Change, Indira Paryavaran Bhawan, New Delhi, India

J. Padhiar Department of Biogas Research and Microbiology, Gujarat Vidyapith, Gandhinagar, Gujarat, India

Ragavan Pandisamy Ministry of Environment, Forest and Climate Change, Indira Paryavaran Bhawan, New Delhi, India

Himakshi Parmar Department of Environmental Studies, Faculty of Science, The Maharaja Sayajirao University of Baroda, Vadodara, India

Bhawana Pathak School of Environment and Sustainable Development, Central University of Gujarat, Gandhinagar, India

Sunil Kumar Patidar Department of Chemistry, Shri Neelkantheshwar Govt. Post Graduate College, Khandawa, Madhya Pradesh, India

Satish Piplode Department of Chemistry, SBS Govt. P G College, Hoshangabad, Madhya Pradesh, India

Bryan Nelson Raveen Institute of Tropical Biodiversity and Sustainable Development, University Malaysia Terengganu, Kuala Nerus, Terengganu, Malaysia

Priyanka Sah G.B. Pant National Institute of Himalayan Environment (NIHE), Ladakh Regional Centre, Leh, Ladakh (UT), India

Paulami Sahu School of Environment and Sustainable Development, Central University of Gujarat, Gandhinagar, Gujarat, India

Prakash Samnani Department of Chemistry, Faculty of Science, The Maharaja Sayajirao University of Baroda, Vadodara, India

Purabi Sarkar Tirambhapur Askote Raj Ecosystem Private Limited (TARE), Pithoragarh, Uttarakhand, India

Rehana Shaik International Institute of Information Technology, Hyderabad, Telangana, India

Aditya Sharma Department of Atmospheric Science, Central University of Rajasthan, Ajmer, India

Subrat Sharma G.B. Pant National Institute of Himalayan Environment (NIHE), Ladakh Regional Centre, Leh, Ladakh (UT), India

Vandna Sharma Swami Vivekanand Government College, Ghumarwin, Himachal Pradesh, India

Hassan I. Sheikh Faculty of Fisheries and Food Science, University Malaysia Terengganu, Kuala Nerus, Terengganu, Malaysia

Kannan Sivakumar Centre of Advanced Studies in Marine Biology (CASMB), Annamalai University, Parangipettai (Portnovo), India

Balaji Vedharajan OMCAR Foundation, Velivayal, Tamil Nadu, India

Priya Chandulal Vithalani Department of Biogas Research and Microbiology, Gujarat Vidyapith, Gandhinagar, Gujarat, India

Saleem Ahmad Yatoo Government Degree College, Baramulla, Jammu and Kashmir, India

Chapter 1
Climate Change and Urban Environment Sustainability: Issues and Challenges

Paulami Sahu and Chitrangada Debsarma

1.1 Introduction

1.1.1 Sustainable Urbanization and Its Significance

The global urban population has intensified from 30% in 1950 to 55% by today while the urban stretches just 0.4–0.9% of world's terrestrial surfaces and with this current rate of urbanization, it is projected to increase 70% by 2050 (Marvuglia et al. 2020). Figure 1.1 shows region-wise increasing trend of urban population distribution (in millions) from statistical data of United Nations, 2012. It is assessed that urban population is the cause for the use of three-quarters of the world's (direct) final energy utilization and carbon dioxide emissions. For instance, in EU, the urban water and final energy consumption are 30% and 40%, respectively, while the urban CO_2 emission is 30% and 30% of waste production. As per Global Energy Assessment, the urban sustainability and energy utilization policies should concentrate on urban built environment, its density, design, energy, and transport systems, which has significant environmental impacts. It is assumed that if the present and future city area spread is continued with consumption of resources with the similar rate, there would be critical environmental, social, and economic effects around the globe (Millennium Ecosystem Assessment, MA 2003). The key priorities to meet the global goals for sustainable development, it is essential to modify cities in such a manner that they should be more inexpensive, resource-efficient, inclusive, resilient, mitigated and adaptive to climate change (Marvuglia et al. 2020). International organizations such as European Commission, Organization for Economic

P. Sahu (✉) · C. Debsarma
School of Environment and Sustainable Development, Central University of Gujarat, Gandhinagar, Gujarat, India
e-mail: paulami_sahu@cug.ac.in

B. Pathak, R. S. Dubey (eds.), *Climate Change and Urban Environment Sustainability*, Disaster Resilience and Green Growth,
https://doi.org/10.1007/978-981-19-7618-6_1

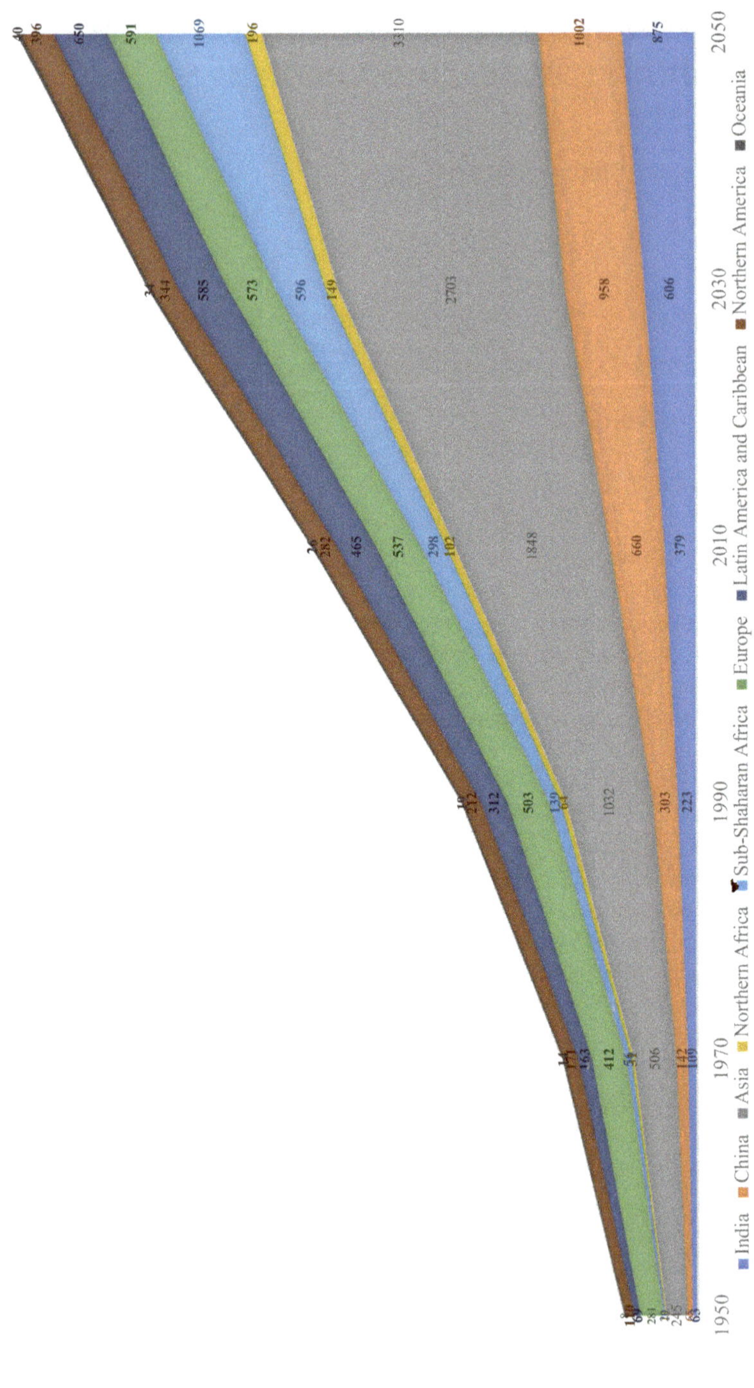

Fig. 1.1 Graph showing region-wise urban population distribution in millions (*Data collected from Statistics of United Nations, 2012*)

Co-operation and Development (OECD), World Bank, and UN Habitat are making appreciable efforts for encouraging the sustainable urbanization practices (Shen et al. 2011). For integration of social well-being, economic equity, and environmental protection to urban development plans, it is necessary to develop, implement, and monitor policies and promote incentives to achieve the desired state of sustainable urbanization (Choguill 1993; Marcotullio 2001; Holden et al. 2008).

1.1.2 Sustainable Urban Development Plans: International and National Scenarios

There is a requirement of quantifiable indicators for measuring the performance of urban sustainability in the method of urban sustainability assessment. The local and national authorities around the globe have established some urban sustainability indicators as per their needs since early 1990s and there is no specific set of indicators that suits alike to all towns or communities (Shen et al. 2011). Figure 1.2 shows some examples of sustainable development plans implemented in city level worldwide.

The sustainable city plan of Melbourne (Australia) titled "Melbourne City Plan 2010" had objectives of improving the quality of environment, maintaining social equality and increasing economic prosperity. "The HK2030 Study" plan of Hong Kong (China) has focal area to maintain quality environment, economy, health, and hygiene. "Sustainable Barcelona", an initiative taken up for urban sustainability of Barcelona, has given the emphasis on the conservation of biodiversity, effective use of resources, maintaining inhabitant's health, gender equality, etc. Mexico City's "Green Plan" aiming toward "Greenest city in Latin America" by focusing on different sectors like water, air, energy, waste management, land conservation, transport, etc. Urban sustainability in Taipei (Taiwan) focused on efficient water resource management, environmental sustainability management, transportation, comfort living, education, etc. Singapore's "Green Plan" is the transformation of the city into a green city by improving air quality, public health, waste management, carbon intensity and reducing per person water consumption and environmental awareness. In India, the City Development Plans of Chandigarh and Pune have the vision toward "greenest city of India" and "economically vibrant and sustainable city," respectively. They have some common focus area of economic development, urban governance, transportation, waste management, basic services, land and river conservation, etc. (Rosenzweig et al. 2018).

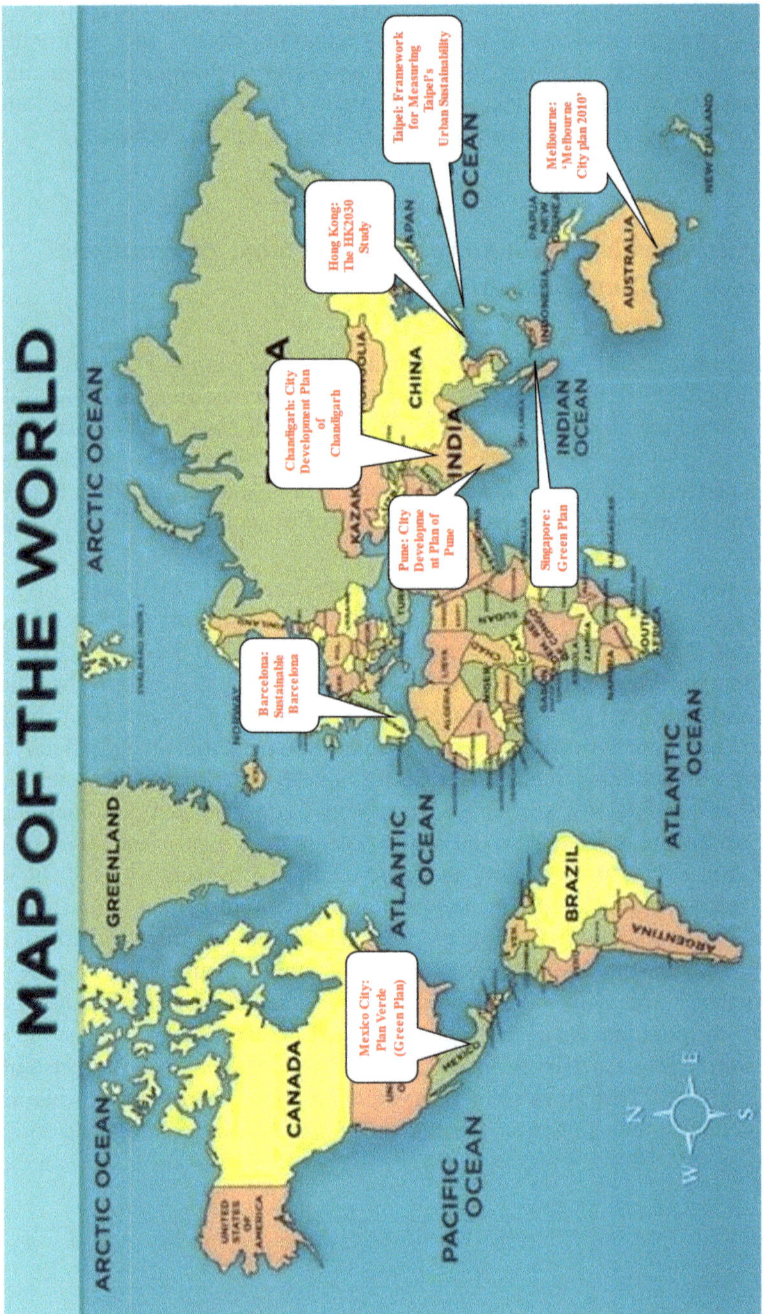

Fig. 1.2 Some examples of sustainable development plans implemented in different cities of the world

1.2 Sustainable Urbanization: Challenges Imposed by Climate Change

The alteration in the climate pattern of global and regional scale, which was evident from mid-20th onwards, primarily allied with an elevated level of greenhouse gases mainly CO_2 owing to consumption of fossil fuel, is known as climate change. According to Intergovernmental Panel on Climate Change (IPCC) Report of 2021

> "Climate change is intensifying the water cycle. This brings more intense rainfall and associated flooding, as well as more intense drought in many regions."
>
> "Coastal areas will see continued sea level rise throughout the twenty-first century, contributing to more frequent and severe coastal flooding in low-lying areas and coastal erosion. Extreme sea level events that previously occurred once in 100 years could happen every year by the end of this century."
>
> "For cities, some aspects of climate change may be amplified, including heat (since urban areas are usually warmer than their surroundings), flooding from heavy precipitation events and sea level rise in coastal cities." (IPCC 2021)

Urban areas are prone to extreme climatic shocks like heatwaves, hurricanes, changes in precipitation pattern, pollutions, diseases, etc. due to its higher density of population. For instance, heat waves, wildfires, urban drought/flood, and extreme climatic events caused by climate change are experienced in northern and southern hemisphere during 2018 and 2019, mostly in the towns and cities of Australia, California, and Chile. Climate change models projected that, in the cities of Europe, Africa, and South America, the mean maximum temperature will increase by 2–8 °C in upcoming decades, which lead to frequent droughts in the areas (Lin et al. 2021).

The extreme weather events and unfamiliar changes in the patterns of climatic factors are reported by the scientists around the globe. These changes took place over several decades of changing climatic conditions. The frequency and intensity of extreme climatic events including storm surge, hurricanes, tropical cyclones, floods, cold–heat waves, drought, etc. are projected to increase in the upcoming years. Other climate change driven challenges include air pollution, water scarcity, sea-level rise, epidemics, etc. Some climate-change-imposed challenges faced by various cities around the world are summarized in Table 1.1.

1.3 Major Urban Solutions for Climate Impact

A large volume of literatures regarding the adaptation measures against the climate challenges for sustainable urbanization including the technological, nature-based, and social solutions are published. But to maximize the benefits of the strategies, there is a need of integration of these solutions (Lin et al. 2021).

Table 1.1 Climate-change-imposed challenges to cities

Climate change driven challenges	Urban Impacts	Affecting zones	Vulnerable Groups	Examples
1. Major climate extremes				
a. Strom surge, hurricanes and tropical cyclones	Infrastructural damage, economic lose, transportation and energy issues, communication system problems	Coastal areas, settlements with fixed infrastructures	Populations with limited resources	*Hurricane Katrina* in 2005 in New Orleans, *Hurricane Sandy* in New York City in 2012
b. Floods and changes in rainfall patterns	Soil-erosion-landslides, disruption in transportation and energy supplies, infrastructural damage, communication systems loses, tourism effects, salt water intrusion, water scarcity in some areas	Various effects depending on the region, mainly the low-lying coastal and flood-prone areas	Same as tropical cyclones	*Mumbai flood* in 2005, *Jakarta* (Indonesia) *flood* in 2007
c. Cold and heat waves, urban heat islands	Negative health effects on human, decrease in labor productivity, changes in requirements of water, food and energy, increase GHGs, economic effects	Mostly mid-latitude areas, high density cities	Economically poor, children and elderly populations	European heatwave in 2003, heatwave in Andhra Pradesh, India

(continued)

Table 1.1 (continued)

d. Drought	Shifts in land-use and economic activities, challenges for water resources	Mostly arid and semi-arid regions, already water scarce areas	Poor populations	About 40% of US is experiencing drought due to warmer temperatures (US Drought Monitor report, August, 2021)
2. Changes in patterns of temperature	Decrease in urban the quality of urban air, increase in energy costs and demands, tourism effects, thawing of permafrost	Diverse effects, mostly the regions with limited resources for adaptations	Largely dense populations with limited resources for adaptation	Global average temperature has increased by 2°F during last 20th century (NASA Global Climate Change Data)
3. Air pollution	Decrease in urban air quality, health effects	Global impacts	Economically poor and vulnerable groups of society	Natural and man-made aerosols altering the albedo of polar ice caps, contributing
				global climate change and melting of polar-ice (IPCC, 2018)
4. Sea-level rise	Challenges for infrastructures, water logging in coastal areas, flood risks, saltwater intrusion	Low-lying coastal areas	Poor populations are more vulnerable	Cities like Mumbai, Shanghai and Dhaka are vulnerable to sea level rise

(continued)

Table 1.1 (continued)

5. Municipal wastes	Methane emission from landfills, surface-groundwater pollution, soil pollution, negative impact on public health	Global impacts	Poor communities with limited resources	Citizens of developed nations produce more waste than the developing and under-developed countries (Vergara and Tchobanoglous, 2012)
6. Water scarcity	Negative impacts on human health, changes in surface and groundwater	Low income countries are more vulnerable	Informal settlements	5 under 700 children die from diarrhea due to insufficient water, sanitation ang hygiene conditions (UNICEF, 2021)
7. Diseases, epidemics and pandemics	Increase in the spread of many vector-borne diseases like Malaria, Dengue, cholera, typhoid, COVID-19	High density cities are more vulnerable	Slum dwellers with low sanitation infrastructure and high density	38-Cholera, Dengue and other vector-borne diseases are sensitive to urban crowding, improper sanitation and hygiene systems (WHO).
8. Urban governance	Gaps between commitments and their effective actions, lack of funding, insufficient stuffs etc.	Global impacts	Poor communities with limited resources	Around 78% of towns and cities have lack of funds for the implementation of projects and programs (Aylett, 2014).

1.3.1 Urban Technological Solutions

Nowadays, emerging and innovative technologies are introducing for increasing the efficiency of the urban infrastructures. For instance, low carbon/impact infrastructures, changing in building materials to increase albedo, permeable pavements, automated systems, and sensors are frequently introducing under smart city agendas. Sustainable urban designing for adaptation to climate change needs to increase the efficiency of buildings, industries, transportation to reduce the emissions of GHG and urban waste heat. Modifications of conventional buildings to well-ventilated buildings provide cooling effects and reduce energy utilization to manage the high temperatures. To manage urban heat island, high albedo construction materials can be used to improve surface reflectance. In the low-lying coastal cities, sustainable infrastructures need to be prioritized as the land-use planning for that area as a measure of coastal protection. In case of informal settlements, retrofitting to the housing systems have multiple co-assistances like more decent quality of living and enhancements to public health. Low carbon/impact infrastructures may manage the increasing trend of urban energy demands and reducing the GHGs emission. Investments in water management, practice of anaerobic reactors for low-carbon energy production, biogas recovering from wastewater, drainage and treatment of wastewater, and exclusion of inter-basin water transfers as high-energy options, etc. are needed to meet the urban water demands. Waste management strategies in context of climate change include reduce, reuse, recycle and waste-to-energy approaches. Sustainable transportation system by improving vehicle propulsion design for fuel use reduction, introducing high quality and capacity mass transport systems and low-carbon based technologies, which are more sustainable and affordable, should be added (Rosenzweig et al. 2018).

1.3.2 Urban Nature-Based Solutions

The use of cost-effective blue–green infrastructures in the urban areas can provide multiple co-benefits of health and ecosystem services to the city dwellers like improving water and air quality, controlling extreme heat, soil erosion, and flooding, and improving health and well-being. Integrate urban plantations into urban design also regulates the local hydrological cycle as well as carbon sequestration of that area. Construction of dunes and restoration of wetlands are the key adaptive nature-based options against coastal flooding and sea-level rise (Rosenzweig et al. 2018).

But the nature-based adaptations are vulnerable to extreme climatic challenges. For instance, green infrastructures are sensitive to extreme temperature and precipitation. Adequate maintenance is required in case of nature-based adaptations to get desired results and performances (Rosenzweig et al. 2018).

1.3.3 Urban Social Solutions

Social adaptations against climate challenges for sustainable urbanization are based on changes in practices, values, and behavior of the city inhabitants. The susceptibility of climate change is magnified especially for the economically poor and vulnerable group of the society. That is why, environmental justice and social equity should be incorporated in the urban climatic policies and programs as key long-term goals and need to be assessed and adjusted time to time in order to maintain equity, resilience, and sustainability. Awareness programs among the economically poor communities and slum dwellers regarding climate risks should be increased in order to grow the self-protection ability of them. To lessen the expenses of climatic influences, economic policies should be aided to the local governments to initiate actions. Disaster prepared trainings, early warning systems, and improving public health care systems are also key factors in handling pressures on human health against climate challenges (Rosenzweig et al. 2018).

1.3.4 Integration of Technological Nature-Based Social Solutions for Climate Change Mitigation

Due to the wide range and complexity of the climate challenges, the individual types of solutions are unable to mitigate the problems. However, the integration of the solutions can provide desired level of results and functions to make the urban environment more sustainable against recent climatic conditions. For instance, in Freiberg (Germany), grassed and pervious surfaces were used to design urban stormwater drainage around the tram corridors at multiple scales, where technological solutions were integrated with ecological solutions. In addition, the costs of public transport were subsidized to discourage the use of private cars, as a social solution. Another example is from Government of Singapore, adopted various interdisciplinary projects between local government and universities, Cooling Singapore, technologies for built-up design (e.g., green roofs and walls) and island-wide digital urban climate twin (DUCT) of Singapore as an eco-technological solution. Figure 1.3 summarizes the sustainable urban solutions for adapting climate change (Rosenzweig et al. 2018).

1.4 Climate Change Adaptation: Case Studies of Various Cities Around the Globe

Several towns and cities have already designed significant plans and strategies for the management of risks and improving city's resilience to combat with climate change and achieve sustainable urbanization. These strategies include innovative

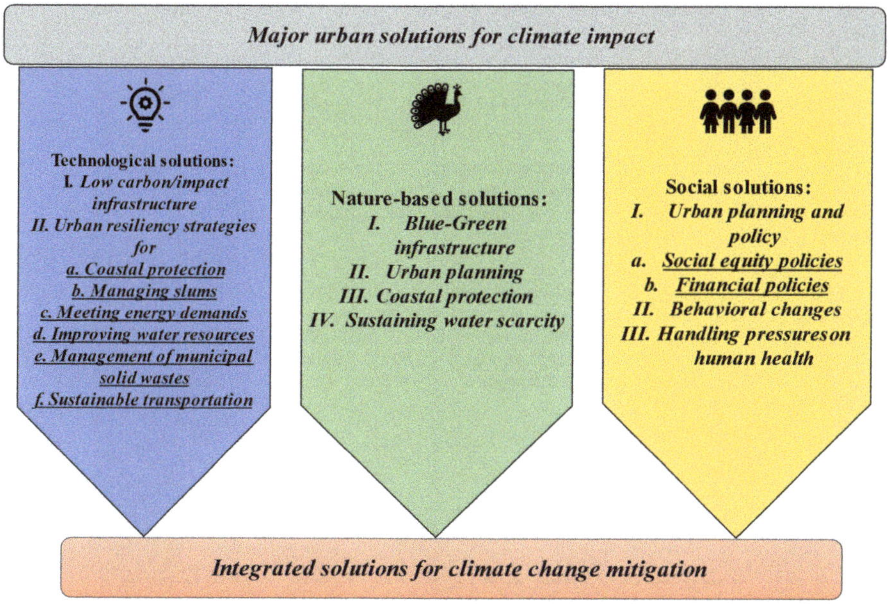

Fig. 1.3 Sustainable urban solutions for climate change adaptation

design for urban infrastructure, modify land-use planning, enhance urban ecological services, integrate disaster risk management, and develop partnerships of public–private sectors and investments, etc. Table 1.2 provides some case studies of sustainable urban planning around the globe (Rosenzweig et al. 2018).

1.5 Concluding Remark

Integrating approaches for adaptation of climate change provides boundless openings and possibilities to address several sustainable urbanization goals. It also increases resilience of the city to climate change across different socioeconomic settings. However, they are context sensitive. Several challenges can be faced for constructing and executing the integrated solutions like lack of resources, competing priorities, and planning under risky or uncertain conditions. Further systematic research for effective solutions to overcome these barriers is required. These include modifications of urban environments suitable for integration in different contexts like technological, social and nature-based solutions, and as a consequence, the integration can work together to improve the overall urban sustainability. Finally, to achieve urban sustainable development in the light of climate change, the urban decision-makers and planners need to promote adaptation along with mitigation strategies as a fruitful solution.

Table 1.2 Case studies of climate change adaptation form various cities around the globe

Case study	Target	Adaptation measures	Adaptation type	Outcomes
Bluebelt concept, Staten Island, New York City, 1990	Storm-water management	Management of natural wetlands and land depressions to accommodate and slow flood water	Integrated ecosystem based and reduction in disaster risk	New York Sustainability Plan or "PlaNYC 2030"
City's Green belt, Medellín City, Aburrá Valley, Colombia	Sustainable and innovative urban planning along with conservation of natural habitats and water resource management	Affordable transport, increase green spaces, good governance and equitable sharing of benefits among the city dwellers	Integrated natural-social adaptations	Lee Kuan Yew World City Prize Special Mention, 2014
Urban Nature Park, Gazelle Valley, City of Jerusalem	Water conservation and rainwater management, combat weather extremes and desertification	Introduction of runoff collection pools and its filtration systems for rechange groundwater, conservation efforts to protect native flora and fauna, park for recreational purposes, etc.	Integrated natural-social management	Gazelle Valley Conservation Program
Integrated water resource management (IWRM), Cape Town	Integrated water resource management and urban fire management	Construction of resilient infrastructure and retention ponds, recanalizing of rivers, restoration of riparian flora, frequent drain cleaning, Usage of fire regime during periods of drought, enhanced systems of disaster warning etc.	Integrated Engineering and Ecological solutions	Launching of "bionet" – a network of green open spaces
NEWater for Singapore	Water resource management	Construct a strong and expanded water supply system including the desalinized water, local catchment water, imported water and treated recycled water (NEWater)	Integrated technological and ecological solution	New surface water drainage regulation
Thornton Creek Water Quality Channel, Thornton Creek, Washington	Integrated water treatment and management	Integrate private development, community services, open public space, resilient design for stormwater drainage etc.	Integrated ecological and social approaches	A community process that supports environment friendly development by private sectors and socioeconomic sustainability for the public in a vastly challenged urban system

References

Aylett A (2014) Progress and challenges in the urban governance of climate change: results of a global survey. MIT Press, Cambridge

Choguill CL (1993) Sustainable cities: urban policies for the future. Pergamon Press, Oxford

Holden M, Roseland M, Ferguson K, Perl A (2008) Seeking urban sustainability on the world stage. Habitat Int 32(3):305–317

IPCC (2018) IPCC report. https://www.unep.org/news-and-stories/story/air-pollution-and-climate-change-two-sides-same-coin

IPCC (2021) IPCC report. https://www.ipcc.ch/2021/08/09/ar6-wg1-20210809-pr/

Lin BB, Ossola A, Alberti M, Andersson E, Bai X, Dobbs C, Elmqvist T, Evans KL, Frantzeskaki N, Fuller RA, Gaston KJ, Haase D, Jim CY, Konijnendijk C, Nagendra H, Neimela J, McPhearson T, Moomaw WR, Parnell S, Pataki D, Ripple WJ, Tan PY (2021) Integrating solutions to adapt cities for climate change. Lancet Planet Health 5(7):479–486

Marcotullio PJ (2001) Asian urban sustainability in the era of globalization. Habitat Int 25(4): 577–598

Marvuglia A, Havinga L, Heidrich O, Fonseca J, Gaitani N, Reckien D (2020) Advances and challenges in assessing urban sustainability: an advanced bibliometric review. Renew Sustain Energy Rev 124:109788

Millennium Ecosystem Assessment, MA (2003) Ecosystems and human well-being: a framework for assessment. In: Report of the conceptual framework working group of the Millennium Ecosystem Assessment. Island Press, Washington, DC

Rosenzweig C, Solecki WD, Romero-Lankao P, Mehrotra S, Dhakal S, Ibrahim SA (eds) (2018) Climate change and cities: second assessment report of the urban climate change research network. Cambridge University Press, Cambridge

Satterthwaite D (2007) Adapting to climate change in urban areas: the possibilities and constraints in low-and middle-income nations, vol 1. IIED, London

Shen LY, Ochoa JJ, Shah MN, Zhang X (2011) The application of urban sustainability indicators—a comparison between various practices. Habitat Int 35(1):17–29

U.S. Drought Monitor (2021) Drought.gov. https://climate.nasa.gov/effects/

UNICEF (2021) UNICEF report. https://www.unicef.org/stories/water-and-climate-change-10-things-you-should-know

Vergara SE, Tchobanoglous G (2012) Municipal solid waste and the environment: a global perspective. Annu Rev Environ Resour 37:277–309

Chapter 2
Infrastructure, Energy Needs and Waste Management for Sustainable Urban Infrastructure

J. Padhiar, P. Vithalani, and N. Bhatt

2.1 Introduction

In 1987, World Commission on Environment and Development of United Nations gave the term sustainable development, in their report "Our Common Future" (Our Common Future 1987). It was stated as "development that meets the needs of the present without compromising the ability of future generations to meet their own needs" for the world to achieve its ultimate growth in various aspects such as cultural, social, economic, and environmental; people and families being the centre of society must function optimally. For a society to prosper or collapse is measured by the sustainability of lifestyle, for example decline of Western Roman Empire and Polynesians on Easter Island by the depletion of resources which can feed them and lead to societal collapse in 1600s (Alvarez-Risco et al. 2020).

The hypothetical concept of sustainability is "possibility to flourish forever", which is completely impractical when implemented. Essentially period of 50–100 years (i.e. 2–4 generations) seems to be practical and appropriate when it comes to sustainably implement the concept. Range of sustainability concern is very wide including pollution, climate change, sanitation, waste management, industrial, housing, etc. To swing towards sustainable urban regions, role of government policies along with their implementation and implication plays a vital role. When conversing about sustainable urban regions, it must have sustainable economic growth which will contribute to increase in social development of citizens which ultimately can mould to attain environment conservation.

J. Padhiar · P. Vithalani · N. Bhatt (✉)
Department of Biogas Research and Microbiology, Gujarat Vidyapith, Gandhinagar, Gujarat, India
e-mail: bhatt@gujaratvidyapith.org

B. Pathak, R. S. Dubey (eds.), *Climate Change and Urban Environment Sustainability*, Disaster Resilience and Green Growth,
https://doi.org/10.1007/978-981-19-7618-6_2

With the increasing desire of modern life, more and more individuals are attracted to urban areas and wish to live there. According to Ferrer et al. (2018), around 66% of world population might live in urban areas by 2050. To fulfil their requirements for infrastructure, energy and management of waste generated by such a large population are becoming a tiresome task. Recent report of Intergovernmental Panel on Climate Change (IPCC) confirms that urban regions use about 67–76% of global energy produced and emit 80% of total greenhouse gas, it is also estimated that energy requirement of urban regions will be 730 EJ by 2050 (Martos et al. 2016; Creutzig et al. 2015). The waste generated by various activities of humans contains harmful gases and non-biodegradable products, negatively affecting climate. Globally 1.3 million tons of municipal solid waste and 1.3 billion metric tons of food waste are generated annually which is estimated to be double by 2050 (Das et al. 2019; Hao et al. 2015).

2.2 Infrastructure for Sustainable Urban Region

Urban infrastructure system (UIS) is nowadays considered a combination of various components and is beyond engineered construction and facilities. UIS is multifaceted in which six major components are shared: (1) natural environment, (2) land use, (3) energy systems, (4) drinking water, storm water and wastewater infrastructure, (5) transportation infrastructure and (6) socio-economics (Pandit et al. 2015). UIS bridges these components and allows the flow of capital into, within and out of urban areas. To construct these kind of infrastructure systems, it requires dedicated input from government as well as citizens.

2.2.1 Bottlenecks in Infrastructure

Various bottlenecks in infrastructure are present which holds back the way of development.

2.2.1.1 Land Acquisition

The compensatory amount offered to land lords is always less than the original value of the land; this creates disagreement in land lords and ultimately leads to disputes before construction.

2.2.1.2 Clearance from Various Agencies

Clearance of the proposed plan take a long time in clearance due to inefficient regulation and co-ordination of various agencies; this in-turn unnecessarily delays the project. Agencies also demand money besides the actual charge for clearance (Rao 2019).

2.2.1.3 Financing

Funding is the most crucial part of construction. Large infrastructures are constructed with the expectation of high revenue generation but the capital investment done for it is low leading to weak construction and affecting the urban development.

2.2.1.4 Environmental Impact Assessment

Infrastructure regulations made to protect environment must be mandatory for new projects but the revised guidelines are also made compulsory for the projects which are under construction; these new changes in the design according to revised guidelines create problems in the midst of construction.

2.2.1.5 Poor Pre-construction Planning

Financial closure, land disputes, delayed clearance take up a very long period of time, resulting in poor pre-construction history. Planning is severely affected due to these issues and poor prior planning causes project to be in pathetic condition.

2.2.2 Dimensions for Sustainable Infrastructure

Many dimensions must be considered while designing new projects; to make it co-ordinate with surrounding environment, social life and economic aspects, resulting in sustainable infrastructure.

2.2.2.1 Environmental Dimension

Natural resources surrounding the project area should be given priority while construction. The land around it should be used to obtain local construction materials, but these resources should not be exploited. Water required for construction should be taken from local region, and water saving and rain water harvesting plans should be installed in the project design to save water and reuse it. Natural ventilation systems must be given in infrastructure to use the surrounding air to its maximum, rather than installing an air conditioner and heater in the rooms. To save electricity, use of solar energy must be emphasized for fulfilling electricity demand, also, natural lightening should be preferred through big windows and not the bulbs. To save the biodiversity, local vegetation must be planted around the building as well as in balconies.

2.2.2.2 Social Dimension

Relation between culture and nature is very ancient. Therefore, along with natural dimension, social life must also be given importance. Opinions and suggestions of local people must be kept in mind while designing the project, contract for constructing the project should be given to local agencies and labours to provide them employment. Infrastructure must be designed in such a way that basic necessities of people can be fulfilled and there must also be easy access to rudimentary requirements like food, transport, etc.

2.2.2.3 Economic Dimension

For the economic growth, connectivity with surrounding is very crucial. Hence, internet and telephonic facility should be provided. Besides this new job, opportunities and activities should not cause harm to ecology, local economics must be given a chance to present and well-organized as well as low maintenance life should be encouraged (Alvarez-Risco et al. 2020).

2.2.3 Approaches for Sustainable Infrastructure

Various ways are nowadays been thought upon to construct an infrastructure that will provide with all the basic needs as well as will be sustainable for many years.

2.2.3.1 Infrastructure Ecology

As there is an ecological niche consisting of various components of nature; similarly, here infrastructure is considered as a niche of six components of UIS. All the components are connected which each other and must also be thought as an integrated system, rather than focusing individually on them efforts should be made to correlate them. Conventional engineering systems had not focused on the

Table 2.1 Interactions between different components of UIS (Pandit et al. 2015)

S. no.	Interactions
1.	Water ↔ energy
2.	Energy ↔ transportation
3.	Land use ↔ transportation
4.	Water ↔ energy ↔ transportation
5.	Land use ↔ energy ↔ transportation
6.	Land use ↔ water ↔ energy

interaction and only on the individual component, resulting in a sub-optimal urban region.

In recent times, water and energy interaction is being promoted as '*Water-Energy Nexus*'. Energy generation requires water and distribution of that water requires energy, but this becomes grave question in hot summer times where demand for both the component increases. This can be solved by installing air-cooled microturbines which simultaneously produces power; it reduces the load on central power house and '*water for energy production*' by 60% (Frankland 2013). Besides water and energy, many other interactions are listed in Table 2.1.

2.2.3.2 Urban Ecology

Urban ecology emphasizes on collaboration of ecological (environmental) and urban (human) systems. In this integration of blue, green and grey land is considered. Blue land consists of ponds, lakes, rivers, etc., green land has all the plantation and grey land comprised of infrastructure like roads, buildings, etc. Integration of these components will make the infrastructure sustainable for longer period of time. Linking the components of blue land with each other will increase communication between aquatic creatures and materials present in water, this will offer habitat to organisms, conserve biodiversity of wetlands and restore urban blue land. Integration of these can also resolve quality by preventing eutrophication, manage flood disasters and add to the beauty of urban regions (Pamukcu-Albers et al. 2021).

In recent times, emphases have been made to look the place beautiful by planting ornamental alien species and compromise with the quality of vegetation. Local species are ignored which causes imbalance in the entire biodiversity of certain region, organisms feeding upon local species are also affected. Instead of this, vegetation growing in locality should be preferred, city gardens, street side vegetation, green roofs and vertical plantation should be encouraged to keep the urban areas green, which will in-turn decrease the temperature of the region and make it sustainable. Grey land is impervious; therefore, it stops the exchange of nutrients, water and air between atmosphere and soil, upsetting urban ecosystem. Infrastructure should be constructed in such a way that it provides room for exchange between soil and atmosphere, this can offer habitat for organisms, pollution control, electricity savings, etc. (Li et al. 2016).

2.3 Energy for Sustainable Urban Regions

Energy occurs in various forms, it can never be produced or destroyed but can be converted from one form to other. Many energy sources are listed in Table 2.2. Present scenario shows that energy requirement is increasing day by day because besides basic equipment, many other new gadgets are also working upon energy. To make urban regions sustainable in terms of energy, stress must be laid on the entire process of energy harvesting and distribution which includes finding energy sources, converting it to desirable form, energy storage and distribution, these processes should also be made sustainable to achieve the goal for urban regions. Energy in buildings must be conserved to fulfil the demand (Chel and Kaushik 2017).

2.3.1 Issues with Energy

Problems associated with energy blocks the path of sustainable energy efficiency.

2.3.1.1 Growing Energy Demand

With the rapidly growing industrial sector, demand of energy is increasing day by day. Most industries require energy in one form or the other to process raw materials, pack finished products, transport and distribute them. Demand of electronic gadgets like mobiles, laptops, power banks, etc. is also increasing which uses electrical energy. Besides industrialization, desire for high standard living leads people to buy air conditioners, electric heaters, mobiles, laptops, etc. also increases use of energy and carelessness of people to turn off electrical items when not in use; wastes energy.

Table 2.2 Energy sources (Alvarez-Risco et al. 2020)

Non-renewable energy sources	Renewable energy sources
Fossil fuels (oil, natural gas, coal)	Solar radiation
Non-fossil fuels	Solar related
Biomass (use rate > replenishment rate)	Hydraulic
Uranium	Wind
Fusion material (e.g. deuterium)	Wave
Wastes (direct or indirect use)	Ocean thermal
	Biomass (use rate < replenishment rate)
	Non-solar related
	Geothermal
	Tidal

2.3.1.2 Environmental Threat

Energy obtained from non-renewable sources generally contribute to pollute the environment. Energy supply and use cause emission of carbon dioxide, sulphur oxides, nitrogen oxide, etc. resulting in increased temperature of atmosphere, eutrophication and acidification of water bodies, in 2015, 80% and 30% of CO_2 and CH_4 emission, respectively, were noted by energy supply and use. Many parts of energy chain also demand large piece of land and huge amount of freshwater which also causes issues (Dincer and Acar 2016).

2.3.2 Urban Energy Service Drivers

There are many factors which affect energy services in urban regions.

2.3.2.1 Environmental Drivers

Climatic variation is one of the important factors which affects energy services both directly and indirectly. Increased or decreased temperature, change in agricultural practices, etc. contribute to variation in demand of energy leading to change in its service.

2.3.2.2 Technological Drivers

Technology affects the levels of energy consumption, new gadgets and improved lights will consume less amount of energy, whereas energy demands of growing urban areas increase due to advancement in industrial sector.

2.3.2.3 Economic Drivers

With increment in the financial sources, people invest more in making their life better with different equipment and also they leave the options with moderate energy consumption (cooking with traditional method), leading to increased energy service (Carreón and Worrell 2017).

2.3.3 *Renewable Energy Sources*

Presently, 50% of global energy production is still done by fossil fuels (coal and natural gas). The problem arises when it comes to convert this energy; conversion efficiency of this energy is very low and it also uses high thermal energy to be converted in different forms (Fig. 2.1).

Renewable energy sources include solar energy, hydropower, tidal energy, geothermal energy, bioenergy, etc.

2.3.3.1 Solar Energy

Solar energy is the energy obtained directly from the sun. Concentrating solar power plants can be constructed to meet the requirements of light and fuel which can be used in transportation. Wind and ocean energy also use sun's energy once absorbed and can be converted into other forms. World Energy Council stated that solar radiation coming on earth are 7500 times more than the total energy consumption of earth (Owusu and Asumadu-Sarkodie 2016).

2.3.3.2 Hydropower

Hydropower uses turbines to produce energy; potential energy of water stored at high level is used when water falls on the turbine towards gravity. This is efficient because hydropower doesn't pollute the environment and can store energy for long period of time (Hamann 2015).

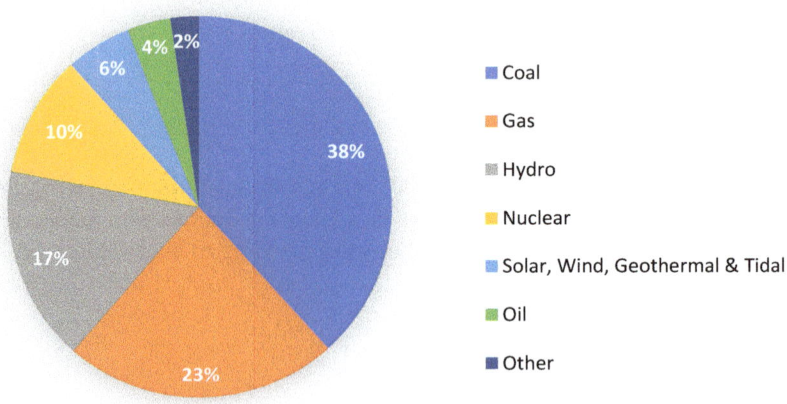

Fig. 2.1 Shows different energy sources and their contribution in producing energy globally, leading by coal followed by gas, hydro and other sources (Nižetić et al. 2019)

2.3.3.3 Tidal Energy

Tidal energy is obtained when wind passes over the water, amount of energy produced is directly proportional to speed, distance and height of the wind. Ocean holds enough energy to fulfil the increasing demand but space for installing plant, differing characteristics of wind, etc. are bottlenecks of this energy (Kammen and Sunter 2016).

2.3.3.4 Geothermal Energy

Earth's crust has huge amount of energy in the form of heat. Natural reservoirs which are sufficiently hot and penetrable are known as hydrothermal reservoirs and those that are hot but requires artificial hydraulic stimulus are called enhanced geothermal systems (Owusu and Asumadu-Sarkodie 2016). Challenges to obtain geothermal energy are heat which is very deep in Earth, distribution is not even and requires extraordinary mechanical support.

2.3.3.5 Bioenergy

This energy is obtained from biodegradable components like agricultural waste (sugar cane waste), animal waste (cow dung), forest waste (shredded leaves, wood residues), etc. Degradation of these wastes provides energy in the form of by-product like gas, biodiesel, etc.; these by-products can be used in cooking, transportation and the left-over stuff can be also used as fertilizer.

2.4 Waste Management for Sustainable Urban Region

With the increasing population of urban region, consumption of goods increases leading to generation of waste. Waste management is becoming a global issue, 40%, 37% and 23% of global waste are generated annually by developed, developing and undeveloped countries, respectively (Das et al. 2019). Huge amount of waste generated is dumped in the natural ecosystems which also gets disturbed due to the entering waste compounds. Jambeck et al. (2015) reported that in 2010, 4.8–12.7 million tons of plastic waste generated entered ocean ecosystems. 2.2 billion tons of municipal solid waste annually will be generated by 2025 (Leal Filho et al. 2016). Waste can be of so many types, different types of waste are mentioned in Table 2.3 and Fig. 2.2.

Table 2.3 Types of waste (Aleluia and Ferrão 2016)

Biodegradable	Non-biodegradable
Agricultural	Plastic
Food	Electronics
Garden and Park	Medical
Paper	Metal

Fig. 2.2 Types of wastes (**a**) and (**b**) solid waste, (**c**) food waste, (**d**) medical waste. Different types of waste disposal on land, waterbodies and in dumping yards. (Source: https://www.caymancompass.com/2018/05/07/untreated-biomedical-waste-dumped-at-brac-landfill/ (n.d.) Accessed on 25 November 2021, https://www.nycfoodpolicy.org/rabbit-hole-measuring-food-waste-confusing/ (n.d.) Accessed on 25 November 2021, Photo credits: rob245 / Fotolia, Nevit Dilmen)

2.4.1 *Threats of Waste*

Waste disposal without care causes many threats to environment as well as health.

2.4.1.1 Environmental Threat

Mismanaged disposal of waste causes soil, water and air pollution. Soil is considered skin of earth which provides minerals for plants to grow, protects ground water and also is the habitat for various organisms. With rapidly growing amount of waste on soil, it is affecting both biotic and abiotic factors of soil, and deposition of heavy metals like Mn, Pb, Zn, Cd, etc. from industries is affecting soil fertility. Wastewater from industries contains many harmful chemicals, high COD, high TDS, chloride, nitrate, fluoride, etc. which when released in open environment without treatment causing ground water pollution, making it unsafe to drink, also when it is released in water bodies aquatic life gets affected, unbalancing the entire water ecosystem. Landfills of waste when undergoes decomposition anaerobically, gases like methane and carbon dioxide are produced which are greenhouse gases and contribute to global warming, besides these gases various sulphur and nitrogen compounds are released from industries as smoke also pollutes air, making it unhealthy to breath.

2.4.1.2 Health Threat

Due to ill practice of people to throw the waste in open areas, health of humans as well as animals is severely affected. Srivastava et al. (2018) stated that people working and residing near waste landfills are at high risk of respiratory diseases, birth defects and cancer. Particularly in developing and undeveloped countries, people have tendency to throw the garbage on road side, rag pickers finding recyclable goods and animals (cow, pig, dog, etc.) finding food dig off these waste heaps; gases produced in these heaps and plastic polyethene which animals consume along with food which deteriorates their health. Alvarado-Esquivel (2013) reported that chances of toxocariasis are more by rag pickers. Scattered waste then blocks the drainage system, leading to stagnant water which gives chance to mosquitoes and insects to breed causing malaria and lymphatic filariasis. Besides this illegal dumping of medical waste increases the chance of pathogenic species to enter environment and gases liberated from open combustion of plastics also cause severe health issues.

2.4.2 Strategies for Waste Management

Various strategies must be adopted to manage the huge amount of waste generated which stops waste to come in path of sustainable urban region.

2.4.2.1 Aerobic Composting

By this method organic compounds in solid waste are subjected to aerobic decomposition under specific pH, humidity and temperature conditions, microorganisms (bacteria and fungi) native to that environment aid the process to form compost. This compost can be used as fertilizer in agriculture and horticulture (Neher et al. 2013). Jain and Sharma (2011) reported that in developing countries, solid waste consists of 60% organic materials, but nowadays these wastes are polluted with paints, artificial pesticides, etc.

2.4.2.2 Anaerobic Digestion

Anaerobic digestion also known as biomethanation is the process by which biodegradable compounds present in waste are decomposed, releasing biogas 55–60% CH_4 which can be used as fuel (Somani et al. 2021).

2.4.2.3 Vermicomposting

Vermicomposting is an integrated action of earthworms and microorganisms (soil and earthworm's gut) to produce valued bioproduct (vermicompost) from organic components of solid waste. Reduction in quantity of carbohydrate, protein, aliphatic and lignocellulosic contents, decrease in volatile solids and C/N ratio along with increase in acid phosphatase activity, humic acid and aromaticity indicate the progress of vermicompost, rich in carbon, nitrogen, phosphorus, sulphur content and enzyme activity and can inhibit plant pathogen when used as biofertilizer (Lv et al. 2013; Bhattacharya et al. 2012).

2.4.2.4 Incineration

Incineration is the thermal process of unprocessed waste at 850 °C in the presence of air, and various by-products like CO_2, SO_2, CO, water vapour, furans, ash are produced (Srivastava et al. 2018). Heat generated in this can be used as electricity. Vergara and Tchobanoglous (2012) stated that to achieve greater efficiency, low moisture content ($<50\%$) and high heating value (>5 MJ/kg) solid waste must be preferred to manage by this method.

2.4.2.5 Pyrolysis and Gasification

In this method, waste is combusted to obtain energy in oxygen scarce environment. Pyrolysis and gasification both thermally degrade waste under pressure and temperature ranging 400–1000 °C in oxygen less environment and 1000–1400 °C with less oxygen, respectively. Syngas is liberated as end product which contains CO, CH_4, H_2, CO_2, N_2 and hydrocarbons (Srivastava et al. 2018).

2.4.2.6 Operated Landfills

Open landfills cause issues of greenhouse gas emission contributing to global warming. Operated landfills are those where waste is partially buried in the soil, and this site is treated with certain strains of microorganisms which are capable of degrading pollutants. These organisms must be capable of degradation of high amount of chemicals because the load of solids deposited over there is ever-changing.

2.4.2.7 Animal Feeding

This approach can be exclusively used for food waste. Food waste produced should not be thrown in dustbins, rather must be given to animals as feed. This way animals will not dig off the waste heaps and their health will also be good (Thi et al. 2015; Mak et al. 2019).

2.5 Conclusion

Sustainability enhances quality of our lives. Sustainability doesn't mean to maintain everything for just one to two decades but it means to sustain it for future generations. With the increasing population of urban regions, it has become ultimate necessity to keep the urban regions sustainable because without proper attention soon these regions will face severe problems of energy, waste management and infrastructure. To keep urban regions sustainable, along with government and various agencies, residents of these regions must also be responsible towards their duty. More and more emphases must be laid on the use of renewable energy and its conservation, use of solar energy must be promoted within peoples. Infrastructure must be constructed in such a way that it co-resides with existing environment without disturbing the ecosystems present in nature. Available commodities should be used in the quantity it is required and waste generation should be minimized, steps must be taken by governmental and non-governmental organizations to manage waste in such a way that it doesn't harm any living organism and also give useful products after processing. Guidelines for discarding waste must followed strictly and violating those guidelines must lead to severe punishments. Sustainable urban regions must have supportable economic growth, contributing to increase social development of citizens which ultimately can led to environment conservation.

2.6 Future Aspects

Sustainability is a key to prosper for every generation on earth. Infrastructure must be designed in such a way that it should not disturb the natural flow of air and water. Native vegetation around the construction should be promoted to protect biodiversity. New constructions should be designed in an eco-friendly way. Energy should be conserved to its maximum possible way and different ideas can be thought to overcome the challenges in the way of renewable energy sources like tidal energy, solar plants can be made affordable and install in building roofs to meet its energy demand. Waste generation should be minimized, many methods exist to manage waste but each has their own pros and cons. Researchers can modify existing methods to attain maximum degradation of waste, reuse and recycle it, emphases can also be laid to manage non-biodegradable waste for sustainable urban regions.

References

Aleluia J, Ferrão P (2016) Characterization of urban waste management practices in developing Asian countries: a new analytical framework based on waste characteristics and urban dimension. Waste Manag 58:415–429. https://doi.org/10.1016/j.wasman.2016.05.008

Alvarado-Esquivel C (2013) Toxocariasis in waste pickers: a case control seroprevalence study. PLoS One 8(1):1–6. https://doi.org/10.1371/journal.pone.0054897

Alvarez-Risco A, Rosen MA, Del-Aguila-Arcentales S, Marinova D (2020) Building sustainable cities: social, economic and environmental factors. In: Building sustainable cities. Springer Nature, Berlin

Bhattacharya SS, Iftikar W, Sahariah B, Chattopadhyay GN (2012) Vermicomposting converts fly ash to enrich soil fertility and sustain crop growth in red and lateritic soils. Resour Conserv Recycl 65:100–106. https://doi.org/10.1016/j.resconrec.2012.05.008

Carreón JR, Worrell E (2017) Urban energy systems within the transition to sustainable development. A research agenda for urban metabolism. Resour Conserv Recycl 132:258–266. https://doi.org/10.1016/j.resconrec.2017.08.004

Chel A, Kaushik G (2017) Renewable energy technologies for sustainable development of energy efficient building. Alex Eng J 57(2):655–669. https://doi.org/10.1016/j.aej.2017.02.027

Creutzig F, Baiocchi G, Bierkandt R, Pichler P, Seto KC (2015) Global typology of urban energy use and potentials for an urbanization mitigation wedge. Proc Natl Acad Sci U S A 112(20): 6283–6288. https://doi.org/10.1073/pnas.1315545112

Das S, Lee S, Kumar P, Kim K, Lee SS, Bhattacharya SS (2019) Solid waste management: scope and the challenge of sustainability. J Clean Prod 228:658–678. https://doi.org/10.1016/j.jclepro.2019.04.323

Dincer I, Acar C (2016) Smart energy systems for a sustainable future. Appl Energy 194:225–235. https://doi.org/10.1016/j.apenergy.2016.12.058

Ferrer ALC, Thomé AMT, Scavarda AJ (2018) Sustainable urban infrastructure: a review. Resour Conserv Recycl 128:360–372. https://doi.org/10.1016/j.resconrec.2016.07.017

Frankland JH (2013) A model-based feasibility study of combined heat and power systems for use in urban environments. Georgia Institute of Technology, Atlanta. https://smartech.gatech.edu/handle/1853/49102

Hamann A (2015) Coordinated predictive control of a hydropower cascade. Dissertation. Carnegie Mellon University, Pittsburgh

Hao HN, Karthikeyan OP, Heimann K (2015) Bio-refining of carbohydrate-rich food waste for biofuels. Energies 8(7):6350–6364. https://doi.org/10.3390/en8076350

https://www.caymancompass.com/2018/05/07/untreated-biomedical-waste-dumped-at-brac-landfill/. Accessed 25 Nov 2021

https://www.nycfoodpolicy.org/rabbit-hole-measuring-food-waste-confusing/. Accessed 25 Nov 2021

Jain S, Sharma MP (2011) Power generation from MSW of Haridwar city: a feasibility study. Renew Sustain Energy Rev 15(1):69–90. https://doi.org/10.1016/j.rser.2010.09.007

Jambeck J, Geyer R, Wilcox C, Siegler TR, Perryman M, Andrady A, Narayan R, Law KL (2015) Plastic waste inputs from land into the ocean. Mar Pollut 347(6223):768. https://science.sciencemag.org/CONTENT/347/6223/768.abstract

Kammen DM, Sunter DA (2016) City-integrated renewable energy for urban sustainability. Science 352(6288):922–928. https://doi.org/10.1126/science.aad9302

Leal Filho W, Brandli L, Moora H, Kruopiene J, Stenmarck Å (2016) Benchmarking approaches and methods in the field of urban waste management. J Clean Prod 112:4377–4386. https://doi.org/10.1016/j.jclepro.2015.09.065

Li F, Liu X, Zhang X, Zhao D, Liu H, Zhou C, Wang R (2016) Urban ecological infrastructure: an integrated network for ecosystem services and sustainable urban systems. J Clean Prod 163: S12–S18. https://doi.org/10.1016/j.jclepro.2016.02.079

Lv B, Xing M, Yang J, Qi W, Lu Y (2013) Chemical and spectroscopic characterization of water extractable organic matter during vermicomposting of cattle dung. Bioresour Technol 132:320–326. https://doi.org/10.1016/j.biortech.2013.01.006

Mak TMW, Xiong X, Tsang DCW, Yu IKM, Poon CS (2019) Sustainable food waste management towards circular bioeconomy: policy review, limitations and opportunities. Bioresour Technol 297:122497. https://doi.org/10.1016/j.biortech.2019.122497

Martos A, Pacheco-torres R, Ordóñez J, Jadraque-gago E (2016) Towards successful environmental performance of sustainable cities: intervening sectors. A review. Renew Sustain Energy Rev 57: 479–495. https://doi.org/10.1016/j.rser.2015.12.095

Neher DA, Weicht TR, Bates ST, Leff JW, Fierer N (2013) Changes in bacterial and fungal communities across compost recipes, preparation methods, and composting times. PLoS One 8(11):79512. https://doi.org/10.1371/journal.pone.0079512

Nižetić S, Djilali N, Papadopoulos A, Rodrigues JJC (2019) Smart technologies for promotion of energy efficiency, utilization of sustainable resources and waste management. J Clean Prod 231: 565–591. https://doi.org/10.1016/j.jclepro.2019.04.397

Our Common Future (1987) United Nations, vol 4. Taylor & Francis, Abingdon. https://doi.org/10.1080/07488008808408783

Owusu PA, Asumadu-Sarkodie S (2016) A review of renewable energy sources, sustainability issues and climate change mitigation. Cognet Eng 3(1):1167990. https://doi.org/10.1080/23311916.2016.1167990

Pamukcu-Albers P, Ugolini F, La Rosa D, Grădinaru SR, Azevedo JC, Wu J (2021) Building green infrastructure to enhance urban resilience to climate change and pandemics. Landsc Ecol 36(3): 665–673. https://doi.org/10.1007/s10980-021-01212-y

Pandit A, Minné EA, Li F, Brown H, Jeong H, James JAC, Newell JP, Weissburg M, Chang ME, Xu M, Yang P, Wang R, Thomas VM, Yu X, Lu Z, Crittenden JC (2015) Infrastructure ecology: an evolving paradigm for sustainable urban development. J Clean Prod 163:S19–S27. https://doi.org/10.1016/j.jclepro.2015.09.010

Rao DCR (2019) Issues and challenges to urban infrastructure in India. Academia, San Francisco

Somani P, Navaneethan RD, Thangaiyan S (2021) Integrated solid waste management in urban India: a mini review. J Phys Conf Ser 1913(1):2084. https://doi.org/10.1088/1742-6596/1913/1/012084

Srivastava V, Ismail SA, Singh P, Singh RP (2018) Urban solid waste management in the developing world with emphasis on India: challenges and opportunities. Rev Environ Sci Biotechnol 14(2):317–337. https://doi.org/10.1007/s11157-014-9352-4

Thi NBD, Kumar G, Lin CY (2015) An overview of food waste management in developing countries: current status and future perspective. J Environ Manag 157:220–229. https://doi.org/10.1016/j.jenvman.2015.04.022

Vergara SE, Tchobanoglous G (2012) Municipal solid waste and the environment: a global perspective. Annu Rev Environ Resour 37(1):277–309. https://doi.org/10.1146/annurev-environ-050511-122532

Chapter 3
Climate Change and Sustainable Urban Transport Environment

Ashwani Luthra

3.1 Introduction

Climate change refers to long-term change in the temperature of the earth that leads to weather changes. It is directly associated with the phenomenon of global warming to indicate gradual increase in the earth's temperature due to increased levels of carbon dioxide, chlorofluorocarbons, and other pollutants (EPA 2008, Web update: April 2021). The change may be natural. But since the beginning of the industrial revolution in 1800s, human activities such as burning of fossil fuels producing heat-trapping gases have led to unprecedented changes in the chemical composition of the earth's atmosphere. Consequently, the concentration of carbon dioxide in the air has increased from a pre-industrial value of about 280 ppm in 1700s to 410 ppm in 2019, an increase of 46% (Morain and Budge 2012). Similarly, the concentration of methane and nitrous oxide has reached to over 1800 ppb and 331 ppb in 2019, respectively. This increase is predominantly due to the use of fossil fuel (EPA 2008, Web update: April 2021). A study conducted by U.S. Environmental Protection Agency reveals that electric power, transport and industry contribute about 88% of the carbon emissions in the air. Transport sector alone contributes to about 24% of the total carbon emissions.

Global insane urbanization can be seen as the major responsible factor for raised the demand for land to house the rising urbanites, and increased the supply of various agricultural, industrial and other products. Thus, unprecedented spatial spread of urban settlements, whimsical increment of industrial units and high reliance on private transport have been the major contributors to changes in chemical composition in the atmosphere. The green lands have turned into heat islands due to

A. Luthra (✉)
Guru Ramdas School of Planning, Guru Nanak Dev University, Amritsar, India
e-mail: ashwani.planning@gndu.ac.in

B. Pathak, R. S. Dubey (eds.), *Climate Change and Urban Environment Sustainability*, Disaster Resilience and Green Growth,
https://doi.org/10.1007/978-981-19-7618-6_3

construction of buildings; unimaginable number of private vehicles are plying on the urban roads to meet the increased travel needs; high traffic densities have degraded the air quality in large volumes. The combined impact of all this has been the raised temperatures and environmental degradation, especially in the urban settlements. Nationally Appropriate Mitigation Actions (NAMA) for climate change have proposed for application of land-use planning principles to reduce energy consumption and limit urban sprawl. Consequently, efforts to plan sustainable human communities are strived for as counter measures reduce the impacts of climate change.

3.2 Climate Change: Concept and Contributors

Climate change is recognized as a significant manmade global environmental challenge. 'Global Climate Change', 'Global Warming', 'Green House Gas Emissions' and the like terms have caught attractions of policy makers, technocrats, scientists and researchers in the middle of twentieth century. Increment in greenhouse gases (GHGs) in the atmosphere is responsible for global climate change. Climate change refers to changes in long-term trends in the average temperature. Man's efforts to cater his needs are primarily responsible for climate change.

Since the past few decades, research has noticed that the global temperature has been rising continuously. It has been agreed in general that the global temperature has increased to the tune of 2 °C (While, 2008) and it is expected to rise by 1.5–3.5 ° C by 2100 (EPA 2008, Web update: April 2021). According to International Penal on Climate Change AR4, the rise in temperature by the end of the century with respect to 1980–1999 levels would range from 0.6 to 4.0 °C and the sea level may rise by 0.18–0.59 m during the same period. The all-India maximum temperatures show an increase in temperature by 0.71 °C per 100 year and all-India mean annual minimum temperature has significantly increased by 0.27 °C per 100 years during the period 1901–2007.

Climate change has been an important global environmental challenge and is part of larger challenge of sustainable development as well. Increased greenhouse gas (GHG) emissions are responsible for climate change. Primary GHGs in earth's atmosphere are water vapours, carbon dioxide (CO_2), methane, nitrous oxide and ozone. Total world CO_2 emissions have been estimated to be to the tune of 27.1 (Billion Tons, 2005). As per the global studies, man's creations, i.e. power generation, transport and industry are the prime accuses of increasing GHGs (Fig. 3.1). They contribute about 46%, 19%, and 23%, respectively, to the carbon emissions globally. The growth of these sectors is the outcome of unprecedented growth of global population in general and urban population in particular and steeply rising automobile population. Therefore, the major brunt of climate change has been faced by the urban settlements. According to working paper on 'Cities, Climate Change and Multilevel Governance by Corfee-Morlot et al. (2009), 'Climate change is expected to impact the cities and their regions through increased intensity of heat waves, directly affecting the human health; increases in intense rainfall events,

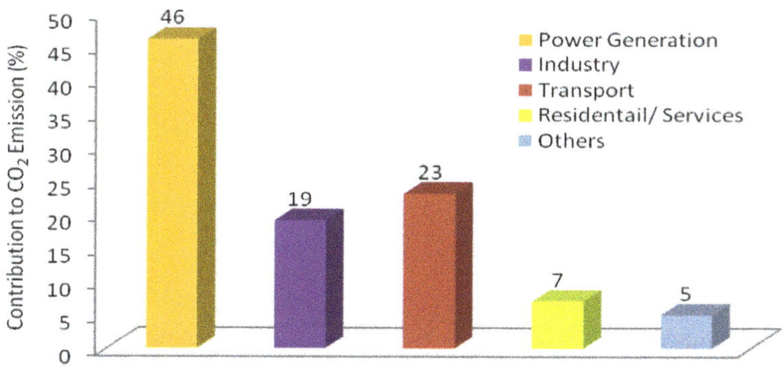

Fig. 3.1 Contribution to CO_2 emissions by sectors

increasing the risk of inland flooding; retreat of mountain glaciers, impacting water availability and its quality; and an increased risk of drought and water shortage in already dry regions. Urban centres may also be particularly vulnerable to some of the distributive impacts of climate change. Climate change is expected to have physical and economic consequences across numerous and diverse human activities. An explanation of rising population and automobiles as major contributors to air pollution will help in understanding the phenomenon of climate change.'

3.2.1 Rising Population and Insane Urbanization

Human souls on the global earth land are increasing at a rapid pace, and urban population increment is considered as the key player in the climate change. 'Global human population growth amounts to around 83 million annually, or 1.1% per year. The global population has grown from 1 billion in 1800 to 7.9 billion in 2020. The UN projected population to keep growing, and estimates have put the total population at 8.6 billion by mid-2030, 9.8 billion by mid-2050 and 11.2 billion by 2100' (Wikipedia 2021). The population trends also reveal that a major proportion of the total population lives in urban settlements. As per UNCTAD (2020), 'All over the world, a growing proportion of the population lives in cities. In 2009, 51.1% lived in urban areas. By 2019, the share of urban population increased to 55.7%. It is generally higher in the developed (80.5% in 2019) than in the developing world (51.1%), with transition economies in between the two (65.4%). During 2009–2019, urbanization has been most pronounced in developing economies, especially in developing Asia and Oceania, which saw the urbanization rate increase from 42.3 in 2009 to 49.1% in 2019'. But concentration of urban population in few bigger cities has a major trend of urbanization in different countries of the world (Fig. 3.2).

India has emerged as the second largest urban system of the world with 377 million (31.16%) of its total population inhabiting in 7935 urban settlements (Census of India, 2011). But the major brunt of this urbanization is taken by the 53 million plus

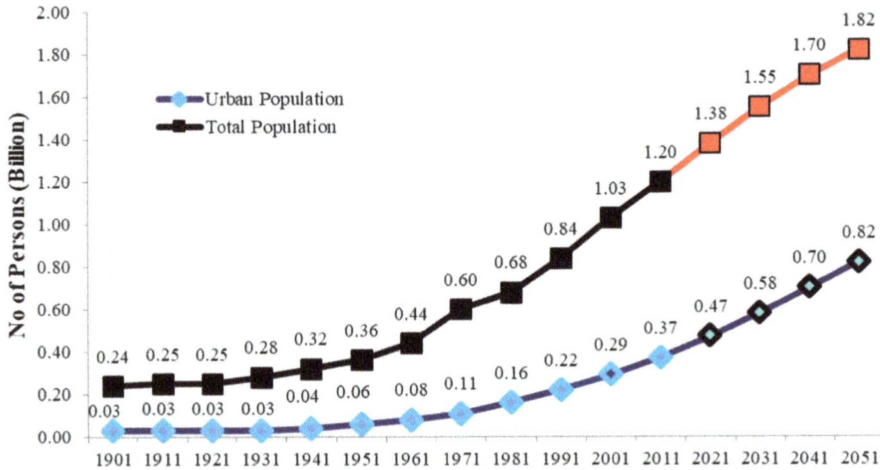

Fig. 3.2 Urban population in India (1941–2011)

cities by inhabiting 42.16% of the total urban population, more than half of which (23.77%) lives in 9 mega cities having million plus persons each. Only 3 (ten million plus) cities of India inhabit about 12.94% of the total urban population of India in 2011.

As per the estimates of UNCTAD (2019), India has 34% urban population in 2018 and McKinsey Global Institute (2010) has anticipated it to be 40% by 2030. It projects that 0.82 billion persons will be living in urban settlements of India by 2051. But the apathy is that the skewness of urbanization still remains towards bigger cities, where 22% and 27% of future urban population will live in mega and metropolitan cities, respectively, by 2051 (McKinsey Global Institute 2010). Cities as engines of economic growth have contributed about 63% to the gross domestic product of the nation and are expected to raise their contribution to 70% by 2031 (MoUD 2014). Rapidly rising total and urban population tend to invite different sectors, particularly transport, to grow to meet the needs of the living souls, which directly or indirectly contribute to environmental degradation and hence climate change.

3.2.2 Rising Number and Densities of Automobiles

Rising migration to urban settlements due to higher job opportunities and better infrastructure has always remained the prime reason for higher urbanization, resulting into increased spatial sprawl and ever rising trip lengths. But absence and inadequacy of public transport facilities tend the urbanites to rely on their self-owned automobiles to cater to their travel needs. A case of India reveals that merely 4.52 million registered automobiles in 1980 have increased to 297.19 million in 2019, registering a compounded annual growth rate of 9.8% during 2010–2019 (MoRTH 2020). Two-wheelers constitute the major share of this vehicular fleet, which has always remained more than 70% since 2001, 73% being in 2016 (MoSPI 2019). Another interesting fact to notice is the mismatched growth of vehicles and road length for their seamless flow. Whereas vehicles have risen by 9.9% in 2001–2011, the road length has risen by only 3.4%. Resultantly, traffic density has increased on the Indian roads. Mumbai tops in traffic density with 530 vehicles per kilometre in 2019 followed by Pune (259), Kolkata (319), Bangalore (149) and Delhi (108) (Sen 2019). Increased number of automobiles on the one hand and rising traffic densities on the other tend to pose serious threats to air pollution leading to climate change apart from energy losses.

3.3 Air Pollution in India

Air pollution is a key global mandate and serious concern in the present time of global warming, ozone depletion, greenhouse effect, acid rain, etc., whose effects are felt in cities far away from the sources. Alarmingly, WHO Global Air Pollution Database released in 2018 reveals that 14 out of 15 most polluted cities in the world are in India, the number being only 2 in 2010. Another study by IQAir Air Visual and Greenpeace has identified that 7 out of the 10 worst polluted cities and 22 of the 30 worst polluted cities of the world are in India (Thornton 2019). Data of National Ambient Air Quality Standards reveal that most of the urban areas of Indo-Gangetic plain exceed the standard limits of particulate matters (NCAP 2019). City size has a significant bearing on the pollution load and it increases geometrically with the growth of the city. NCAP (2019) has viewed that primary contribution of vehicles to $PM_{2.5}$ is higher in winters (20–23%) in Delhi and the NCR regions than in summers (18–20%). Over a 730 day period between 2013 and 2015, the Indian capital's air met 'healthy' standards for the particulate matter on only 7 days (Awaz 2017). Unprecedented growth urban population and vehicles, ever rising traffic densities, crawling speeds, bad road conditions, encroached roads, poor engine and fuel technology, inadequate transport infrastructure, and ad hoc traffic management measures have been the important causes for the rising air pollution loads in the cities.

Transport sector has always remained the major contributor to air pollution of the cities. As per Sinha et al. (1989), automobiles are responsible for 22% of manmade CO_2 emission in the world and 75–95% of emissions from mobile sources stem from road transport (Faiz 1993). In another study (Shrivastava et al. 2013) confirmed that road transport sector is responsible for 94.5% of CO_2 and 60% of the greenhouse gas emissions. In 2016, study of Sharma and Dikshit (2016) reveals that vehicular emissions are responsible for 25% to the pollution during summer and these emissions remain concentrated in the urban centres. Sharma (2018) study states that India is the fourth biggest CO_2 emitter in the world and 16% of the man-made CO_2 emissions stem from transport sector.

Air pollution from traffic in metro cities accounts for 60–80% of total air pollution. Pollution from motor vehicles produces about one-fifth of the incremental CO_2 in the atmosphere arising from human activity (which potentially contributes to global warming), one third of the chlorofluorocarbons (CFCs) (that contributes to the depletion of the ozone layer), and half of the NO_x (which contribute to continental scale acidification and ecological damage). Delhi alone produces about 1.6 times and 3.5 times more pollution to that of Mumbai and Calcutta, respectively, and that is why Delhi is identified as the fourth most polluted city of the world. Since over 70% of the vehicular fleet is constituted of two-wheelers, therefore, they are the main culprits in the air pollution in the cities, especially the bigger cities.

Contribution of two-wheelers in the total pollution load has been about 79% whereas cars, three wheelers and buses contribute about 8%, 7% and 1.85%, respectively. But in mega cities, percentage share of two-wheeler, cars, three wheelers and buses has been 29%, 47%, 27% and 53%, respectively. Since the composition of vehicles in urban areas has remained the same over the years; therefore, percentage pollution load has remained almost the same at present even.

3.4 Effects of Air Pollution

Vehicular pollution is not restricted to India only but has impacted all the nations across the globe. Hence, United Nations (2015), while setting 17 sustainable development goals (SDG) to address various issues, has set SDG 3 as 'good health and well-being for the people' with an objective to ensure healthy lives and promote well-being for all at all ages. Apart from reducing or ending deaths from various deceases, SDG 3 categorically focused on reducing the global deaths and injuries from road accidents to half by 2020, it also stressed on substantially reduce the number of deaths and illnesses from hazardous chemicals and air, water and soil pollution and contamination by strengthening the capacity of all countries, in particular developing countries.

According to the State of Global Air 2019 report of Health Effects Institute (HEI) and The Institute for Health Metrics and Evaluation (Health Effects Institute 2019; IHME 2016), air pollution is the fifth highest mortality risk factor globally, which causes people to die younger as a result of cardiovascular and respiratory diseases, and also exacerbates chronic diseases such as asthma, causing people to miss school or work and eroding quality of life. The report states that 92% people worldwide are living in areas exceeding the World Health Organization (WHO) guideline for healthy air and more than half lived in areas that do not even meet WHO's least-stringent air quality target in 2017, which led to about 4.9 million deaths and 147 million years of healthy life lost (ibid). The report profoundly highlighted that South Asians, including India, are dying early as their life expectancy has reduced by over 2.6 years (ibid).

Nearly two decades ago, eighth report on 'World Resources' had viewed that air pollution alone takes the lives of four million children the world over every year

Table 3.1 Vehicular emissions and health effects

Pollutant	Health effects
Carbon monoxide	• Difficulty in breathing (due to lower oxygen concentration) • Effects on central nervous system • Changes occur in cardiac and pulmonary functions • Concentration more than 500 ppm has suffocation effects like headache, giddiness
Hydrocarbons	• Severe throat and eye irritation. Some of them are carcinogenic • Induces some cancer producing substances • Causes abnormal leaf and bad development • Causes visibility reduction
Nitrogen oxides	• Increased susceptibility to respiratory infection • Adverse changes in cell structure of lung well
Carbon particles	• Associated with wide range of respiratory symptoms • Long-term exposure may increase risk of heart and lung disease
Sulphur dioxide	• Increased asthmatic attacks • Reduced lung capability
Lead	• Impair normal intellectual development and learning ability in children • Abnormalities in fertility and pregnancy

(United Nations Children's Fund (UNICEF) 2016). WHO's report on 'Air Pollution and Child Health: Prescribing Clean Air' reveals that over 90% of the world's children (1.8 billion) breathe toxic air every day (World Health Organization 2018). The report also reveals that 98% and 52% of children under 5 years are exposed to PM 2.5 pollution in low- and middle-income, and high-income countries, respectively. World Bank report states that more than 40,000 people die prematurely every year in India because of air pollution related health problems. Another study by IHME (2016) reveals that premature deaths due to air pollution in 2013 cost the global economy about $225 billion in lost labour income, or about $5.11 trillion in welfare losses worldwide.

Vehicular emissions have been the dominant source of air pollution through their contribution in rising hazardous gases like carbon monoxide (CO), carbon dioxide (CO_2), volatile organic compounds (VOCs) or hydrocarbons (HCs), nitrogen oxides (NO_x) and particulate matter (PM). The impacts of rising emissions are reflected through rising number of cases of various diseases in different cities. The effects of these pollutants are extremely deadly, and careful update and awareness are required for sustainable health, living and well-being of urban inhabitants. Some of the important are mentioned in Table 3.1.

Different ranges of air quality indexes reflect different health hazards. Therefore, the global air quality indexes have been localized by India, and their health impacts and health advisory are mentioned in Table 3.2.

Global reports on air pollution index reveal that mega and metropolitan cities of India are among the most polluted cities of the world. Following facts about impacts of vehicular pollution in India on the health of its residents are worth noticing.

Table 3.2 India's air quality index

Air quality index	Health impact	Health advisory
Good + satisfactory 0–100	Minimal impact	No cautionary action required
Moderate 101–200	May cause breathing discomfort to the people with lung disease (asthma), heart disease, children and older adults	Unusually sensitive people should consider reducing prolonged or heavy exertion and heavy outdoor work
Poor 201–300	May precipitate severe attack on short-term exposure in high-risk individuals and respiratory symptoms (breathing discomfort) in normal individual on long-term exposure	Children and adult with heart or lung disease should reduce prolonged or heavy exertion and limit outdoor activity
Very poor 301–400	May cause mild respiratory problems in normal individual/more pronounced in people with lung and heart disease	Everyone should reduce prolonged or heavy exertion. More caution for children or adult with heart or lung disease
Severe 401–500	May cause respiratory effect even on healthy people and serious health impacts on people with lung and heart diseases. The health impacts may be experienced even during light physical activity	Everyone should avoid all outdoor physical activity. Sensitive individuals should remain indoor with minimal activity

Source: Mukta (2017)

- Almost 44% of school children in Delhi had reduced lung function compared to 25.7% in the other cities studied (Awaz 2017).
- Delhi's PM levels are linked to six million asthma attacks per year in Delhi.
- As per a study based on 2016 data, at least 140 million people in India breathe air that is 10 times or more over the WHO safe limit (Bernard and Kazmin 2018).
- Compared with other countries, India ranked second after China in the number of deaths attributable to transportation emissions in 2015. An estimated 740,001 premature deaths were attributable to transportation emissions in India 2015. This represents a 28% increase in annual transportation-attributable deaths in India compared with 2010 (Anenberg 2019).
- About 1.2 million Indians died prematurely due to exposure to unsafe air in 2017 (Health Effects Institute 2019; IHME 2016).
- Indian cities have experienced 6,73,100 and 4,81,700 deaths due to exposure to outdoor and household pollution in 2017.
- As much as two thirds of deaths (3,85,000) from air pollution in India can be attributed to exhaust emissions from diesel vehicles, which were responsible for nearly 385,000 deaths in 2015 (The Economic Times 2019).
- According to the Greenpeace Report, 1.2 million deaths take place every year in India due to air pollution, which is only a "fraction less" than that caused by tobacco usage, and 3% of the GDP is lost due to air pollution (Hindustan Times 2019).

- A study published in The Lancet has estimated that in 2017 air pollution killed 1.24 million Indians—half of them younger than 70, which lowers the country's average life expectancy by 1.7 years.
- On-road diesel vehicles contributed 60% of the transportation health burden in New Delhi (ICCT 2019).
- Of the 17,000, 1800 premature deaths in New Delhi were attributable to ambient PM 2.5 and ozone from transportation tailpipe emissions, accounting for just over one-tenth (11%) of all deaths from air pollution that year in New Delhi in 2015.
- Compared with other major urban areas in India, New Delhi had the highest number of deaths attributable to transportation emissions in 2015 and the highest mortality rate (9 deaths per 100,000 population).

Though current issues and challenges reflect grim transport emission conditions in the cities, forecasts for urban traffic and travel scenario state that solutions need to be devised for sustainable transport in the cities. As a result of urbanization in India, pressure on urban transport is likely to increase substantially in the next millennium. Following are some of the points of due considerations to forecast the vehicle air pollution levels:

- Rate of urbanization is expected to be rapid, expecting 40% of Indian population to reside in urban settlements and contribute about 70% to country's GDP by 2030 (McKinsey Global Institute 2010). The current imbalanced urban system is expected to continue in future as well where about 49% of the urban population is expected to reside in mega and metropolitan cities by 2051. Thus, increased traffic densities and rising vehicular emissions will increase the health hazards exponentially in nothing done conditions.
- It is estimated that every extra one million population generates an extra 3.5 to four million public transport trips per day.
- Urban transport demand is expected to grow by 2.6 times at the existing model split in the large and medium sized cities. Larger cities are expected to register yet higher demand (Pandey 2012).
- Ministry of Housing and Urban Affairs (2008) report anticipated increased per capita trip rate (PCTR) in all metropolitan cities in next two decades. Metropolitan cities and mega cities are expected to attain PCTR ranging between 1.4 to 1.8 and 1.8 to 2.0, respectively, by 2031, indicating increased travel demand in bigger cities.
- India is expected to have 5–7.5 times growth in vehicle population in 2025 and 2035, respectively. About 63–70% of the vehicles are expected to be two-wheelers. Thus, the present trend of owning private vehicles and their composition in favour of two-wheelers is expected to remain the same in the next two to three decades.
- CAI-Asia News (Clean Air Asia 2009) cited transport related CO_2 emissions to increase to 57% worldwide by 2030, and passenger and freight transport in developing countries to contribute about 80% of this increase. Most of the current GHG emissions in the transport sector and virtually all the expected growth in emissions come from private cars and trucks.

Thus, future transport scene is also bringing alarming air pollution and hence health challenges in the times to come. A comprehensive vision is required to be developed to control of air pollution in Indian cities. Smart sustainable transport options would be needed to reduce vehicle tailpipe emitted fumes remarkably.

3.5 Government Initiatives Against Air Pollution

Since more than three decades, India has been striving to reduce the air pollution levels of its urban settlements towards the green limits. As a commitment to sustainable development goals, for SGD 3 in particular, India under the National Clean *Air* Action Plan has aimed to *reduce* particulate *pollution* by 30–35% by 2024 (NCAP 2019). Hence, past and futuristic policies, programmes and regulatory reforms speak high of India's commitment to reduce the air pollution levels for healthy living. Following are the key initiatives mentoring the air pollution reduction including the vehicular emissions:

- National Air Monitoring Programme (NAMP), 1984.
- Environment Pollution (Prevention and Control) Authority (EPCA), 1998.
- National Action Plan on Climate Change (NAPCC), 2008 to include.
  - National Solar Mission.
  - National Mission for Enhanced Energy Efficiency.
  - National Mission on Sustainable Habitat.
  - National Water Mission.
  - National Mission for Sustaining the Himalayan Eco-system.
  - National Mission for a Green India.
  - National Mission for Sustainable Agriculture.
  - National Mission on Strategic Knowledge for Climate Change.
- National Ambient Air Quality Standards (NAAQS), 2009.
- National Air Quality Index (AQI), 2015.
- Green Highways (Plantation, Transplantation, Beautification and Maintenance) Policy, 2015.
- Solid Waste Management Rules, 2016.
- Pradhan Mantri Ujjwala Yojana, 2016.
- National Urban Transport Policy, 2016.
- Graded Response Action Plan (GRAP), 2017.
- National Clean Air Programme (NCAP), 2019.
- Draft Guidelines for Setting up, Authorization and Operation of Authorized Vehicle Scrapping Facility, 2019.
- National Electric Mobility Mission Plan, 2020.
- Advanced Vehicle Emission and Fuel Quality Standards—BS-IV from 2017 and BS-VI from 2020.

- The India Cooling Action Plan (ICAP), 2021.
- India's Net-zero Target at COP26 Summit, 2021.

Apart from these policies and plans, few more initiatives have been taken by the Indian government to improve the air quality in Indian cities and reduce its contribution to climate change. Following are some of the key initiatives:

- Introducing gas as an automotive fuel in many cities.
- Introduction and enhancement of the metro-rail and bus-based public transport systems in selected cities.
- Introducing an energy-efficiency labelling programme for energy-intensive home appliances such as air conditioners.
- Notifying new stringent standards for diesel generator sets for standby power generation.

Specific to transport pollution National Urban Transport Policy, 2006, Advanced Vehicle Emission and Fuel Quality Standards—BSIV from 2017 and BS-VI from 2020, Plan to introduce a voluntary fleet modernization and an old-vehicle scrappage programme in India, introducing gas as an automotive fuel in many cities, and introduction and enhancement of the metro-rail and bus-based public transport systems in selected cities are notable.

3.6 Smart Urban Transport Options to Reduce Air Pollution

Though many initiatives are ongoing and few more are initiated to reduce the air pollution in Indian cities, but still the achievements are far, especially in the bigger cities. Following transport options are proposed for sustainable healthy living and well-being in Indian cities:

- Strive for **'walkable cities'** by developing self-contained communities and neighbourhoods and decentralized distribution of activities reduces the travel needs and trip lengths of their inhabitants. In the existing set up, there is an urgent requirement to develop high speed bulk capacity transit corridors following the **transit-oriented development (ToD)** principles. About ½ to 1 km area around the MRT/LRT/BRT stations should be planned for high density mixed land-use development to propagate walkability to the mass transport facilities. Such cities will be the healthy cities of tomorrow for the well-being of all.
- Introduce **world class transport system** to attract the private bikers and car drivers to meet their travel needs. Provision of world class transport infrastructure such as smooth roads with required markings and signs, smart terminals, and application of intelligent transport solutions, supported by reliable, safer, viable, and barrier free transport system shall be the key to modal shift from private to

public transport. This will bring the traffic densities down to a great extent and reduce the air pollution level in the city.

- Promote **green mobility solutions** in the cities by introducing **e-vehicles**, be it private or public. Though the efforts on the same are on but clear-cut policies and awareness programmes need to be brought forward to the public. Also, the necessary infrastructure such as electricity charging stations, preferably the solar panelled, should be evenly distributed in each city. **Non-motorized modes** are usually the most neglected components in the whole city transport scenario. That is why they are the most vulnerable from pollution and safety point of views. Cities are required to prepare green mobility plans that promote and protect the pedestrians and cyclists. Principle of exclusivity within inclusivity needs to be followed while proposing and developing the required infrastructure for them to make them attractive for usability. Violators must be taken to the task heavily under the tag of 'polluter to pay'.
- Introduce **innovative transport system management (TSM) measures** for different components of city transport, i.e. roads, terminals, modes, and sings and markings. Such an effort will improve the performance, efficiency, reliability and seamlessness of the transport system. Following are some of the suggested TSM measures proposed for the Indian cities:

  - **Introduce intelligent transport system** to control and regulate the traffic on city roads to bring in seamless flow conditions so that the congestion, jams and frequent delayed are minimized. Such an effort will lower down the air pollution level to a great extent.
  - **Introduce vehicle quota system,** i.e. restricted production of number of automobiles in a year to restrict rising traffic densities and hence the vehicular pollution load.
  - Introduce **off-peak car and week-end car scheme** to allow more cars primarily during off timings to ensure less congestion during working hours/days and more use of public transport to reduce air pollution levels. **Area Licensing Scheme** is another such mechanism to reduce traffic congestion and hence air pollution in the central areas of the cities, especially during peak hours.
  - Curb **double transfer of private vehicles** to remove old high fuming cars from the city roads.
  - High **road tax, excise duty and cess** on petrol or diesel to discourage the car/bike use.
  - **Road pricing scheme** in the congested areas and crucial arterial roads for augmenting the capacity of the existing roads and seamless flow of traffic. **Electronic road pricing** (ERP) should be introduced for faster clearance at the toll plazas to reduce vehicle emissions and energy losses.
  - **Encourage car-pool** wherever possible so that the number of private vehicles can be reduced, which in turn will reduce the vehicular emissions drastically.

- **Restricted core area and peak hour on street parking, multi-storied parking, and progressive parking charges** should be introduced to decongest the major roads, which will improve the area/city air quality.
- Introduction of **staggering of office hours** is expected to decongest the traffic conditions and hence reduce the vehicular emissions.

3.7 Conclusion

Rapid urbanization, that too skewed towards bigger cities, is the driving force to increased traffic densities, which in turn tend to raise the air pollution levels in the cities. Unprecedented growth of vehicles, particularly the two-wheelers, and use of poor quality fuel and indigenous engine technology have made the Indian cities the 'gas chambers', causing various health hazards. Hence, the residents of the bigger cities are living unhealthy quality of life. Different policies and programmes initiated by the government of India from time to time are the efforts to reduce country's contribution to air pollution and hence the climate change. Though all-round efforts are required from all the sectors contributing to the air pollution, but smart innovative actions in the transport sector are expected to bring down its contribution to air pollution level. Transit oriented development, world class public transport system, innovative transport system management and green fuel/engine technologies are expected to reduce the air pollution levels to a great extent to make the city life healthy.

References

Anenberg S (2019) Health impacts of air pollution from transportation sources in Delhi. ICCT: fact sheet India. https://theicct.org/sites/default/files/ICCT_factsheet_health_impact_airpollution_Delhi_ 20190705.pdf. Accessed 23 Dec 2019

Awaz YK (2017) 8 Facts about pollution in India that will leave you gasping for breath. https://www.youthkiawaaz.com

Bernard S, Kazmin A (2018) Dirty air: how India became the most polluted country on earth. https://ig.ft.com/indiapollution/. Accessed on 4 Dec. 2021

Clean Air Asia (2009) Sustainable urban transport in Asia 2009: a year in review. World Resources Institute, Washington, DC

Corfee-Morlot J, Kamal-Chaoui L, Donovan MG, Cochran I, Robert A, Teasdale PJ (2009) Cities, climate change and multilevel governance. In: OECD environmental working papers no. 14, 2009. OECD Publishing, Paris

EPA (2008) Inventory of greenhouse gas emissions and sinks: 1990–6006, U.S. Environmental Protection Agency. https://www.epa.gov/ghgemissions/inventory-us-greenhouse-gas-emissions-and-sinks-1990-2006. Accessed 4 Dec 2021

Faiz A (1993) Automotive emissions in developing countries-relative implications for global warming, acidification and urban air quality. Transp Res A Policy Pract 27:167–186. https://doi.org/10.1016/0965-8564(93)90057-R

Health Effects Institute (2019) State of global air 2019. In: Special report on global exposure to air pollution and its disease burden. Health Effects Institute, Boston. https://www.stateofglobalair.org
Hindustan Times (2019) Air pollution causes 12L deaths in India annually. Delhi worst off: Greenpeace, New Delhi. https://www.hindustantimes.com/india-news/air-pollution-causes
IHME (2016) The cost of air pollution—strengthening the economic case for action. University of Washington, Seattle
McKinsey Global Institute (2010) Urban world: mapping the economic power of cities. McKinsey & Company, New York. www.mckinsey.com/mgi
Ministry of Housing and Urban Affairs (2008) Report-01-May-2008—Urban Local Body. UMTA. Unified Metropolitan Transport Authority. V/C ratio. Volume–Capacity ratio. WSA. Wilbur Smith Associates. https://mohua.gov.in
Morain SA, Budge A (2012) Environmental tracking for public health surveillance, 1st edn. CRC Press, Boca Raton. http://www.enotes.com/public-health-encyclopedia/climate-change-human-health
MoRTH (2020) Road accidents in India-2019. Ministry of Road Transport and Highways, Government of India, New Delhi. https://morth.nic.in/sites/default/files/RA
MoSPI (2019) Statistical Year Book India. Ministry of Statistics and Programme Implementation—Reports. https://mospi.gov.in
MOUD (2014) Ministry of Urban Development annual report 2014–2015. https://mohua.gov.in
Mukta P (2017) India's first early warning system in Ahmedabad aims to reduce health impact of air pollution. Hindustan Times, New Delhi
NCAP (2019) National clean air programme. Ministry of Environment, Forest and Climate Change, Government of India. NCAP_Report.pdf (https://moef.gov.in)
Pandey VC (2012) Population education. Isha Books, New Delhi
Sen S (2019) Worst car density in India worsens, Mumbai now has 530 per km. Times of India. https://timesofindia.indiatimes.com/city/mumbai/worst-car-density-in-india-worsens-mumbai-now-has-530-perkm/articleshow/72117782.cms
Sharma A (2018) New trends in mobility and implications for OEMs. Auto Tech Review https://autotechreview.com
Sharma M, Dikshit O (2016) Comprehensive study on air pollution and green house gases (GHGs) in Delhi. Indian Institute of Technology Kanpur, Kanpur. https://vindhyabachao.org/embeds/thermal/IIT_Kanpur_Delhi_Air_Pollution_2016.pdf
Shrivastava RK, Saxena N, Gautam G (2013) Air pollution due to road transportation in India: a review on assessment and reduction strategies. J Environ Res Dev 8(1):69
Sinha K, Varma A, Souba J, Faiz A (1989) Environmental and ecological considerations in land transport: a resource guide. In: The World Bank Policy Planning and Research Staff, Infrastructure and Urban Development Department. Technical paper (Report INU 41). The World Bank, Washington, DC
The Economic Times (2019) Majority of air pollution deaths in India linked to diesel vehicle emissions: study. https://auto.economictimes.indiatimes.com/news/industry/majority-of-air-pollution-deaths-in-india-linked-to-diesel-vehicle-emissions-study/68186796
Thornton A (2019) 7 of the World's 10 most polluted cities are in India. World economic forum. https://www.weforum.org/agenda/2019/03/7-of-the-world-s-10-most-polluted-cities-are-in-india/

UNCTAD (2019) E-handbook of statistics, development statistics and information branch. UNCTAD, Cham

United Nations (2015) Transforming our world: the 2030 agenda for sustainable development. United Nations development program. Transforming our world: the 2030 Agenda for Sustainable Development | Department of Economic and Social Affairs (un.org)

United Nations Children's Fund (UNICEF) (2016) Clear the air for children the impact of air pollution on children. www.unicef.org/environment

World Health Organization (2018) Air pollution and child health: prescribing clean air. World Health Organization, Geneva

Chapter 4
Importance of Climatological Contributions in the Green Infrastructure Designs, Sustainable City Planning Towards Better Urban Settlement

Vinars Dawane, Saleem Ahmad Yatoo, Satish Piplode, Sunil Kumar Patidar, Vaishali Joshi, Adil Muhammad, Surabhi Agarwal, and Pankaj Kumar

V. Dawane (✉)
Department of Microbiology and Biotechnology, Sardar Vallabh Bhai Patel College, Mandleshwar, India

S. A. Yatoo
Department of Environmental Science, Faculty of Sciences, Government Degree College Baramulla, Baramulla, India

S. Piplode
Department of Chemistry, SBS Govt. P G College Pipariya, Narmadapuram, Madhya Pradesh, India

S. K. Patidar
Department of Chemistry, Shri Neelkantheshwar Govt. Post Graduate College, Khandwa, M.P., India

V. Joshi
Department of Chemistry, Govt. P G College, Manawar, Dhar, Madhya Pradesh, India

A. Muhammad
Department of Chemical Engineering, Government Engineering College Kozhikode, Kozhikode, Kerala, India

S. Agarwal
Indian Institute of Management (IIM), Bengaluru, Bengaluru, Karnataka, India

P. Kumar
Integrated Regional Office, Ministry of Environment, Forest and Climate Change (MoEFCC), Govt. of India Saifabad, Hyderabad, Telangana, India

Department of Environmental Science, Parul Institute of Applied Sciences, Parul University, Vadodara, India

B. Pathak, R. S. Dubey (eds.), *Climate Change and Urban Environment Sustainability*, Disaster Resilience and Green Growth,
https://doi.org/10.1007/978-981-19-7618-6_4

4.1 Introduction

Environment has always remained a critical criterion for the human lifestyle/standards (Živković 2020), cultures (DeMenocal 2001), and population settlements (Tomita-Yokotani et al. 2009) since the civilization of human societies and before (Hannon and Bradshaw 2000; Zhang et al. 2005; Bodley 2011), thus provided opportunities and restrictions for settlement issues (Knottnerus 2005; Nolon 2021; Wu et al. 2021). Very diverse kind of variation of settlements are present in worldwide today, according to the human requirements, capabilities, and objectives, domains of social, political, and economic associations (Mutizwa-Mangiza et al. 2011) as well as environmental domains (Mcbean and Ajibade 2009). The surrounding environment, climate conditions, and landscape have been one of the most important issues and directly associated factors with the human settlement problems and solutions (Hales et al. 2007). These factors, thus, influence human settlement and their way of living. Gradually, as human population increased, they changed their surroundings according to need and requirement which gradually and consequently changed the climate at local, regional, and global level, and significantly accelerated with the formation of more urban areas and cities. According to reports, in urban areas percentage of people will rise from 50% in the year of 2010 to nearly 70% by year of 2150 (United Nations 2018; Sturiale and Scuderi 2019). It was the first time in history, the population living in cities has exceeded the population that lives in the villages/countryside or away from these populated centres. Increasing human population have certainly led to increase in settlements and thus consequently results in environmental degradation, pollution, and decrease in green spaces in urban areas (Yatoo et al. 2020).

The 2030 Agenda for Sustainable Development "Transforming our world" which also means our cities including the places associated with the urban regions should be well-organized and sustainable in design (United Nations 2015). This 2030 agenda establishes specific 17 eye-catching goals and 169 target to achieve within next 15 years. Among 17 goals, one goal, which is the goal number 11, is solely focuses on about "sustainable cities and communities" and its aim is to "make settlements inclusive, nonviolent, resilient and justifiable (Sturiale and Scuderi 2019)".

Moreover, proper maintenance of green spaces and approaches for green infrastructure development again suggested by UN Agenda to 2030 (United Nations 2015). In this context of sustainable development and UN agenda 2030, climate change remediation and reduction in green spaces need to be addressed and enhanced through adaptation of green infrastructures (GIs) on sustainable basis in urban areas. Besides this, protecting, conserving, and enhancing forest patches and green spaces like parks, gardens, and street trees (Yatoo et al. 2020) through the improvements of the infrastructures and thus actually being "demand areas for Ecosystem Services". As per IPCC 2014 report, care of green spaces is among exceptionally good tactics (IPCC 2014).

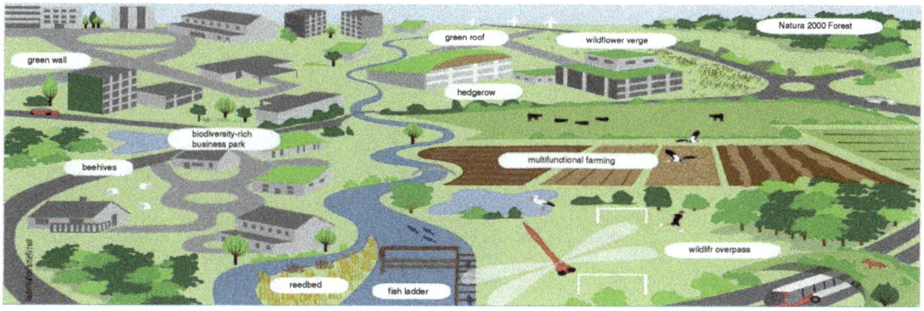

Fig. 4.1 The green infrastructures in different forms [adapted from EU 2016 (European Commission 2016). *Image Courtesy: European Union 2013*]

Fig. 4.2 Green storm water infrastructure in Philadelphia (USA). (*Image Courtesy: Philadelphia Water Department*)

Adaptation of green infrastructures (GIs), preservation and renovation of green roofs, green walls, urban forest cover, rain gardens, urban agriculture (like gardens, community gardening and green spaces, and agricultural parks), river bank parks, artificial/constructed wetlands, small alternative/renewable energy farms, maintenance of nature conservation reserves (Fig. 4.1) are among the most better inputs in sustainable city and better urban arrangements (European Commission 2016; Sturiale and Scuderi 2019). Moreover, while framing of urban plans and guidelines for sustainable city, such ideas and recommendations can be efficiently inculcated around buildings/institutions.

GIs are spreading across the world for at least 10 years due to their benefits and have capacity to resist climate change in urban spaces. Some of the interesting examples of GIs designs adopted at global level by some countries are in USA, most cities have adopted specific GIs plans including New York, Chicago, Washington, etc. Moreover, Greenworks Philadelphia have planned to convert Philadelphia into "the greenest city in USA" and have made a comprehensive list of GIs (Fig. 4.2) measures to be undertaken (Nowak et al. 2006; Economides 2014). Even in India, Bhopal, Madurai, Bangalore, and Delhi, city development plannings are having remarkable visions towards the green urbanization and various plans are already initiated. Other eco-designs adopted in urban infrastructures are "the vertical forest in Milan (Fig. 4.3); presence of Green Belt in Turin; green ring of the municipality of Mirandola (Modena); gardens in Catania (Sturiale et al. 2020)."

Fig. 4.3 Vertical forest in Milan (Italy). (*Image Courtesy: Boeri Studio (June 13, 2016)*)

Fig. 4.4 Germany's green roofs. (*Image Courtesy: Treibhaus Landschaftsarchitektur and Mathias Friedel*)

Besides above projects and managements of GIs designs, green planning in cities has been initiated in Europe and Asia, in Copenhagen (Caspersen and Olafsson 2010), Berlin (Kabisch 2015) (Fig. 4.4), Hong Kong (Jim 2002), Beijing (Yang et al. 2005), and in Nanjing of district Pukuo (Wei et al. 2018) and is preferred to be adopted in other countries.

Thus, the eco-friendly structures, sustainable settlements, and infrastructures have been increasing day-by-day and gaining popularity among various metro city plainings. Here, the purpose of this chapter is to highlight these shifts and prove a strong understating towards the climatological inputs in present/future infrastructures.

4.2 Understanding the Rural/Urban Human Settlement and Climatological Inputs

The human settlement can be defined as a place where human habitation occurs thus being an important aspect of each and very policy related to the stable human inhabitability and associated environment. Therefore, it has been referred as to the entirety of human civic with all the societal, physical, administrative, spiritual, and ethnic elements which sustain the human settlement. Any form of human residence,

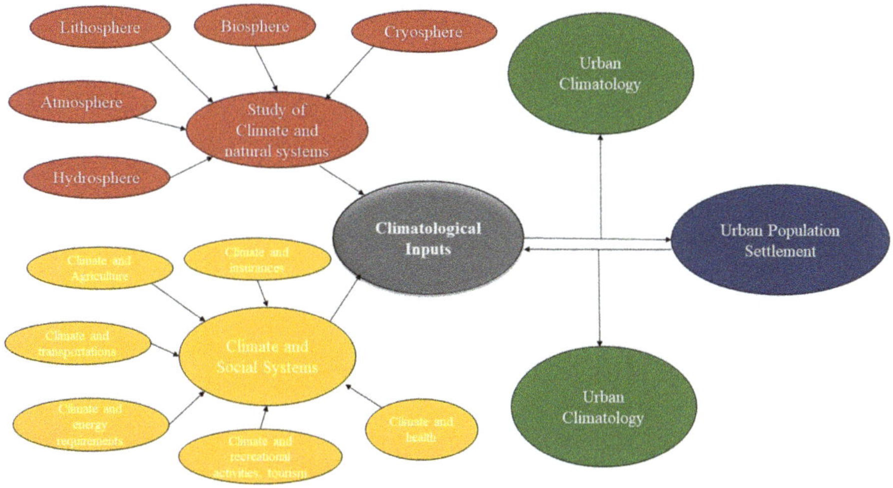

Fig. 4.5 Urban climatology and various climatological inputs

from the smallest household to the major city, where the community of humans exist in and follow their life objectives, can be assumed as settlement.

Human settlements originate in numerous arrangements (i.e. structures such as three-dimensional organization and alignment of elements of buildings) and/or forms (i.e. physical forms including built-up zones, including the figure, magnitude, compactness, and configuration of settlements) and it can be everlasting and impermanent, rural and metropolitan, moveable and inactive/stationary, dispersed and agglomerated (Fig. 4.5). The habitable land has been powerfully interlinked with the influences of climates and extremities of climate existing nearby and thus beforehand living there or making a prearranged population condensed settlement one should recognize to accomplish the environmental issues and climatological contributions to gain a better habitability and future strategies for their managements.

Diverse countries have their own standards to distinguish settlements as rural or urban, which costume their own circumstances and necessities. Some of the common bases for sorting are size of the population, work-related configuration, and administration. Understating this classification is important to know the policy and a working plan for a settlement along with to implant the geographical and environmental influences. In this context, the rural settlements have been closely and/or directly connected to the indigenous natural resources. The size and the occupational point of views have been significant criteria for the habitability in the quantitative sense along with the management of environmental influences like various primary actions such as agriculture, animal husbandry, fishing, mining, and forestry.

The place of settlement where some one wants to live is important such as sea coastal or river line peoples along settlement of fisheries, farmers settlements in the fields, and plains nearby riverine banks, which has been rural prospects, while urban-dense settlements are types of settlement that has been nodal in personality and non-agricultural prospects. The industries, trades, and inter-connected services have been associated with the economy and occupation. Thus, they have been dense as compared with rural settlements due to polycentric urbanization and metropolitan development (Wei and Knox 2015). In this sense, the climatological inputs may vary but more required in the urban places like big towns, city, or bigger new platforms like metro cities or megacities.

The climatological inputs for planned or eco-friendly—intentional construction development have been a complex matter and planning this at a larger platform is a critical challenge. The understanding of a suitable climatological input has always been comparative and place dependent or conditional aspect to analyse and understand. Climatological inputs thus, an interdisciplinary or combined input, have come as amelioration from various domains such as the archaeological prospect, chronological dynamics of geochemistry and/or geology, geography natural resources sciences along with life sciences, sedimentological and paleontological inputs, remote sensing, GIS, and other interconnected issues.

Thus, the field of urban climatology is associated with the various domains of environmental aspects such as the use of interlinked natural resources, their effects on natural resources, their sustainable uses, their missuses, scale of uses–misuses, generated environmental complications such as pollution, etc. (Coutts et al. 2010). Thus, the idea highlights comprehensive faces of the impacts of urbanization and/or urban areas and links meteorology with physical, chemical, and biology frontiers of associated climate conditions and their associated issues/impacts (Mills 2007).

4.3 Social, Economic, and Environmental Benefits of Climatological Inputs in City Design

Enormous rapid rate of increase in urbanization and increasing population living in urban settlements have been created the need for these settlements to be sustainable cannot be neglected further. The confluence of climatology and city designs found to be essential as it has to bear the benefits across the spheres of society, economy as well as environment, disqualifying the usual debate wherein development and environment have conflicted against each other as trade-offs. From such issues, green infrastructures (GIs) and sustainable infrastructure contribute in accruing these benefits. The integrated approach has been considered to get effective results from the climatological inputs towards the city planning (Heidt and Neef 2008). The important benefits can be social (Zhou and Rana 2012), economic, environmental, and/or interdisciplinary (De la Barrera et al. 2016; Fallmann and Emeis 2020).

4.3.1 Social Benefits

While some social benefits accrue directly, some are nested. These occur in the form of ecological services through various functions that help adapt cities to be more habitable and suitable for healthy living (Demuzere et al. 2014; Hegetschweiler et al. 2017). This happens in the form of purification of air, moderation of temperature, mitigation of run-off from water bodies, relieving of heat waves, protection from solar radiation, and even noise reduction. Many of these functions in turn ensure better health and well-being of urban dwellers (Kourtit et al. 2020). Physical health improvements happen in the form of better respiratory systems, better capability to indulge in physical activities, and an increase in life expectancy. It improves health recoveries in the hospital and at home, with overall enhanced health levels. It can help alleviate the growing trend of non-communicable or lifestyle diseases caused by sedentary lifestyles (Leon 2008; Sugar and Kennedy 2021). The impact is not only limited to the current population but also extent to incoming generations in the form of better childhood development—both physically and mentally. Provision for healthy in-person interaction in green open spaces between children helps spur holistic development (Crane et al. 2021).

This has also been helpful to reduce prevalent health inequality existing due to socio-economic differences. Even in the 2020 pandemic, slums and informal settlements were emerging as hotspots for COVID-19 with billions of people facing severe risk due to sub-standard living conditions and overcrowding (Corburn et al. 2020; Patel and Shah 2020). Once addressed with inputs in city design, the ripple effect of this is multi-fold, leading to a step towards social equality and social inclusion. This also helps tackle issues pertaining to poverty and aids in lowering crime rates in the city (Vardoulakis and Kinney 2019; Chigbu and Onyebueke 2021).

City designs with climatological inputs account for spaces, which facilitate social interaction and building of strong communities (Chu et al. 2016). There has been engagement and belongingness within the community and with nature. Community building along with physiological well-being also has been influenced the psychological health. There has been improved levels of mental health, reduction in stress, and a subsequent decline in self-induced crimes and suicide rates. Incorporating climatology has been facilitated clean, safe, and affordable mobility solutions with efficient routes and sustainable transportation. An improvement has been seen in water quality—a growing urban concern with population explosion and ad-hoc settlements along with a strong management of drainage and solid waste controlling in a sustainable manner (Vardoulakis and Kinney 2019). All of these benefits mentioned above point to a better, hassle-free living experience and an overall improvement in quality of life for individuals, communities, and the society.

4.3.2 Economic Benefits

A city offering better quality of life, planned use of resources, and numerous other social benefits directly translates into increments in land value (Stevenson et al. 2016). Property rates of residential and commercial properties witness a huge surge (Anttiroiko and Caves 2014). This also attracts massive investment in the city.

Tourists are drawn to safe, efficient, climatologically planned cities, as there is better quality of tourism opportunities, recreational services, and easy mobility (Straupe and Liepa 2018). An efficient well-functioning public transportation system benefits not only tourists but helps improve unit economics and benefits to public dwelling in the city (Pan et al. 2018). Moreover, efficiency in many aspects of living provides economic benefits like smart cities and smart homes, which enable low energy use. It helps to avoid costs which are otherwise additional burdens like that of solid waste management and air, water purification (Straupe and Liepa 2018).

Community spaces are also a source of additional revenue and provide scope for economic value addition through the likes of parks and playgrounds. There is scope for economic value through biodiversity preservation and use of innovative alternatives instead of coarse destruction like vertical forests and sanctuaries (Admiraal et al. 2013). There is huge economic loss when resources, which provide value, are left unutilised—like rainwater in the absence of rainwater harvesting systems or solar and wind energy. Apart from the immense social loss, the direct economic cost of natural disasters was US$268 billion. This can be mitigated and salvaged with careful city planning with a foundation of climatological inputs.

Incorporation of climatological inputs ensures expedited planning approvals with reduced red tapism. This provides the way forward to increase ease of doing business and invites more investment into the city. There is also benefit from enhanced job creation in urban climatological planning. This also extends to implementation of such inputs which most often require interventions of latest technology and hence human capital (Valenzuela-Levi and Abreu 2021).

4.3.3 Environmental Benefits

Urban climatology is an essential tool for climate action. It helps and efficiently slow down the impact of climate change. The 2021 Emissions Gap report shows the alarming global temperature upsurge of 2.7 °C by the finish of the century, leading to catastrophic variations to the planet's climate. The world needs to split yearly greenhouse gas (GHG) discharges in the next 8 years. Green practices with green roofs and walls reduce GHG emissions, help regulate Urban Heat Islands (UHI), and subsequently keep global warming in check. Asia has seen a loss of thousands of people and hundreds of billions of dollars while heavily burdening the ecosystem and infrastructure because of extreme weather conditions/events. Climatology and its inputs enable better, stable and more predictable weather while also equipping the city better to respond to such events (Cities Alliance 2007).

There is merit in regeneration of land that has undergone development and concretisation to regain some of its lost value. This also leads to restoration of ground water and improves the water table (Ronchi et al. 2020). Similarly, forest cover can be restored in urban cities and be a source of a multitude of other benefits. Another regulating service of such urban ecosystems is that providing protection from natural disasters. Preserving wetlands and mangrove forests ensures buffers for flood protection and facilitates high carbon sequestration. This also fosters a rich biodiversity and harbours flora and fauna. Climatological inputs can help increase the population of endangered species and create an environment more suited for "coexistence".

Thus, urban climatology and its inputs positively impact the interlinked social, economic, and environmental domains in urban city design. There is also an overall enhancement of "ease of living" and a sense of collaboration and ownership in the citizens of the city—facilitating higher sustainability.

4.4 Selected Case Studies on Climatological Inputs in Sustainable City Design and Urbanization

4.4.1 Indian Case Studies

India is the seventh largest country on this planet along with approximate of 1,269,219.34 square miles area and very diverse environmental conditions and geographical landscape. The Indian Ocean in southern part, Lakshadweep around west followed by Arabian Sea in the west, Bay of Bengal in the east as well as northern parts conjugated with Middle East of Asian including desert and cold regions have been created a significant importance of climate variability studies. The important geographic regions including Indo-Gangetic plans, Thar Desert, various Islands, and The Himalayas and associated regions, coastal plans, and Ghats as well Peninsular Plateau have been important merger of settlement planning's and city building strategies and critical factors for present/future urban city planning under the shed of climatological point of view (Fig. 4.6).

Very important cities in India including Bhopal, Madurai, Bangalore, and Delhi have been implementing the green-blue infrastructures, eco-friendly urban settlement strategies, and action plans by their respective Municipal Corporations and Smart City managements to maintain the green covers and develop environmentally sustainable city, and improve the overall health. Various recent studies have been reported to highlight the significant attention on the issues regarding climatological inputs towards the urbanization in India. In this regard, Sheikh (2018) worked on city planning with special reference to Jaipur City and discussed the proper city planning issues. Jaipur situated at the central part of India and capital of Rajasthan

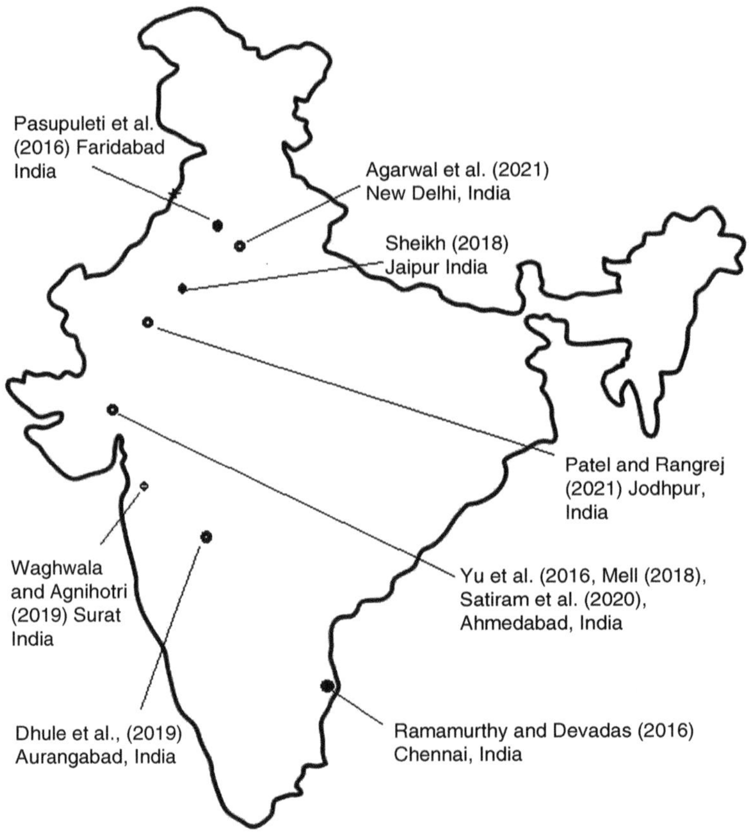

Fig. 4.6 Some locations of selected case studied and the urban climatological projects cited here with the understanding of the geographical position in the map of India

state. It is known as "Pink City". They reported that improved electricity and water quantity, better quality sanitation, and reutilizing of waste, appropriate transportation, and vehicle management schemes are the main green infrastructure component for the sustainable city planning towards better urban settlement. Combination of these components makes a green and clean city. In the present study, Jaipur city selected as model to evaluate the important of these green infrastructure components for urban development. Challenges, problems, and solutions of these components have been discussed with special reference to Jaipur City. They reported that water shortage, air pollution, water pollution, loss of biodiversity, environment degradation, and migration are the main problem in the Rajasthan state. Various types of green infrastructure strategies have been provided the solutions of these problems. Smart buildings, intelligent transport systems, smart water system, smart grids, smart governance, smart health care, smart education, and smart security systems are the solution of above problems. It also deals the climate change and natural disaster management through green infrastructure. They reported that problem of solid waste can be solved through proper collection, identification, and application/disposal of solid waste. Urban environment or economy problem can be solved using proper development of industrial and residential area. Population problem can be solved using proper development of residential area. Greening Commercial Corridors and City Parks Revitalization may be used for green infrastructure. Operative drainage structures, sewage schemes, transportations, aquatic lines, rooftops, raised porches, and solider practicalities are the important point for slum redevelopment in Jaipur city (Sheikh 2018).

In another study related to the Rajasthan state area, Patel and Rangrej (2021) worked on urban city planning with special reference to Jodhpur city. Jodhpur belongs to the Rajasthan state. It is situated on the Thar Desert of India and northwest of India. In the present study city planning of Jodhpur city has been considered with keep in mind climate change adoption and disaster risk management. They reported importance of green infrastructure for sustainable city development. Study concluded that green infrastructure could improve the healthy life as well as increase the beauty of the city. In the present study, campus of IIT Jodhpur has been used as model. They highlight some important feature of green campus to avoid climate change and natural disaster including smart buildings in campus, plantation, rainwater harvesting, and sewage recycle treatment. These suggestions already applied in the IIT Jodhpur campus to prevent from climate change and natural disaster (Patel and Rangrej 2021).

In an important study associated with Gujarat state, Mell (2018) worked on green infrastructure development in Ahmedabad city. Ahmedabad belongs to the state of Gujarat and largest city of Gujarat. It is located in the western part of India. Ahmedabad city identified as largest urban area and commercial centre of Gujarat. Ahmedabad Urban Development Authority (AUDA) and Ahmedabad Municipal Corporation (AMC) two most important agency worked on the planning of the Ahmedabad city for sustainable development. All the discussions and suggestion were related to increase eminence and effectivity for the city's environment of Ahmedabad city for the development of Ahmedabad city have been briefly reported. Various important suggestions including improvement of Sabarmati Riverfront, Kankaria and Vastrapur Lakes in Gujarat have been identified by these two working agencies. Detailed planning of these water bodies with limitation has been also discussed in present investigation. Many other suggestions related to including street tree, urban grooves, and meanwhile space have been also discussed. They reported that these suggestions have been easily adopted for Ahmedabad city. The city planning easily prevents climate change in the study area (Mell 2018). Another study with same place, Satiram et al. (2020) also worked on green infrastructure implementation with special reference to Ahmedabad city. In the present study, importance of green infrastructure (GI) has been highlighted with special reference to ecological sustainable development. They reported that development of Vastrapur Lake, plantations along Nehrunagar—Shivranjini road, recreational space and social hub, and lake conservation project at Kankaria are the main projects which have been used for the sustainable city planning. These projects can prevent the climate change and natural disaster in the areas of Ahmedabad city. Various key challenges to apply these suggestions in the Ahmedabad city also discussed in the present investigation. They also reported the key factors to improve these suggestions (Satiram et al. 2020). Similarly, Yu et al. (2016) worked on green infrastructure and disaster resilience in Ahmedabad. In the present study slum, upgrading strategies have been discussed with special reference to Ahmedabad. Current study reported infrastructural strategies for the prevention on natural disaster. Protection from water leakages, individuals cleaning community drainage problems, storage of livelihood goods, use of fans and shade, structural adaptations for earthquakes are the main point discussed in the present study. They reported that effective management of the systems and cycles needed for disaster management in slum area of Ahmedabad (Yu et al. 2016). Again, in a case study related to the climates and their influence in Gujarat state region city planning, Waghwala and Agnihotri (2019) worked on flood risk management in Surat city using green infrastructure. Surat belongs to the Gujarat state and situated at the western part of India. It is placed at the mouth of Tapi river and having large seaport. It is identified as one of the largest urban areas of western India. In 2006, flood accrued in the Surat city and more than 80% Surat city was affected by flood. After that various green infrastructure development has been done to prevent the Surat city from flood. They reported that there is a dramatically change in percentage of urban area, water bodies area; vegetation area and open space area have been observed from 1968 to 2006. They reported that percentage of urban area and vegetation area have been increased from 1968 to 2006. In addition, percentage of water bodies' area and open space area have been deceased from 1968 to 2006. These are the main reason behind 2006 flood in Surat city. After 2006, a lot of work have been done in the field of flood risk management. Various planning have been done to reduce the flood risk. Surat city planning has been considered in the direction of sustainable development. Various suggestions for flood mitigation and risk transfer have been discussed in the present investigation. They reported that under green infrastructure planning Khar land, wasteland and existing pond were used for flood resilience in Surat city. New ponds also developed to divert the floodwater. Various bank creeks present at the left and right of Tapi river also have been used for the diversion of floodwater. These suggestions and planning 50% reduce the risk of flood and resiliency of Surat city (Waghwala and Agnihotri 2019).

In a significant study on northern part of India and city plaining, Pasupuleti et al. (2016) worked on green infrastructure for Faridabad city. Faridabad belongs to the Haryana state of India. Faridabad is the part of National Capital Region of Delhi and Delhi Metropolitan Area. It is most populated city of Haryana and situated at the North West region of India. This study deals with the relation between green infrastructure and smart city. Study examined the possibility of various component of green infrastructure including urban transportation, solid waste management, and urban environment for the climate change and disaster management in the Faridabad city along with convert it into smart city. Paper discussed various problems and their solutions related to urban transportation, solid waste management, and urban environment in brief. They reported that problem of solid waste can be solved through proper collection, identification, and application/disposal of solid waste. Urban environment or economy problem can be solved using proper development of industrial and residential area (Pasupuleti et al. 2016).

In a study associated with southern part of India and city plannings, Ramamurthy and Devadas (2016) worked on sustainable urban development strategy of Chennai City. Chennai is situated at the eastern part of India. It is the capital of Tamil Nadu state. It is also one of the metropolitan cities in India. They reported main factors of the city planning including clean energy, mobility, buildings, natural resources management, pollution prevention, population growth, and green services. These factors can support the climatically, socially, and economically sustainable urban development. Various aspects of these factors have been discussed in brief. They also discussed the problem and solution for city planning related to these factors. They reported that green and clean energy including solar or wind energy may be used for sustainable development. Various types of natural resources may be used to deal with climate change and natural disaster management. Problem of solid waste can be solved through proper collection, identification, and application/disposal of solid waste. Population problem can be solved using proper development of residential area. Greening Commercial Corridors and City Parks Revitalization may be used for green infrastructure (Ramamurthy and Devadas 2016).

In a very recent study in the geographical heart of India, Agarwal et al. (2021) worked on sustainable city planning with reference to New Delhi. New Delhi is the capital of India. It is situated at the north part of India. This paper deals with the green infrastructure problem and solution in New Delhi. Various principles of green urbanization and sustainable development have been also discussed. Finally, it represents the sustainable city model for city planning towards better urban settlement. Urban sprawl, housing, squatter settlements, environmental concern, transport, water crisis, trash disposal are the main problems/issue/challenges of the urban green infrastructure. Due to this issue, climate changes and natural disaster have been observed in the particular region. All the problem mentioned above resolved by the green urbanism. Green urbanism worked on triple-zero frameworks principle. "Zero" fossil-fuel energy, "zero" waste and "zero" gas emissions are the main feature of green urbanism (Agarwal et al. 2021). Similarly Dhule et al. (2019) worked on sustainable city planning strategies in Aurangabad city. Aurangabad city is situated at the western part in Maharashtra state of India. They reported that green infrastructure could provide digital technology for the development of sustainable city. Due to the use of green infrastructure city, become smart city. Quality and performance of services have been enhanced using green infrastructure in city. Smart buildings, intelligent transport systems, smart water system, smart grids, smart governance, smart health care, smart education, and smart security systems are the main features of any sustainable or smart city. In Aurangabad, city excellent works have been done for the management of issue related social and economic development. Water supply system has been improved using various water resources including Jayakwadi dam, Neher-e-Ambari, and Harsool dam. Domestic water effluent also treated on sewage treatment plant situated at Salim Ali Lake. Study concluded that overall technological, economic, and environmental changes used for

Aurangabad city planning have been excellent examples of sustainable city planning towards better urban settlement (Dhule et al. 2019).

Thus, the above studies highlighted the importance of climatological inputs in the urban climatology and their benefits in India.

4.4.2 United States of America (USA)-Based Case Studies

USA is the third largest country on this planet along with approximate of 3,794,100.4288 square miles area and very diverse environmental diversities. Varied topographies and different geographical regions made a very dynamic climatological equations in USA thus as a reflection the climatological insides are also very different and dynamic. The major regions included various interior highlands and plans, intermontane plains and plateaus, mountain regions including Pacific and Rocky Mountains, Appalachian Highlands, Atlantic plans, and Laurentian Upland made a diver habitability in USA. That have been reflected in the urbanization patterns, types, and the settlement issues as well as critical factors for present/future urban city planning under the shed of climatological point of view (Fig. 4.7).

Various recent studies have been reported to highlight the significant attention on the issues regarding climatological inputs toward the urbanization in the USA. In this regard, Shandas (2015) has studied the impact of environmental management and neighbourhood change on green infrastructure for storm water in Portland city, Oregon, USA. Portland city is a city of Northwest US Pacific. Portland is one of the rainiest cities in Oregon, with 94 cm/year of rainfall, and thus there are huge storm water challenges for Portland's landscapes. In many cities, wastewater and storm water infrastructure has become practically pathetic and needs to be improved. In this study, Tabor to River (T2R) program, has been added which was a green infrastructure project. It has been applied on 5.6 km^2 in the southeast quadrant of the metropolitan. This project has focused on many physical fluctuations to the land comprising the arrangement network of 500 vegetated storm water facilities. Project

Fig. 4.7 Some locations of selected case studied and the urban climatological projects cited here with the understanding of the geographical position in the map of USA

selected apart from more than 100 private storm water facilities including plantation of 3500 trees and restoration or supersede the 266 km of sewer pipes. This project also related to the behavioural study of residents of the area and a relation between how citizens got engage in stewardship of environment and green infrastructure and their workability. The results of this study revealed that there is a lack of willingness for behavioural changes among respondents irrespective of the biophysical alteration occurring in their neighbourhood, but physical changes can lead to motivate the social behaviour on urban environmental management, this reflection may help in planning to construct better approaches for green infrastructure development for storm water challenges. It has been reported that to conquer the environmental challenges and for green infrastructure and challenges due to climate change, it is essential to develop linking between human and environment (Shandas 2015).

In another and comprehensive, work by Lovell and Taylor (2013) on providing urban ecosystem facilities through poly-functional green infrastructure in the United States. This work has contributed to a multiscale approach in US for participatory green infrastructure planning process from the neighbourhood to the city and region to get a proper lead on green infrastructure. In this piece of work, they reported that there has been a critical need for developing the infrastructure, which can help human and nature from the uncertain and frightening future calamities, changing climate, water scarcity, food insecurity, and energy crisis. Work cited to scrutinize the probabilities to develop this eco-sustained infrastructure in cities that could allow citizens to adapt and transform to more anticipated expansion trajectories. They observed that landscape multi-functionalities have been initially applied on agricultural settings and have found opportunities to develop green infrastructures on urban landscapes. It was found that in many cases just a small-scale grass root project might lead to a broader transformation. To embolden the communities to work on green infrastructure projects and involving the residents in planning the green infrastructure, multifunctional landscape may help in promising outcomes with visible benefits to city health (Lovell and Taylor 2013).

In a remarkable work by Meerow (2020) recently has studied the important dynamics of multifunctional green infrastructure plannings in New York City. This program for New York City manly focuses on water quality but other feasible benefits have also been targeted. This paper suggested that how the possible benefits can be prioritized, articulated and what would be their implications. The local stakeholders have been explored for their opinion about the comparative significance of all the benefits of green infrastructure. Furthermore, spatial multi-criteria analysis has also been performed. Trade-off and synergies between different criteria have been studied and set side by side with model priorities and existing green infrastructure locations. The results clearly suggest that for setting green infrastructure, priority areas may differ which is greatly affected by different potential planning criteria, for example, storm water, urban heat island. Among all the criteria, the most compelling positive correlation has been observed between storm water, city heat islet and air value. Another positive correlation has been observed between contact to green spaces and city heat island, apart from these, more socially accessible areas are positively correlated with those having greater air quality threats. These synergies give an appealing idea about which area should be chosen for implementing green infrastructure and what type of green infrastructure should be chosen, for an instance, large park is beneficial for a civic than a minor bioswale or tree pit. In short, planning criteria and resident opinions greatly affect the selection of area and type of green infrastructure, which is to serve the city health in many ways (Meerow 2020).

In a significant finding, Tilt and Ries (2021) have investigated limits and promoters influencing green infrastructure projects in minor towns, they have chosen Coos History Museum located in Coos Bay and George S. Birniein Park in city of LaGrande as case study sites, both are minor communities in the state of Oregon, USA. Coos Bay is a city located along Oregon coast with 64 inches per year of average rainfall. It has been developed from a fishing community to a tourist destination now with workshops, cafeterias, and other attractions; one of them is Coos History Museum. The utmost priority of city is to manage stormwater and specifically prevent the entry of polluted stormwater in bay and estuaries. LaGrande is the city located in eastern Oregon, USA. This city collects a regular precipitation of 16.54 inches per year. The city faces a speedy snowmelt with weighty spring showers, which has followed by seasonal summer. Such climatic condition flips challenges to handle stormwater run-offs before it combines with water bodies. Since the city is at higher elevation, its nearby areas are prone to face stormwater entry. In addition, city also faces occasional flooding due to high water table and natural springs. In this study, it has been reported that in smaller communities, green infrastructure projects and policies are being potentially affected with changing regulatory structure and climate. The implementation and sustenance of green infrastructure for small city have specific set of restraints like both unwieldy regulations and a lack of regulation restraint the implementation of green infrastructure. The small city which has to follow federal regulations was restricted by regulations specially limited staff. It can be concluded from the present study that close relationship with staff, property-owners, strong public relations, and a devoted finance authority are the catalysts that can promote an easy implementation and sustenance of green infrastructure (Tilt and Ries 2021).

In a comprehensive work, Cheng (2016) has studied Spatial Climate Justice and Green Infrastructure Assessment: A case study for the Huron river watershed, USA. The Huron river covers 7 countries in southeast Michigan USA and its watershed drains more than 2300 km^2 area with a population of 500,000 and includes cities like Ann Arbor and Ypsilanti. One of the focuses of this work is on the hydrological conjugation of green infrastructure for resolving flooding hazards and water pollution caused due to climate change in the cities of Waxon, Ann Arbor, and Ypsilanti in Huron river watershed. This paper reported that various important information could be drawn if climate justice hotspots can be examined for varied climatic conditions. This information play a vital role in city planning, to rank green infrastructure for climate change variation in the areas, which seek greater social benefits (Cheng 2016).

Another case by Mell (2014) has studied aligning disjointed development structures through a GI approach to metropolitan expansion in the UK and USA. Author reported that GI has willingly flourished with a theoretical and employment viewpoint, these result in a dual plot of understanding, one deals with the set of conceptual principles within wide global debate and the other focusses on the localized understanding of this principles within divergent delivery approaches. Due to this dual plot, a level of complexity arises in developing green infrastructure policy and succeeding investments. Changing green infrastructure policies, practice, and funding largely affects this process. This paper further reports that even though conceptual understanding of green infrastructure has been easily available in literature, but still a greater variability is obvious for applying these principles on a local scale (i.e. national, regional, and sub-regional) planning. This paper suggests that green infrastructure development needs the conceptual expansion as well as implementation. It has been reported that the dual plot of evaluation has both positive and negative features of green infrastructure planning, which highlights the difficulty in applying it on global and local development narratives for green infrastructures (Mell 2014).

4.4.3 UK and European Countries Location-Based Case Studies

The UK's climate changing day-by-day due to the growing of greenhouse gases discharge. The recorded hottest summer of England (in 2018) and the recent years flooding pattern indicate the impact of climate transformation in the cities of UK. The climate change continuously disrupting the everyday life of UK's people in the form of flooding, overheating, draught, polluted air, and contaminated water. The climate change is unavoidable, and the role of socio-economical is crucial in adopting the new climate changes. The appropriate mitigations are needed to increase adaptation capacities in a period. Herein some case studies have been reviewed from different cities of UK's to illustrate the importance of climatological inputs in the designing of sustainable/green infrastructure and city planning (Fig. 4.8). In this regard, Hathway and Sharples (2012) presented the role of City Rivers interactions in mitigating the Urban Heat Island (UHI) properties in their case study on the river Don, Sheffield, UK. The variation in weather conditions was also considered in their study. They concluded that in the ambient temperature (200 °C) condition, the average 10 °C cooling is reported over the river. However, during the hot days of May, this cooling enhanced to 20 °C over the river. They reported that the temperature changes only during the daytime rather than the night. They also highlighted the role of surrounding materials of river in air temperature changes. The highly vegetated banks are cooler than the well-engineered hard banks. They suggest to enhance the cooling effects of river to mitigate the effects of UHI as close consideration of urban designs. Therefore, the consideration of urban rivers in the urban designs play a crucial role to reduce the UHI effects (Hathway and Sharples 2012).

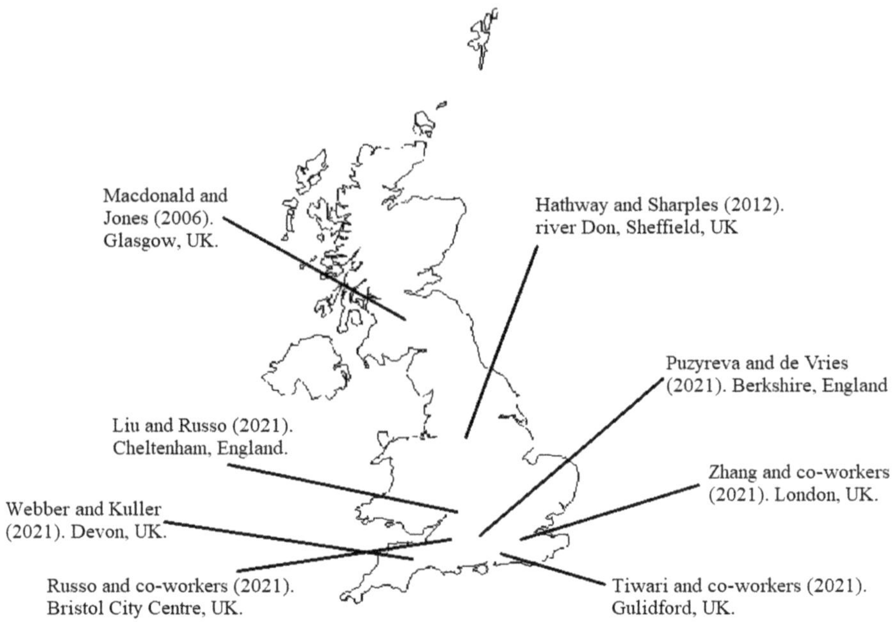

Fig. 4.8 Some locations of selected case studied and the urban climatological projects cited here with the understanding of the geographical position in the map of UK

In another noteworthy study, Liu and Russo (2021) explored the influence of Urban Green Space (UGS) in implementation of green infrastructure (GI) approach for urban Ecosystem Conditions (EC) and Ecosystem Services (ES). In their remarkable work, they used UGS components to implement GI through a case study of urban town Cheltenham, England. They used Mapping and Assessment of Ecosystems and their Services (MAES) approach for their studies and the results of urban ecosystem conditions (UEC) and urban ecosystem services (UES) were evaluated. They showed that for securing citizens health, maintaining of ecosystem service is a main option for adaptation as well as in sustainable city planning. They reported that the different GI strategies could be implemented by altering the arrangement and three-dimensional arrangements of city green space (Liu and Russo 2021).

In a very recent study, Tiwari et al. (2021) explored the role of green infrastructure in mitigation of urban heat island. A small-town area of UK (Guildford) was selected for a case study. They used ADMS-TH (temperature and humidity) model, existing green infrastructures [BASE-Green Infrastructure (BGI)] like trees, grassland, green roofs, and four hypothetical scenarios, namely (1) Hypothetical-No Green Infrastructure (HGI-No), (2) Hypothetical-Maximum Green Roofs (HGR-Max), (3) Hypothetical-Maximum Grassland (HG-Max) and Hypothetical-Maximum Trees (HT-Max) in their study. They also evaluate the effectiveness of various GIs in mitigation of UHI. The average temperature perturbation was highest for HGI-No and lowest for HT-Max and follow the order HGI-NO $>$ HGI-Max $>$

BGI > HGR-Max > HT-Max. They reported that the trees are best tools to mitigate the UHI among all employed GIs. The canopy and evaporations of trees enhance the effectiveness than the other GIs. However, if the large areas has been covered by the green roofs then they can be useful as a supplementary apparatus tool in extenuation of UHI (Tiwari et al. 2021).

Another fresh work by Russo et al. (2021) employed two web-based freely available i-Tree Canopy and the United Kingdom Office for National Statistics tools to evaluate the ecosystem services by taking an example of Bristol city centre. They emphasize on the role of such tools for calculation of ecosystem services and in the tree planting strategies. Both tools require little to no expertise knowledge to use it and having features of simplicity, fast, and trackability. The i-Tree Canopy tools consider only green coverage (trees), whereas the United Kingdom Office for National Statistics tools access all the environmental surfaces such as trees, water, and vegetation. The Bristol city is located on the Avon river; therefore, both the tools show different results due to the difference in calculation methodologies. They also explore the strength and weakness of such tools in planning of GIs and sustainable cities (Russo et al. 2021).

Another recent study by Zhang et al. (2021) investigated the role of city green space in mitigation of urban flooding in a case study of London, UK. The ecohydrological model along with 2018 UK climate projection UKCP18 was employed to investigate the capacities of urban green infrastructures for natural flood management under changeable climatic conditions. They found that the green infrastructure plays a crucial role in stormwater control and the run-off water is highest when vegetations are inactive. They reported that, for effective flood control, an integrated approach of grey infrastructure drainage system and the plant water stress should be employed. They suggested that, for enhancement of future flood management and storm water control system in London, there should be increased plantation by 50% along with the proper drainage system and adopting of green infrastructure management system (Zhang et al. 2021). Similarly, in this same year, Puzyreva and de Vries (2021) performed an ethnographic case study to examine the role of local communities and flood history in flood management in the county of Berkshire, England. The data of this case study is based on semi-structure interviews, observations, and it was qualitatively analysed using ATLAS program package using constructive strain of grounded theory. They reported that the local groups play a significant part in flood danger management. The local knowledge, practices, and expertise of communities help in flood management (Puzyreva and de Vries 2021).

In a noteworthy work by Macdonald and Jones (2006) showed in a remarkable work on sustainable drainage systems through a case study of Glasgow, UK. The flood management and storm water management especially in urban industrial city are challenging tasks. They reported that the Glasgow Strategic Drainage Plan (GSDP) which depends on the strategies of sustainable drainage systems (SuDS) helps to effectively resolve the city drainage problems. To abort the surface water entry in drainage system is a major task in this process. They highlighted that the

GSDP not only helps in flood management but also shows a key part in mitigation of water pollution from various sources (Macdonald and Jones 2006).

Again a fresh work by Casares et al. (2021) explored a case study on retrofitting of sustainable drainage systems (SuDS) based on the project Counters Creek Flood Alleviation Scheme, implemented in London Borough of Hammersmith and Fulham, and Royal Borough of Kensington and Chelsea in the UK. The adaptation of SuDS to resolve the flooding problems with cost effective manner and to protect the environment and communities is a challenging task. They noted that for flood risk management in selected sites of UK, the flips (drainage pumping device), local schemes for enhancing sewer capacity, and implementation of the SuDS play a major role. Rethinking about the implementation of four pillars of SuDS (quantity and quality of water, biodiversity, and facility) with innovative tools is helpful for flood management and beneficial for communities and environments (Casares et al. 2021).

In a recent as well as mind-catching work by Webber and Kuller (2021) used preliminary screening tools, SSANTO (Spatial Suitability Analysis Tool) and GIS-MCDA (Geographic Information System-Multiple Criteria Decision Analysis) for suitability of SuDS in surface water management system. They performed a case study on Devon, which is a largest county in England and located in South West of the UK. They noted that the screening tools are helpful for pre-identifications of SuDS opportunities and for the effectively implementation of water management strategies. They reported that in Devon the SuDS were successfully installed and benefited using these pre-screening tools (Webber and Kuller 2021).

A significant study by Emmanuel and Loconsole (2015) represented a significant role of green infrastructure adaptation for mitigation of urban overheating problem in the Glasgow Clyde Valley (GCV) region of UK. The case study of GCV was carried out using Local Climate Zone (LCZ) system. The Computational Fluid Dynamics (CFD) simulations were also performed to evaluate the utility of GI adaptation strategies. They reported that increasing 20% of the green infrastructure cover in the GCV leads to reduce the local surface temperature up to 20 °C. In order to achieve the 20% green cover goal in their study, they used Green Area Ratio (GAR) methods and suggested the use of roof garden, street tree, green walls, mini-parks for practical implementations. They highlighted the role of CFD simulations of green cover in mitigation of urban overheating and climate change. The thermal comfort and air temperature pattern were also discussed in the case study of GCV. They also concluded that the 20% increase of green surface could help to decrease the extra city heat island effect in 2050 by one-third to one-half (Emmanuel and Loconsole 2015).

The above-discussed case studies from the cities of UK have provided the significant information and important facts, which can be helpful in developing the green infrastructure models and in designing of new sustainable infrastructure systems in both established and emerging countries. The above case studies are based on the importance of GI in urban adaptation to climate transformation and the role of communities and technologies in achieving the societal, economic, and environmental benefits. The use of adaptation tools in mitigation of climate change has explored. The reducing greenhouse gas emission, reducing urban heat islands,

Fig. 4.9 Some locations of selected urban climatological projects cited here with the understanding of the geographical position in the map of European countries

and flood managements have been a current need of UK for development of sustainable cities in well-being of people.

The chapter has also discussed the selected case studies and working plans of selected European cities and specific green-blue infrastructure and the projects that influenced by climates (Fig. 4.9).

In a very comprehensive report work conducted by De Bellis et al. (2015) associated with Almada city project have carried out as among the first project case studies of related areas. The city named Almada, is a dense city of Lisbon Metropolitan Area, Portugal and among the one of tourist spot of Portugal attracts around eight million tourists to its coastline. Almada faces many natures' challenges like landslide, drought, rising sea levels, salination, flash floods, biodiversity loss, and forest fires due to climate change. An adaptive strategy has been developed for safer, healthier, and more resilient city. The Department of Sustainable Environmental Strategy and Management (DEGAS) has established in 1999 by the Almada municipality for the mitigation of environmental problems existed in the city. The "EU Cities Adapt" project provided additional funds in 2012, this helped to integrate the adaptive strategies further into the various other departments, projects, and plans. The development of strategies and its implementation are led by DEGAS, and selected stakeholders, including research groups in the universities and the Energy Agency of Almada, are invited for collaboration. They monitored impacts of climate change, assessed vulnerabilities, and modelled scenarios. The DEGAS has clear focus; its multi-proficient team become a key success for Almada. A multifunctional and multiscale approach was taken, the "urban resilience" and "ecosystem services" concepts are used for the climate change assessment and reducing its vulnerabilities.

This involves mapping and featuring various range of functions (potential and existing) to reduce the environmental impact, which can be achieved by providing green spaces throughout the city. In addition to this, funding, knowledge exchange, and capacity building by participating in EU Cities Adapt and other EU projects highly catalysed the efforts. A range of projects implemented based on these strategies resulted in incorporating adaptive measures across urban agriculture, mobility, reducing the heat island effect, and coastal restoration. Despite these, capital and social resource are measured as the important factors for the further development. Lowland seaside zones cities are at the danger of intensifying sea levels, which is associated with the chance of submergence, coastal erosion, and flooding. The possible solutions associated with this involve the preservation and reestablishment of seaside ecosystems, aggregate vegetation to manage the sand dunes (De Bellis et al. 2015).

Another remarkable and comprehensive study associated with Edible Edinburgh, Scotland city project. The Edible Edinburgh Sustainable Food City Plan was presented in 2014 by the City Council of Edinburgh, Scotland, with a vision for fresh food for everyone there, healthful, vibrant communities, and a better environment. A cross-sector collaboration resulted in the proposal. The goals have been decided to attain sustainable and resilient food gaining in the city's three chief public segment organizations, among other things, to develop the city's indigenous, self-governing food subdivision, support skills training, increase land obtainable for food creation, reduce the city's ecological outline, expand healthiness and security, and support communities and their affairs with food. Food was identified as a major societal issue during a survey for another strategic plan, "Sustainable Edinburgh 2020", in 2011, which provided the drive for the plan. To examine the prospects for a local food policy, a consortium of 12 organizations from the governmental, corporate, and public society subdivisions (including philanthropic, academics, eateries, and business groups) shaped. For the first year, monthly seminars were held where the goals, goals, activities, and techniques of functioning together were discussed. The alliance concluded at the end of the meeting that there were abundant synergies to officially approve "Edible Edinburgh". By the autumn of 2013, a shared visualization had been developed and was available for public comment. The final version took into account almost 400 suggestions. The consortium has self-governing property, but because it has headed by a councillor, it has tied to prevailing governance systems. Project had two major challenges: first, reaching a consensus on the initiative's goals and objectives with wide support from the diverse players and interests involved, and second, gaining imports from legislators and important organizations in endeavouring to encourage the administrative improvements. Both of these obstacles have already been surmounted; for example, the organization was successful in petitioning the leaders of Edinburgh and Glasgow City Councils to announce a combined pledge on food deficiency. These accomplishments were aided by a regulatory framework that encourages local governments to engage with other ministries and the non-profit sector, as well as the Community Empowerment Act's (approved in 2015) goal of increasing community input over expansion conclusions (De Bellis et al. 2015).

Another comprehensive study on De Bellis et al. (2015) has associated with Copenhagen city Cloudburst management plan. On July second of 2011 within 2 h, Cloudburst hits Copenhagen, resulted in 150 mm of rain drowned the majority areas under water up to 1 m, caused an approximately $one million damage in Copenhagen. This resulted in solution to mitigate climate change became most priority in Copenhagen. The consequences due to flood outdistanced the jurisdictional boundaries, demanding an assuredly collaborative effort should be established between various stakeholders like planners, engineers, utility providers, economists, politicians, investors, and citizens to integrate strategies for climate adaptation within regulatory planning. The case highlighted the Copenhagen Cloudburst Management Plan (2012) that showed how a combined approach to storage of stormwater in highways and transferring it to central sewage channels could save money in the long turn run. By incorporating trees and other plants, redesigned streets can improve their visual and recreational qualities while also promoting biodiversity. According to a cost-benefit analysis, executing these procedures in the internal urban places between 2013 and 2033 would cost around €500 million, paralleled to the €800 million in flooding destructions triggered by a solo chief rainstorm in 2011 (De Bellis et al. 2015).

In a noteworthy work by Jongman et al. (2004) as an important case study on Berlin city plain provided a comprehensive model as green infrastructure. In Berlin, Germany, the citywide Landscape Program (LaPro) has been a key intentional tool for enhancing societal and ecological connectedness against a backdrop of fast population increase. The LaPro has been a legally mandated plan for civil servants that was closely tied to the city's land management system. The goals were divided into four categories: natural environment, which includes urban climate, ecosystem and biodiversity protection, leisure, and beautification of landscape. The "General Urban Compensation Plan" (GAK) backs them up. The GAK appraised breaches in the urban green linkage and recommends action plans, with financing for implementation coming from development projects' mandated effect mitigation and restitution requirements. Developers pay for measures taken to compensate in other places if environmental effect mitigation is not practicable inside a specified location. Traditional surveying methodologies can be improved by using public participatory geographic information systems (PPGIS), which can capture the social benefits of green spaces. Traditionally, resident comfort has been monitored by simple postal investigations enquiring about residential use of it and experience with gardens in general, through one-on-one questionnaires conducted on-site. Cultural ecosystem functioning such as recreational, appealing appreciation, educational and social prospects, and motivation have all been assessed using the same methodology.

While both systems can give useful data, PPGIS allows communities to collect it crossways whole districts or the entire city. The outcomes of a PPGIS survey can substantially improve ability to understand how gardens are used, park-goers' requirements and inclinations, the advantages that such spaces offer, and potential encounters. An electronic PPGIS research was done in Berlin as part of GREEN SURGE to investigate how green spaces are used and how the cultural ecosystem services that are always perceived. Although the PPGIS survey delivered much more aspect about precise parks than the city's last urban green feedback form in terms of all-purpose uses and perspectives of desires, the results were comparable to the city's last green space fulfilment inspection in terms of universal usages and awareness of needs (Jongman et al. 2004).

Thus, the above selected case studies highlighted the working plans and their contemporary outputs with the respect of desired expectations in a very comprehensive way. The results of these above studies would become a strong base for future city planning and urbanization in under the lens of climatological point of view and environmental angle.

4.5 The Key Challenges and Future Prospects

In urban spaces or in the cities, the climate change has been strongly linked with complementary socio-economic variabilities: demographic progresses, advanced percentage of variable aged or sensitive aged individuals, challenging requests for natural resources utilization. These variations strengthen the vulnerability of societies, property, and systems of ecology and ecosystems underneath surviving the climate environments until unless key challenges have been taken to the attentions.

The influential effects (whether positive or negative) of climate change in the urban regiments such as changes in heat profiles, temperature fluctuations, precipitation abnormalities, sea level rises, pollution, and other linked patterns have been required significant amount of actions and understandings. Thus, strong foundation has been required to put the urban climatological agenda in to various minds, policies, and interconnected frameworks to get, solution focused, and action-oriented plans as well as desired outcomes. In this regard, the major key challenges can be assigned as sincere attempts towards eco-friendly economic frameworks, time management, and fast-tracking towards the outputs, focus on multi-stakeholder and their sectoral participation, urban planning and policies and sustainable governance/ management as well as their interconnected boundary points.

4.5.1 Sincere Attempts Towards Eco-Friendly Economic Frameworks

The uncertainty about how to plan, develop, implement, and manage green infrastructure is reflected in the design standards problem (Sinnett et al. 2018). Design-related technological constraints include a lack of data on presentation features and absence of practical awareness (NRC 2009). The application of design, process, care and assessment of urban green infrastructure, design standards that are suited to cities environments and react to precise hazards and resources accessibility are required for the fruitful planning (Li et al. 2017). In India, significant attempts are also being made to address this problem. "Green Infrastructure: A Practitioner's Guide", issued by Centre for Science and the Environment in mid-2017, presents approaches for water-sensitive city scheme and development. This resource is intended for specialists functioning at a variety of scales in the Global South, from specific parcels to cities (Rohilla et al. 2017).

4.5.2 Loops of Control and Regulatory Paths

Finding a proper regulatory environment to further mainstream urban green infrastructure is a challenge. Few authorities established methods for green infrastructure and its advantages, unlike, say, fire protection. Even in the Netherlands, which has been a green infrastructure initiator, there are no legislative structures that provide water specialists with adequate assurance in long-term profits of flexibility methods, such as green infrastructure (Gersonius et al. 2016). Legal agreements must provide clarity in the new division of obligations as well as a long-term commitment to maintenance (Gersonius et al. 2016). A major obstacle is the relatively inadequate combination of controlling agencies into a system that entirely realizes the multi-faceted profits of green infrastructure (Kremer et al. 2016). Separate regulatory authorities in United Kingdom view even well-planned green infrastructure exclusively in terms of respective monitoring concerns; for example, the environment organization views green infrastructure in view of water profits in the environment, the Health and Safety Supervisory in terms of social fitness implications, and so on. In the United Kingdom, this has hindered the ecological drainage structures as fragment of urban green infrastructure (Ellis and Lundy 2016). National storm water infrastructure laws in China, too, solely consider grey infrastructure schemes (Li et al. 2017).

4.5.3 Challenges Regarding Societal and Economical Nexus

Equity is one of the ideologies for building resilience according to the Urban Land Institute. Though most resilience efforts include social justice promotion, the reality is rather different. Flood vulnerability is frequently higher in lower areas of towns; however, geography does not always influence risk dispersal. Instead, it is common in many cities to discover poor areas with low access to green place as compared to rich areas (Hoang and Fenner 2016). As the critical ecological facilities provided by natural surroundings, this unequal access is an issue of ecological integrity (Smiley et al. 2016). Green space is linked to positive benefits on individuals' welfare, including their health (Zuniga-Teran 2020). Eccentricity can develop due to poverty, education, background and religious conviction in various parts of the world, and in all circumstances, eccentricity generally entails a deficiency of admittance to green infrastructure profits. When evaluating access to the wellness advantages of green infrastructure, the public conception of green space, as well as its eminence, are key factors to evaluate (Fernández-Alvarez 2017).

Community involvement, as well as citizen-led knowledge, is gaining popularity. A key component of the Rockefeller Foundation's examination of hundred Resilient Cities' policies is public participation. With a specific reference to green infrastructure, the City of Bristol's flexibility policy expressly intends to strengthen ability in susceptible and marginalized populations (Bristol City Council 2016). At both the organizational and community levels, public participation is critical. Although necessary for efficient GI implementation, low-income individuals may find it more difficult to participate in community initiatives due to a lack of leisure time (Furlong et al. 2018).

4.5.4 Challenges Regarding Money, Capital, and Finances

Due to the harm to city structure, cities confront substantial economic issues following a flood occurrence. Annually, the United States uses roughly \$2 billion dollars on floods (Subramanian 2016), whereas China grieved the damages of about \$15.77 billion USD (Li et al. 2017). As a result, green infrastructure's flood-reduction benefit could be a financial incentive for communities. However, it is still unknown how to accurately assess the costs and advantages of green infrastructure technology, as well as how to convert these benefit estimates into finance simulations for functioning expenditures.

Fines and economic incentives may be used to help cities implement green infrastructure at the household level. Centralized penalties for non-obedience with water quality requirements may be considerable in the United States; therefore evading such fees is a solid reason to appliance green infrastructure. As a result, several towns in the United States have introduced financial inducements to support green infrastructure schemes (Vogel et al. 2015). Residents in Cleveland can get rain barrels for free or at a reduced cost, as well as technical assistance in properly installing those (Baptiste et al. 2015). Alternative profitable tool to stimulate green infrastructure is subsidies. Furthermore, secondary financial inducements may aid in the widespread adoption of green infrastructure. Chicago announced a \$100,000 USD allowance programme to encourage the adoption of green roofs since they reduce both run-off and energy expenditures (Tayouga and Gagné 2016). Even if the maintenance funding issue is resolved, the accountable party must be identified, which may necessitate more funding.

4.5.5 Challenges Regarding Grass Root Innovation and Resilience Policies

By leveraging infrastructure innovation, resilience can be improved (ULI 2018). Since, the green infrastructure is an innovative idea for urban flexibility, creative methods are needed to offer an extensive range of ecological amenities to urban people. Sutton-Grier et al. (2015) discovered key potential for improvement in mix methods in a study of coastal towns in the United States. Several states have embraced and even controlled "living shorelines", which are renovated ecosystems sheltered from destruction by manmade infrastructure.

It is critical to learn from pilot initiatives in order to successfully deploy green infrastructure on a broader scale. Cities may learn from one another, reduce risk, and uncover future prospects through innovation (Li et al. 2017). The successful placement of green infrastructure requires handy co-operation among organizers, creators, legislators, and researchers. They may involve shareholders and develop awareness that will lead to widespread green infrastructure implementation by working together (Tayouga and Gagné 2016). It is also crucial to recognize that urban ecosystem services don't exist in a vacuum and are intertwined with the countryside. As a result, GI in cities should be seen as part of a larger system of green space essential for biological and communal structures.

In terms of regulatory problem, cities require regulatory paths that capture the multifaceted benefits of GI. Because integrating several levels of government is difficult, we propose forming a new spatially defined regulatory body that will take precedence over other parts of the complicated regulatory system. In circumstances where regulatory approaches are not feasible, stakeholder participation and afforestation projects can aid in the implementation of green infrastructure. Local government certification programmes can help improve performance on a variety of levels. Finally, urban planning strategies must protect green space from enlargement in order to permit biological practices to take place.

In terms of the socio-economic difficulty, green infrastructure is grappling with serious problems. Most resilience efforts include an equality component, and there is a growing awareness of the need to address disparity. Gentrification, a time lapse between the installation of grey infrastructure and the transition to green infrastructure, and a lack of public participation, however, continue to exacerbate disparities.

The cost of inaction presents a compelling case for investing in green infrastructure and other resilience efforts when it comes to the finance issue. Green infrastructure is less expensive than grey infrastructure, and economic mechanisms can help with green infrastructure implementation. However, it is unclear who would pay for this expenditure, who will maintain green infrastructure, and how all green infrastructure advantages will be quantified. Residents who reside near green space may approve the use of property taxes to fund maintenance.

4.5.5.1 Sustainable Governance and Management

The term "lack of political will" explains why urban specialists provide basic service expansion, promotion of easy clearances, climate change extenuation, catastrophe hazard lessening, and other concerns of comprehensive sustainable development, a low priority in practise. Whereas governmental and financial interests clearly influence significances, the widespread use of the term "political will" fails to explain the limitations, constraints, and variety of current governance systems, as well as how they define development possibilities (Stepputat et al. 2016).

The functions of administration are unclear, and there is little collaboration between them. Metropolitan governments provide a single answer for handling rural–urban transportation, water, and energy networks, but they typically lack the jurisdiction and capacity to do so.

Cities and local governments frequently lack the power, resources, and competence to carry out the tasks entrusted to them by central governments in basic facilities and additional areas.

Governments have inadequate resources for negotiating agreements with commercial service providers and overseeing public–private partnerships. The expansion and efficiency of fundamental services have been hampered as a result. The scarcity of disaggregated information at the city level complicates urban development and the measurement of progress in important pointers for SDG 11 and other targets. Cities' capacity to acquire financing for growth, whether through indigenous levies, governments, or global aid, is restricted. In terms of subsidy for climate change extenuation, the "New Climate Economy" has made a resilient financial instance for implementing low-carbon measures in cities. Cities, on the other hand, must have the financial resources and affluence to support low-carbon initiatives.

While those authority issues are widespread, the governance gap is particularly visible in smaller cities, which house the majority of the city residents and will experience the greatest population growth over the next 15 years. The "fragile cities", expand extraordinarily quickly and where facilities are equally pressured due to the inflow of people expatriate by armed clash, are a particularly challenging set of cities.

4.6 Conclusion

The urbanization and the nexus of climatological aspect as input for city development and future city planning is relatively a new concept. The many of global and economically important places in the world and related cities have already begun this evolution. The impacts of global climate change on social, economic, and environmental domains are very aggressive; thus, the climatological solutions towards the present/future population settlements are the need of the hour. It is important to combine these sustainable transitions while planning for adaptation and resilience to gain a better control on the future visions towards the sustainable development and goals. The societal, economic, and environmental benefits of climatological inputs in urban settlement plannings are remarkable. The acceptance, sincere attempts, and strategic plannings are required to gain the desired output from this remarkable eco-friendly concept.

References

Admiraal JF, Wossink A, de Groot WT, de Snoo GR (2013) More than total economic value: how to combine economic valuation of biodiversity with ecological resilience. Ecol Econ 89:115–122

Agarwal M, Ruchi, Fakhr FA, Choudhary K (2021) A sustainable model of urbanization for Indian cities, a case study of New Delhi. Int J Eng Res Technol 10(3):82–90

Anttiroiko AV, Caves RW (2014) Urban planning 3.0: impact of recent developments of the web on urban planning. In: IT in the public sphere: applications in administration, government, politics, and planning. IGI Global, Hershey, pp 233–257

Baptiste AK, Foley C, Smardon R (2015) Understanding urban neighborhood differences in willingness to implement green infrastructure measures: a case study of Syracuse, NY. Landsc Urban Plan 136:1–12

Bodley JH (2011) Cultural anthropology: tribes, states, and the global system. Rowman Altamira, Lanham

Casares PS, Benton L, Bull J (2021) Sustainable drainage retrofit case study: Counters Creek Flood Alleviation Scheme, UK. J Environ Eng Sci 16:166–173. https://doi.org/10.1680/jenes.21.00021

Caspersen OH, Olafsson AS (2010) Recreational mapping and planning for enlargement of the green structure in greater Copenhagen. Urban Urban For Green 9:101–112

Cheng C (2016) Spatial climate justice and green infrastructure assessment: a case study for the Huron River watershed, Michigan, USA. GI_Forum 4(1):176–190. https://doi.org/10.1553/giscience2016_01_s176

Chigbu UE, Onyebueke VU (2021) The COVID-19 pandemic in informal settlements: (re)-considering urban planning interventions. Town Plan Rev 92(1):115–121

Chu E, Anguelovski I, Carmin J (2016) Inclusive approaches to urban climate adaptation planning and implementation in the global south. Clim Pol 16(3):372–392

Cities Alliance (2007) Liveable cities: the benefits of urban environmental planning—a cities alliance study on good practices and useful tools. Cities Alliance, Brussels

Corburn J, Vlahov D, Mberu B, Riley L, Caiaffa WT, Rashid SF, Ayad H (2020) Slum health: arresting COVID-19 and improving well-being in urban informal settlements. J Urban Health 97(3):348–357

Coutts A, Beringer J, Tapper N (2010) Changing urban climate and CO_2 emissions: implications for the development of policies for sustainable cities. Urban Policy Res 28(1):27–47

Crane M, Lloyd S, Haines A, Ding D, Hutchinson E, Belesova K, Turcu C (2021) Transforming cities for sustainability: a health perspective. Environ Int 147:106366

De Bellis Y, Santos A, Tosics I, Davies C, Hansen R, Rall E, Pauleit S, Lafortezza R (2015) Green infrastructure planning and implementation

De la Barrera F, Reyes-Paecke S, Harris J, Bascuñán D, Farías JM (2016) People's perception influences on the use of green spaces in socio-economically differentiated neighborhoods. Urban For Urban Green 20:254–264

DeMenocal PB (2001) Cultural responses to climate change during the late Holocene. Science 292(5517):667–673

Demuzere M, Orru K, Heidrich O, Olazabal E, Geneletti D, Orru H, Faehnle M (2014) Mitigating and adapting to climate change: multi-functional and multi-scale assessment of green urban infrastructure. J Environ Manage 146:107–115
Dhule SB, Ambhore G, Paikrao R, Ingle R, Gawale D, Meher V (2019) Case study on smart city. Int Res J Eng Technol 6(5):7744–7750
Economides C (2014) Green infrastructure: sustainable solutions in 11 cities across the United States. Columbia University Water Centre, New York
Ellis JB, Lundy L (2016) Implementing sustainable drainage systems for urban surface water management within the regulatory framework in England and Wales. J Environ Manag 183:630–636
Emmanuel R, Loconsole A (2015) Green infrastructure as an adaptation approach to tackling urban overheating in the Glasgow Clyde Valley Region, UK. Landsc Urban Plan 138:71–86. https://doi.org/10.1016/j.landurbplan.2015.02.012
European Commission (2016) The forms and the functions of the green infrastructures. Brussels, European Commission/Environmental
Fallmann J, Emeis S (2020) How to bring urban and global climate studies together with urban planning and architecture? Dev Built Environ 4:100023
Fernández-Álvarez R (2017) Inequitable distribution of green public space in Mexico City: an environmental injustice case. Economía, sociedad y territorio 17(54):399–428
Furlong C, Phelan K, Dodson J (2018) The role of water utilities in urban greening: a case study of Melbourne, Australia. Util Policy 53:25–31
Gersonius B, Rijke J, Ashley R, Bloemen P, Kelder E, Zevenbergen C (2016) Adaptive delta management for flood risk and resilience in Dordrecht, The Netherlands. Nat Hazards 82 (2):201–216
Hales S, Baker M, Howden-Chapman P, Menne B, Woodruff R, Woodward A (2007) Implications of global climate change for housing, human settlements and public health. Rev Environ Health 22(4):295–302
Hannon GE, Bradshaw RH (2000) Impacts and timing of the first human settlement on vegetation of the Faroe Islands. Quat Res 54(3):404–413
Hathway EA, Sharples S (2012) The interaction of rivers and urban form in mitigating the urban Heat Island effect: a UK case study. Build Environ 58:14–22. https://doi.org/10.1016/j.buildenv.2012.06.013
Hegetschweiler KT, de Vries S, Arnberger A, Bell S, Brennan M, Siter N, Hunziker M (2017) Linking demand and supply factors in identifying cultural ecosystem services of urban green infrastructures: a review of European studies. Urban For Urban Green 21:48–59
Heidt V, Neef M (2008) Benefits of urban green space for improving urban climate. In: Ecology, planning, and management of urban forests. Springer, New York, pp 84–96
Hoang L, Fenner RAA (2016) System interactions of stormwater management using sustainable urban drainage systems and green infrastructure. Urban Water J 13(7):739–758
IPCC (2014) Intergovernmental Panel on Climate Change 2014. In: Core Writing Team, Pachauri RK, Meyer LA (eds) Contribution of working groups I, II and III to the fifth assessment report of the Intergovernmental Panel on Climate Change. IPCC, Geneva, p 151
Jim CY (2002) Planning strategies to overcome constraints on greenspace provision in urban Hong Kong town. Plan Rev 73:127–152
Jongman RH, Külvik M, Kristiansen I (2004) European ecological networks and greenways. Landsc Urban Plan 68(2–3):305–319
Kabisch N (2015) Ecosystem service implementation and governance challenges in "urban green spaces" planning-the case of Berlin, Germany. Land Use Policy 42:557–567
Knottnerus OS (2005) History of human settlement, cultural change and interference with the marine environment. Helgol Mar Res 59(1):2–8
Kourtit K, Nijkamp P, Suzuki S (2020) Are global cities sustainability champions? A double delinking analysis of environmental performance of urban agglomerations. Sci Total Environ 709:134963

Kremer P, Hamstead Z, Haase D, McPhearson T, Frantzeskaki N, Andersson E, Kabisch N, Larondelle N, Rall EL, Voigt A, Baró F (2016) Key insights for the future of urban ecosystem services research. Ecol Soc 21(2)
Leon DA (2008) Cities, urbanization and health. Int J Epidemiol 37(1):4–8
Li F, Liu X, Zhang X, Zhao D, Liu H, Zhou C, Wang R (2017) Urban ecological infrastructure: an integrated network for ecosystem services and sustainable urban systems. J Clean Prod 163: S12–S18
Liu OY, Russo A (2021) Assessing the contribution of urban green spaces in green infrastructure strategy planning for urban ecosystem conditions and services. Sustain Cities Soc 68:102772. https://doi.org/10.1016/j.scs.2021.102772
Lovell ST, Taylor JR (2013) Supplying urban ecosystem services through multifunctional green infrastructure in the United States. Landsc Ecol 28(8):1447–1463. https://doi.org/10.1007/s10980-013-9912-y
Macdonald N, Jones P (2006) The inclusion of sustainable drainage systems in flood management in the post-industrial city: a case study of Glasgow. Scott Geogr J 122(3):233–246. https://doi.org/10.1080/00369220618737268
McBean G, Ajibade I (2009) Climate change, related hazards and human settlements. Curr Opin Environ Sustain 1(2):179–186
Meerow S (2020) The politics of multifunctional green infrastructure planning in New York City. Cities 100:102621. https://www.sciencedirect.com/science/article/abs/pii/S0264275119314179
Mell JC (2014) Aligning fragmented planning structures through a green infrastructure approach to urban development in the UK and USA. Urban For Urban Green 13:612–620
Mell IC (2018) Greening Ahmedabad—creating a resilient Indian city using a green infrastructure approach to investment. Landsc Res 43(3):289–314
Mills G (2007) Cities as agents of global change. Int J Climatol 27(14):1849–1857
Mutizwa-Mangiza ND, Arimah BC, Jensen I (2011) 2013. Global report on human settlements, Cities and climate change
Nolon JR (2021) Local land use power: managing human settlements to mitigate climate change. Envtl L Rep 51:10426
Nowak DJ, Crane DN, Stevens JC (2006) Air pollution removal by urban trees and shrubs in the United States. Urban For Urban Green 4:115–123
Pan SY, Gao M, Kim H, Shah KJ, Pei SL, Chiang PC (2018) Advances and challenges in sustainable tourism toward a green economy. Sci Total Environ 635:452–469
Pasupuleti NS, Lathwal ASS, Jagdish PS (2016) Sustainable smart solutions for city of Faridabad–a case study addressing urban infrastructure problems. Int J Civil Eng Res 7(1):33–40
Patel MS, Rangrej S (2021) Experiences of green infrastructure in urban planning. Int J Res Eng Sci 9(1):11–13
Patel A, Shah P (2020) Rethinking slums, cities, and urban planning: lessons from the COVID-19 pandemic. Cities Health 21:S145–S147
Puzyreva K, de Vries DH (2021) 'A low and watery place': a case study of flood history and sustainable community engagement in flood risk management in the county of Berkshire, England. Int J Disaster Risk Reduct 52:101980. https://doi.org/10.1016/j.ijdrr.2020.101980
Ramamurthy A, Devadas MD (2016) A model of eco-welfare planning: sustainable urban development of Chennai MetropolitanCity, India. Int J Sci Eng Res 7(9):179–203
Rohilla SK, Matto M, Jainer S, Kumar M, Sharda C (2017) Policy paper on water efficiency and conservation in urban India. Centre for Science and Environment, New Delhi
Ronchi S, Salata S, Arcidiacono A (2020) Which urban design parameters provide climate-proof cities? An application of the urban cooling InVEST model in the city of Milan comparing historical planning morphologies. Sustain Cities Soc 63:102459
Russo A, Chan WT, Cirella GT (2021) Estimating air pollution removal and monetary value for urban green infrastructure strategies using web-based applications. Land 10(8):788. https://doi.org/10.3390/land10080788

Satiram BG, Chakraborty I, Banerjee S (2020) Challenges in implementation of green infrastructure in metropolitan cities—case of Ahmedabad. Int J Archit Infrastruct Plan 6(2):7–14
Shandas V (2015) Neighborhood change and the role of environmental stewardship: a case study of green infrastructure for stormwater in the City of Portland, Oregon, USA. Ecol Soc 20(3):16
Sheikh MM (2018) Smart city and urban issues: a case study of Jaipur. Ann Geogr Stud 1(1):23–33
Sinnett D, Calvert T, Smith N, Burgess S, King L (2018) The translation and use of green infrastructure evidence. In: Proceedings of the Institution of Civil Engineers-Water Management, vol 171, no 2. Thomas Telford Ltd., pp 99–109
Smiley KT, Sharma T, Steinberg A, Hodges-Copple S, Jacobson E, Matveeva L (2016) More inclusive parks planning: park quality and preferences for park access and amenities. Environ Just 9(1):1–7
Stepputat F (2016) We need to rethink 'disarmament, demobilization and reintegration'
Stevenson M, Thompson J, de Sá TH, Ewing R, Mohan D, McClure R, Woodcock J (2016) Land use, transport, and population health: estimating the health benefits of compact cities. Lancet 388(10062):2925–2935
Straupe I, Liepa L (2018) The relation of green infrastructure and tourism in urban ecosystem. Res Rural Dev 1:17
Sturiale L, Scuderi A (2019) The role of green infrastructures in urban planning for climate change adaptation. Climate 7(10):119
Sturiale L, Scuderi A, Timpanaro G, Foti VT, Stella G (2020) Social and inclusive "value" generation in metropolitan area with the "urban gardens" planning. In: Values and functions for future cities. Springer, Cham, pp 285–302
Subramanian R (2016) Rained out: problems and solutions for managing urban stormwater runoff. Ecol LQ 43:421
Sutton-Grier AE, Wowk K, Bamford H (2015) Future of our coasts: the potential for natural and hybrid infrastructure to enhance the resilience of our coastal communities, economies and ecosystems. Environ Sci Pol 51:137–148
Sugar L, Kennedy C (2021) Urban scaling and the benefits of living in cities. Sustain Cities Soc 66: 102617
Tayouga SJ, Gagné SA (2016) The socio-ecological factors that influence the adoption of green infrastructure. Sustainability 8(12):1277
Tilt JH, Ries PD (2021) Constraints and catalysts influencing green infrastructure projects in small cities. Urban For Urban Green 63:127138. https://doi.org/10.1016/j.ufug.2021.127138
Tiwari A, Kumar P, Kalaiarasan G, Ottosen TB (2021) The impacts of existing and hypothetical green infrastructure scenarios on urban heat island formation. Environ Pollut 274:115898. https://doi.org/10.1016/j.envpol.2020.115898
Tomita-Yokotani K, Anilir S, Katayama N, Hashimoto H, Yamashita M (2009) Space agriculture for habitation on mars and sustainable civilization on earth. In: In 2009 4th International conference on recent advances in space technologies. IEEE, New York, pp 68–69
United Nations (2015) Transforming our world: the 2030 agenda for sustainable development. United Nations, New York
United Nations (2018) Revision of the world urbanization prospects. United Nations, New York
Valenzuela-Levi N, Abreu M (2021) The income equality dividend in 'green infrastructure': evidence from the transport and waste sectors in the OECD. J Clean Prod 315:128088
Vardoulakis S, Kinney P (2019) Grand challenges in sustainable cities and health. Front Sustain Cities 1:7
Vogel JR, Moore TL, Coffman RR, Rodie SN, Hutchinson SL, McDonough KR, McLemore AJ, McMaine JT (2015) Critical review of technical questions facing low impact development and green infrastructure: a perspective from the Great Plains. Water Environ Res 87(9):849–862
Waghwala RK, Agnihotri PG (2019) Flood risk assessment and resilience strategies for flood risk management: a case study of Surat City. Int J Disaster Risk Reduct 40:101155

Webber JL, Kuller M (2021) Enhancing the visibility of SuDS in strategic planning using preliminary regional opportunity screening. ISPRS Int J Geo Inf 10(11):726. https://doi.org/10.3390/ijgi10110726
Wei F, Knox PL (2015) Spatial transformation of metropolitan cities. Environ Plan A 47(1):50–68
Wei J, Qian J, Tao Y, Hu F, Ou W (2018) Evaluating spatial priority of urban green infrastructure for urban sustainability in areas of rapid urbanization: a case study of Pukou in China. Sustainability 10:327
Wu L, Lu S, Zhu C, Ma C, Sun X, Li X, Guo Q (2021) Holocene environmental archaeology of the Yangtze River Valley in China: a review. Land 10(3):302
Yang J, Mcbride J, Zhou J, Sun Z (2005) The urban forest in Beijing and its role in air pollution reduction. Urban For Urban Green 3:65–78
Yatoo SA, Sahu P, Kalubarme MH, Kansara BB (2020) Monitoring urban land use changes and its future prospects using cellular automata simulation and artificial neural network for Ahmedabad city, India. GeoJournal 87:765–786. https://doi.org/10.1007/s10708-020-10274-5
Yu J, Shannon H, Baumann A, Schwartz L, Bhatt M (2016) Slum upgrading programs and disaster resilience: a case study of an Indian 'Smart City'. Procedia Environ Sci 36:154–161
Zhang Q, Zhu C, Liu CL, Jiang T (2005) Environmental change and its impacts on human settlement in the Yangtze Delta, PR China. Catena 60(3):267–277
Zhang Z, Paschalis A, Mijic A (2021) Planning London's green spaces in an integrated water management approach to enhance future resilience in urban stormwater control. J Hydrol 597: 126126. https://doi.org/10.1016/j.jhydrol.2021.126126
Zhou X, Rana MP (2012) Social benefits of urban green space: a conceptual framework of valuation and accessibility measurements. Manag Environ Qual Int J 23(2):173–189
Živković J (2020) Human settlements and climate change. In: Climate action. Springer, Cham, pp 573–584
Zuniga-Teran AA, Staddon C, de Vito L, Gerlak AK, Ward S, Schoeman Y, Hart A, Booth G (2020) Challenges of mainstreaming green infrastructure in built environment professions. J Environ Plan Manag 63(4):710–732

Chapter 5
Methodology for Sustainable Urban Planning

N. Kapoor and V. K. Bansal

5.1 Urban Planning

5.1.1 Introduction to Urban Planning

The number of persons per unit area, economical condition, and existence of a municipality are the three key factors which designate a location as urban. For a place, to be called as urban, it must have a certain number of persons living there; these persons must not be scattered over large but concentrated in a small area; there must be minimum population density of the area; all persons must be involved in commercial activities or services in addition to agriculture or animal rearing; and there must be a municipality in the locality, various types of town committees for decision-making, or town planning and government and semi-government governing body to administer the facilities or services of the area. The planning refers to a continuous process that entails a variety of activities. It comprised of decisions and choices on how to use the available resources in a normative way to attain specific goals at a later date.

Urban planning spans a wide range of disciplines and aims to achieve a sustainable, ecological, economic, and social organization of all aspects of an urban setting. It is responsible for providing shelter, jobs, restoration, business, water supply,

N. Kapoor
Department of Architecture, National Institute of Technology (NIT) Hamirpur, Hamirpur, Himachal Pradesh, India
e-mail: neetu@nith.ac.in

V. K. Bansal (✉)
Department of Civil Engineering, National Institute of Technology (NIT) Hamirpur, Hamirpur, Himachal Pradesh, India
e-mail: vkb@nith.ac.in

B. Pathak, R. S. Dubey (eds.), *Climate Change and Urban Environment Sustainability*, Disaster Resilience and Green Growth,
https://doi.org/10.1007/978-981-19-7618-6_5

sanitation facility, transportation, and communication as well as the preservation of natural and built heritage of the region. Urban planning involves setting objectives, collecting data, analysis of data, strategic thinking, forecasting, and public participation in both the development of green land and redevelopment an urban setting. It concentrates on location, geography, economical condition, and social consequences of an urban environment. Urban planning involves locating various activities within the available space in order to design and regulate the use of a given space.

The urban population of developed countries is predicted to grow slowly, from 0.9 billion in 2005 to 1.0 billion in 2030; however, there is increasing need of urban planning in the developing countries (UN-Habitat 2009). The urban areas of developing countries will experience significant growth. Growth in developing countries will be very rapid, averaging 2.2% per year from 2005 to 2030, which corresponds to a 30 year of doubling time. In 2030, urban population will be 5.0 billion, accounting for 60% population of the world. The cities of developing countries will account for practically all population growth, with a population of about 4.0 billion by 2030, roughly equal to the whole population of the developing world in 1990. This makes urban planning decisions more important in the developing countries.

5.1.2 Approach of Urban Planning in Practice: Master Plan

The conventional planning systems of developed countries have a significant influence on the planning systems in the developing countries, partially because they are vestiges of a colonial history. While the urban planning system in the developed countries has also grown to include more than just land-use planning. However, developing countries still depend upon *master plans*. The *master plans* are the spatial plans to depict the state and form of an urban space on a map at a future period (Todes et al. 2010). Keeping the developments in the developing countries into account precise and location-specific planning is required. The *master plans* do not adjust changes of the future. The researchers have indicated that *master plans* are incompatible with modern economic realities (Yeh and Wu 1999). In the developed countries, underlying tension among planning professionals and the market requirements always exist, it leads to planning alteration to accommodate market trends or requirements and plans that answer to economic factors.

Master plans are not suitable for dynamic urban environment but they are also thought to be completely malevolent (UN-Habitat 2009; Watson 2009). For example, *master plan* developed by *Shenzhen* dose not include floating or temporary populations, thus rejecting their access to the city (Watson 2009). Several experts have criticized China's master planning for failing to adapt to changing economic realisms on while pushing urban growth. The *master plans* were utilized to impose geographic segregation along racial lines in South Africa during the Apartheid era.

Similarly, Indian metropolises have been transformed into places of exclusionary behaviours. Grandiose infrastructure projects have been built in these places with an attempt to accelerate places towards becoming global cities, despite the fact that they continue to be battlegrounds for the urban poor. This spans the scope of urban development from the official and informal, planned and unplanned, legal and illegal spectrums. Formal development is defined as development that adheres to established structural standards, whereas informal development is defined as development that does not. The acquiescence of real estate development to city plans or byelaws is referred to as planned development. The non-compliance of real estate development to city plans is referred to as unplanned development. The legality or illegality of a real estate development is determined by the type of structure used.

Unfortunately, not much has changed in urban planning in the last 100 years. The *environmental sustainability* and *participation of people* are important issues for planners to consider because of which urban planning will ideally evolve into a more environmentally conscious discipline.

5.1.3 Climate Change and Urban Planning

Climate change is a product of human activities. Climate change and resulting rise in global temperature have severe consequences on the earth's natural system and human society. To avoid severe consequences, mitigation strategies are required. Mitigation strategies help in reducing climate change by reducing greenhouse gas emissions. The changes to accommodate the unavoidable consequences are required. This generates the need for adaptation strategies. In adaptation, various communities prepare themselves to deal with an unexpected climate in the future.

Extensive human intervention at the international, national, state, and local level is required for mitigation as well as adaptation strategies. Most of the mitigation strategies revolve around reducing greenhouse gases (GHG) emissions as the overproduction of GHG makes a layer in the earth's atmosphere and causes rising temperature by trapping the solar heat which results in rising global temperature. This rising global temperature has two main issues over the coming decades and may have massive consequences on the ecosystem:

- Sea-level rise due to melting of ice may lead to submergence of coastal land.
- Changes in weather patterns will result in flooding, wildfires, drought, and landslides.

An amount of increased global temperature is inevitable, which will require adaptation strategies for climate change. These adaptation strategies include reducing human settlement in flood-prone areas, control floods, reducing soil erosion, and improving immunization against infectious diseases.

Most of the catastrophic climate change effects are avoidable through mitigation strategies, which are centred on fuel efficiency, fuel substitution, and energy efficiency of built-up infrastructures. A considerable effect of global warming is taken care through efficient urban planning strategies. The sustainable planning of urban centres reduces the demand of energy to a great extent. Published literature indicates that urban planning strategies reduce the fuel consumption of vehicles by compact planning with a variety of housing typologies, providing safe and efficient public transport, and placing required facilities within the walking distance from the residential area. Ewing and Cervero (2010) established that vehicular movement reduces between 20% and 40% in squeezed urban places compared to loose suburbs which are dependent on others. The strategies that reduce vehicular movement by restraining suburban expansion and encourage more compacted, walkable, and independent suburbs contribute to the mitigation of climate change.

A sustainable land use approach for urban planning reduces GHG emissions by up to 10%. Further investment in an efficient transportation system will result in additional GHG reduction (Ewing and Cervero, 2010). Hill et al. (2006) stated that about 66% of developments in 2050 will be new, efficient land use, and transportation strategies may decrease the emission of GHG. Extensive research has been carried out to study the impact of compact planning on the reduction in transportation cost; however, its effect on energy consumed by built-up infrastructure has been studied less extensively. Although, Norman et al. (2006) reported that GHG emissions and per capita energy consumption are 2–2.5 times less in compact planning.

Concerns of continuous climate changes, requirements for clean air, need of good drinking water, and best spatial planning draw attention towards sustainability, more precisely sustainable urban planning of urban cities. There is the need of developmental plans and routine practices to ensure self-sustaining urban societies over the long duration. Sustainable urban planning contains many areas that make it multidisciplinary area rather than a mono disciplinary. Sustainable urban planning contributes towards welfare of people and their surroundings, shaping urban surroundings strategically into healthier and more efficient places. To accomplish these objectives, planners, in particular urban planners make plans and device strategies to help industries and people in the management of their resources and reduce the overall impact on the environment. This strongly generates the need of the sustainable urban planning.

5.2 Sustainability and Urban Planning

5.2.1 Recent Initiatives for Sustainable Urban Planning: A Global Prospective

Most of the researchers used operational methods and tools, such as participatory planning, GIS-based methods, and scenario planning for sustainable urban planning (Paul and Basu 2016; Sharma and Chandel 2017). These operational methods and tools help in making urban planning decisions more sustainable, rational, logical and provide a scientific base to the solution (Lian 2018). The operational methods use qualitative as well as quantitative data for solving urban planning problems at different levels/scales. Urban planning through participatory planning uses qualitative data for decision-making. GIS has been used for quantitative data analysis and the creation of scenarios. The different levels of urban planning are also integrated together for rational decisions based on qualitative as well as quantitative data.

Various sustainable urban planning initiatives proposed by different researchers based on the context and issues have been addressed in literature (Pickett et al. 2013; Wheeler 2013). The researchers have proposed planning paradigms, like new urbanism, suburban retrofit and regeneration, and smart growth to achieve sustainability by distributing the functions spatially focusing on urban form. However, the outcomes of these initiatives on socio-economic, cultural, and environmental parameters appear to be weak and sometimes questionable. Some researchers have criticized the effectiveness of the proposed planning paradigms in terms of evaluation of sustainability (McCrea and Walters 2012). Morales et al. (2018) developed a collaborative planning framework for supporting sustainable urban planning, and Li et al. (2007) developed a framework for quantifying the environmental impacts of urban development.

The issues of developing countries are complex because of multidimensional challenges of sustainable urban planning which has to ensure interaction of spatial, social, economic, and environmental dimensions with each other. As a result, planners need new sharp instruments to deal with this complexity and fulfil the expanding technical demands in planning procedures. The Planning Support System (PSS) is a collection of innovative tools that can help planners anticipate future land utilization scenarios and develop better plans (Vonk et al. 2007). PSS are intended to assist stakeholders in improving their knowledge base as well as integrating and balancing various demands on space (Geertman and Stillwell 2003).

5.2.2 Planning Support Systems for Sustainable Urban Planning

Although the concept of building tools to support planning activities extends back much further, the name PSS first appeared in the planning industry in the mid-1980s. PSS was first characterized in 1987 as a growing support for the planning process that included the use of digital tools such as GIS. Klosterman (1997) later described PSS as an information framework that incorporates the whole range of existing and future information technologies that are useful for planning. Geertman and Stillwell (2003) elaborated the scope of PSS, by using wide range of geo-information technology capabilities to provide planning solutions at any given spatial scale and inside any specific planning context. Brail (2005) provides the most recent definition of PSS, indicating that its goal is either projection to a future point or estimation of effects of some form of development.

The majorities of PSS are GIS-based and aid in the storage, management, analysis, and display of spatial data. Because GIS is used for a variety of reasons and on a variety of spatial challenges, each PSS is distinct and focuses on assisting a certain planning activity hence contains a specific collection of tools. A typical PSS

Table 5.1 Comparison of existing planning support systems

Scale/ Policy Level / PSS	Scale				Policy Level				
	Block level	Neigh-bourhood level	District	Regional level	Data acquisition	Interpre-tation	Design and strategies	Impleme-ntation	Evaluation
Community Viz			✓	✓	✓	✓	✓		
Development Pattern Approach	✓	✓	✓	✓	✓	✓	✓	✓	
Envision Tomorrow	✓	✓	✓	✓	✓	✓	✓	✓	
INDEX and Cool Spots	✓	✓	✓	✓	✓	✓	✓	✓	✓
I-PLACE's	✓	✓	✓	✓	✓	✓	✓	✓	✓
Metro Quest				✓	✓	✓	✓		
Tool for evaluating Neighbourhood Sustainability		✓			✓	✓	✓		
UPlan				✓	✓	✓			

brings together data, knowledge, methodologies, and tools related to a specific planning problem into a unified framework, with access offered via a common graphical user interface (Geertman and Stillwell 2003). PSS are useful tools for assisting planners in navigating the complexities of producing higher-quality plans while maximizing both resources and time.

PSS works on three levels: data structure for planning problems, optimization methods for planning processes, and virtual scenario development. Some existing PSS have been created to foresee future scenarios and help planners in making better decisions (Vonk and Geertman 2005). A vast number of PSS have been created, but only a small number have grown to the point where they are available for free or as commercial packages. *CommunityViz suite* (Communityviz 2021), *Envision Tomorrow INDEX* (Envision Tomorrow 2021), UPlan (2021), and I-Place's (2021) are just a few examples.

PSS may be classified based on their scope, technique, scalability, and decision-making assistance. PSS may also be classified by the step or stages of the policy-making cycle that they assist, inform, and influence. Different PSS use different methodologies such as spatial or non-spatial, top-down or bottom-up, simulation or end-state, observation-based or process-based. All PSS have been built to operate at various geographical scales ranging from block level, neighbourhood level, district level, and regional level. Most existing PSS function on a single scale, but only a few provide a flexible framework that may be used at several scales. PSS that works fluidly across several geographical scales are the most desirable, but also the most difficult, for sustainable urban planning at all sizes. Table 5.1 provides comparison of existing widely used planning support systems.

PSS are not universally accepted in planning practise. The main barrier to widespread adoption of these existing PSS is that they do not fit into the planner's workflow and operate on a black box paradigm (Vonk and Geertman 2005). Furthermore, these PSS are both difficult to find and use. Others are oversimplified to the point where they lose the ability to model the relationship between urban planning and sustainability. Some require expertise in specific domains, as well as a significant amount of time and resources, while others are oversimplified to the point where they lose the ability to model the relationship between urban planning and sustainability. To develop effective climate change mitigation methods, a better ability to model the link between urban planning and sustainability is required. This shows that a new PSS might be built on the merits of existing technologies. Continued development of PSS has strengthened links between various tools and develops new techniques for analysing urban planning decisions in order to achieve sustainability. An ideal PSS should be comprehensive, multi-scalar, iterative, additive, affordable, easily accessible, and policy relevant. Also it should have three-dimensional capabilities for precise descriptions of the effects of future community design in order to prevent planners from being unable to explain the real-world consequences of suggested activities.

5.3 Methodology for Sustainable Urban Planning

The methodology for sustainable urban planning has three major stages which are *scenario planning*, *public participation*, and *energy-intensive planning* as shown in Fig. 5.1. In the first stage, the context of the planning problem is understood in terms of *environmental aspects*. The data corresponding to the *environmental aspects* are required to be acquired for creating the scenarios of the potential planning solutions.

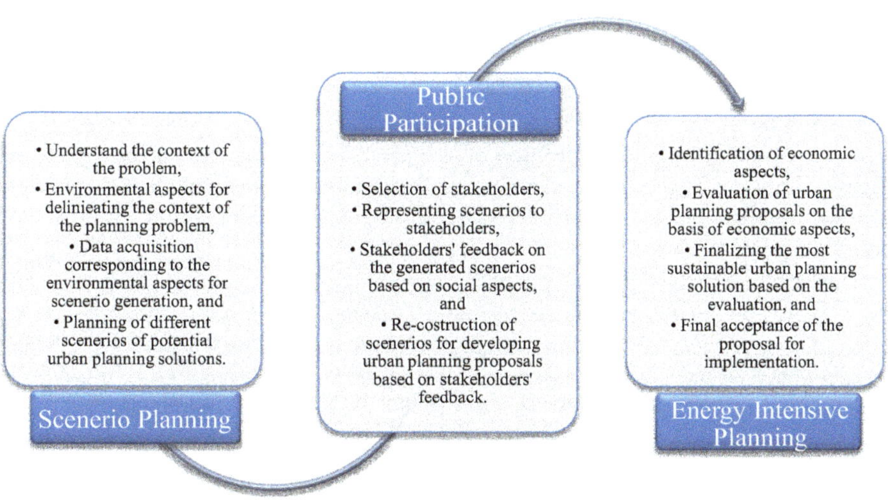

Fig. 5.1 The three major stages of the methodology for sustainable urban planning

The second stage helps in making the urban planning solutions more sustainable. In this, the stakeholders are selected, and the scenarios created in the first stage are presented to the stakeholders for their feedback. The *social aspects* are taken into consideration while taking the stakeholders' input. The urban planning proposals are reconstructed on the basis of feedback received from the stakeholders, which ensures the active participation of public in taking urban planning decisions. The third stage of the methodology is *energy-intensive* planning which makes the urban planning solutions more economical. The first step in this stage is the identification of *economic aspects*. On the basis of *economic aspects,* the planning proposals constructed in the second stage are evaluated, and the most sustainable planning solution is finalized for implementation.

The three stages of sustainable urban planning technique need consideration of *environmental*, *social*, and *economic* aspects. These aspects are identified through a survey of the most widely used methodologies for making sustainable urban planning decisions among professionals (Kumar and Biswas 2013; Dai et al. 2001; Dong et al. 2008; Youssef et al. 2011; Das et al. 2013; Yang et al. 2008; Bathrellos et al. 2012) and are applicable at various scales of site analysis. This includes McHarg's scientific systematic method (McHarg 1969), Lynch and Hack's technical method (Lynch and Hack 1984), Rubenstein's experimental method (Rubenstein 1996), and Beer and Higgins (2000) and LaGro's context-sensitive method (LaGro 2007). For regional urban planning, the scientific systematic method is appropriate. The technical and experimental procedures are primarily focused on the local scale. The context-sensitive technique works on a variety of scales, but mostly at the local scale. As a result, first and foremost, a study of the methodologies was conducted in order to compile a thorough list of aspects. The various aspects used for sustainable urban planning reported in the literature have been compiled in Fig. 5.2.

5.3.1 *Scenario Planning:* Environmental Aspects

The *environmental aspects* are unique for every context and are required for delineation of the existing condition of the site. The *environmental aspects* affect the development of the urban planning proposal in the form of alternate scenarios. The *environmental aspects* considered for sustainable urban planning are discussed below in brief:

Elevation: It is the height of a location above a defined reference point, which is most typically the Earth's mean sea level. Slopes with both a gradient and an orientation are created by spatial change in elevation (aspect). Because of their impacts on numerous other criteria such as drainage pattern, visibility, wind pressure, and so on, site elevations have a significant impact on urban planning decisions (Kraus and Charvátová 2016).

Vegetation: Existing vegetation on a place provides several ecological, economic, and social advantages. Trees perform a variety of services that directly benefit humans. Trees lower the cost of heating and cooling neighbouring buildings,

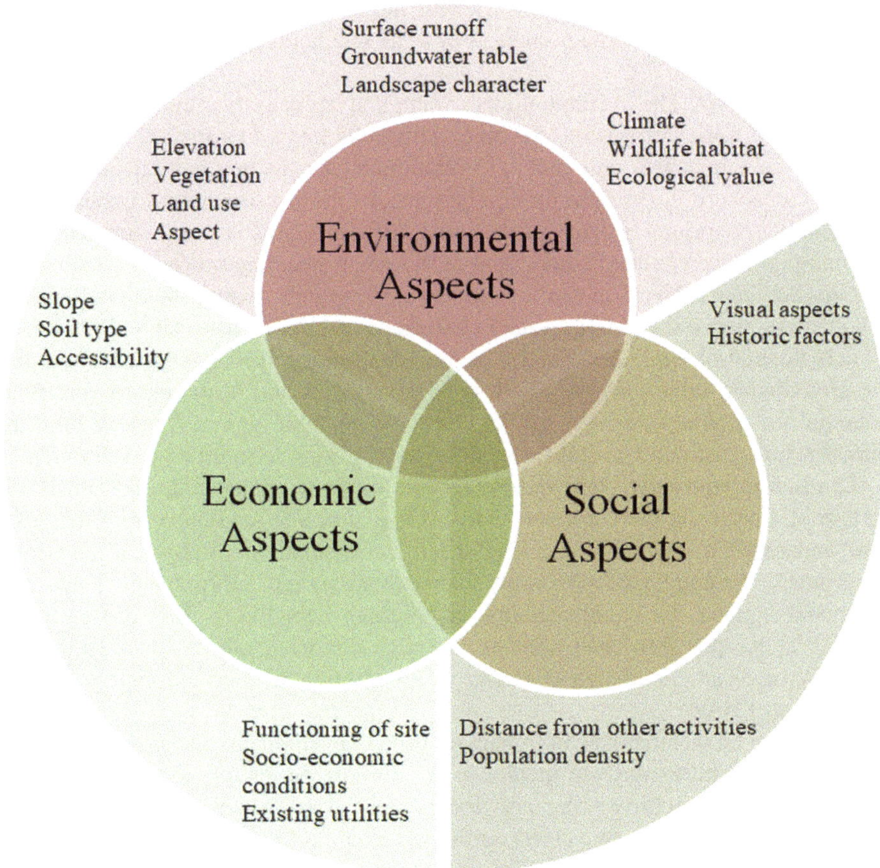

Fig. 5.2 Various aspects used for sustainable urban planning

offer shade, block out unwanted views, and act as windbreaks. Trees are also important for their aesthetic value. To make a plan more sustainable, the development proposal should maintain current vegetation and include new (Petit et al. 2004).

Land use: It relates to how land is used, such as farms, water bodies, woods, and built-up areas. Before making urban planning decisions, current land use on the site should be evaluated for contextual development. The previous land use of a location can have a range of effects on infrastructure development. The projected land uses of the property may be influenced by surrounding land uses in either a good or negative way. Natural areas, such as wetlands, wildlife habitats, lakes, woods, groundwater supplies, and economically productive regions, such as farms and forests, should be protected for sustainable urban planning (Persson 2013).

Aspect: It is the compass direction in which a slope is facing. The consideration of aspect while planning sustainable development proposal is essential, because it impacts an area's microclimate. The north facing side of slopes in the northern hemisphere is frequently shadowed, whereas the southern facing side receives more sun energy in the winter. The proneness of a site to a landslide is also affected by its aspect (Tian et al. 2010).

Surface runoff: The overland flow of water that happens over the Earth's surface as a result of rain, storm water, or other sources is known as surface runoff. This happens when the soil is saturated to its maximum capacity and rain falls faster than the soil can absorb it. Because it frequently entails the construction of impermeable or virtually impervious structures and paved surfaces, surface runoff management is becoming a more regulated component of the urban planning process.

Groundwater table: The top level of an underground surface below which the soil or rocks are saturated with water is known as the groundwater table. Because of climate fluctuations and the quantity of precipitation received or utilized by plants, the groundwater table fluctuates both with the seasons and from year to year. The potential for groundwater rise and its corrosive potential pose a threat to built-up infrastructure, making it a vital factor to address in long-term urban development.

Landscape character: It is described as a distinct, recognized, and consistent pattern of features in the landscape that distinguishes one landscape from the next. The landscape character is created by specific combinations of physical landscape components (geology, landform, soils, and vegetation) and anthropogenic landscape components (land use, field patterns, and human habitation). This distinguishes diverse landscapes from one another and provides each one its own feeling of place. As a result, landscape character is a reflection of the landscape's holistic nature (Jessel 2006).

Climate: The weather has an impact on urban development decisions. Precipitation, air temperature, and sun incidence are all climatic factors that impact urban development. Microclimate changes dramatically across small distances and time periods. Brown and Gillespie (1995) proposed a method for representing the microclimate of the site that considers both solar radiation and wind exposure (Beer and Higgins 2000).

Wildlife habitat: Most wildlife species' populations are inherently discontinuous. The consideration of wildlife habitat is important before locating different infrastructure facilities for the protection of existing species.

Ecological value: It is critical to consider the influence of planned urban design decisions on the site's current ecological value. Geo-climatic factors (slope, altitude, aspect, and climatic), biological features (vegetation and animals), and social significance (cultural, economic, and historical value) are among the aspects of ecological evaluation (Patowary and Sarma 2019). The geo-climatic, ecological, and social relevance traits are chosen based on the location and condition of the place under evaluation.

5.3.2 *Public Participation:* Social Aspects

The *social aspects* are required for understanding the intended purpose of modifying the site under consideration. This requires the stakeholders' input.

Visual aspects: Visual characteristics of a site have an impact on its usability. Visual resources help to establish lasting impressions of a site. Before allocating land for a specific facility, the aesthetic quality of the site as well as visible off-site characteristics must be addressed. Views of historically significant buildings, notable mountains, or other landmarks are crucial site qualities because they communicate a strong identity that residents and tourists can sense.

Historic factors**:** The land allocation for different facilities should be done by keeping in mind the historic resources present on or near the site under consideration. Existing bridges, buildings, walls, signs, and many other noteworthy structures or pieces erected in former times are examples of historic resources. Buildings, bridges, and other structures listed on the list of nationally significant historic resources in India are legally protected by the Archaeological Survey of India. At the state and local levels, there are regulations in place to safeguard historic treasures. A structure's or neighbourhood's historic value is determined by numerous factors, including age, quality, uniqueness, and representativeness (Ames and McClelland 2002).

Distance from other activities**:** Urban planning decisions should be taken in a manner that optimizes the distance between different activities. It should strive to reduce the number and distance of automobile journeys by combining land uses and ensuring that walkers, bicycles, and transit passengers have safe and comfortable alternatives.

Population density**:** The quantity of people living per square kilometre is referred to as population density. It is a crucial aspect while taking urban planning decisions, specifically in areas which are urbanizing at a fast pace. If it is too sparse, then the efficiency is lost, and if it's too dense, then it becomes an impossible and uncomfortable habitat to live in, resulting in squatters and jams.

5.3.3 *Energy-Intensive Planning:* Economic Aspects

The final urban planning proposal is evaluated on the basis of the following *economic aspects*.

Slope: The gradient or incline/decline of the terrain is referred to as a slope. It is calculated by dividing the difference in elevation between two places by the lateral distance between these two sites. The slope value can be stated as a percentage, a ratio, or an angle (Steven et al. 2013). Slope analysis is a crucial method that aids a planner in selecting the parts of a site that may be developed.

Soil type: It is defined as the systematic classification of soils based on their texture, such as sand, silt, and clay. A unique soil type is made up of all soils that share a set of well-defined features. Because of characteristics such as carrying capacity, fertility, and permeability, soil type is vital to consider when making urban design decisions.

Accessibility: It refers to how accessible a location is via roads and other forms of transportation. The transportation of raw materials for construction as well as other vehicle services necessitates the accessibility of a site. Before making any urban design decisions, it is critical to consider the site's accessibility (Curtis 2008).

***Functioning of site*:** It refers to the efficient pattern of various activities associated with different facilities. This is an important aspect to be considered by the planners before locating different infrastructural facilities, so as to optimize the utilization of resources, which ultimately result in efficient functioning of a site. Un-thoughtful allocation of land for various facilities leads to inefficient functioning of the site.

***Socio-economic conditions*:** Socio-economic conditions shape the quantity and quality of spaces and their ability to generate activities by influencing planning decisions. Variation in socio-economic conditions alters people's desires and needs and thus demands different facilities. The impacts of urban planning decisions are required to be assessed on the socio-economic conditions of the existing population. Urban planning is defined as one of the most important strategies for achieving a place's social and economic development. According to Healey et al. (1997), socio-economic factors influence urban planning policies, regulations, and practices. Furthermore, the kind of limitations and possibilities given by planning policies, regulations, and practices affect socio-economic dynamics.

Existing utilities: Water supply networks, overhead electric supply lines, telephone lines, existing roads, and walkways are examples of public utilities that offer daily requirements to users of various facilities constructed, and so influence urban design decisions. Before making urban planning decisions, it's critical to know where current utilities are located in order to identify where the new construction will link to the existing one.

5.4 Geospatial Technologies for Sustainable Urban Planning

5.4.1 Need of Geospatial Technologies

Geospatial technology is the term used for the tools and technologies that provides and assists in analysing the locational and spatial information such as remote sensing (RS), geographic information system (GIS), and global positioning system (GPS). For sustainable urban planning, geospatial technologies have become more important. These technologies help in understanding the biological, environmental, hydrological, geological, and physical features of the Earth surfaces. Satellites with better frequency and resolution pictures in terms of spectral, radiometric, and geographical extents than previous satellites have been launched as part of recent advances in geospatial technology. This advancement in geospatial technology aided in recognizing, comprehending, and solving environmental problems in a very efficient manner.

The most extensively used geospatial technology is GIS, which has undergone significant refining since its inception in the 1960s. Many national and sub-national statistics offices provide GIS data as well as geocoding demographic data to overlay on maps. Planners utilize GIS to show and analyse various types of geographic data. GIS has advanced geographic analysis capabilities, which aids in making better informed, logical, and efficient decisions based on scientific methodologies [90]. Geospatial approaches and frameworks may give answers to urban planning challenges by assuring a full review of many aspects such as topography, existing infrastructure facilities/utilities, and other aspects that need geospatial analysis. The manual method makes it difficult to analyse huge datasets of diverse elements, which is easily handled by overlay techniques in GIS (Steinitz et al. 1976). GIS store information in digital form in place of manually mapping of various aspects for planning. The comparison of alternative scenarios is also made for choosing the optimal strategy for urban planning and development. GIS assists planners in generating different scenarios efficiently and logically (Ministry of Urban Development, 2014). GIS have powerful computing functions which can help to achieve more informed and more sustainable urban planning decisions (Ervin 2012).

Because developments in technology have made RS more precise and pervasive, it has become a new significant basis for defining the existing condition of the urbanizing world. Advanced satellites are launched with higher recording capabilities for more frequent and higher resolution images, better algorithms and computing capacity for refining interpretation processes, and enhanced memory and power to handle the large amounts of data. The data obtained by the RS and GPS is combined with other spatial and non-spatial data in the GIS for supporting urban planners and decision-makers in achieving long-term sustainability. The use of geospatial technologies can help in making wiser policy decisions for more sustainable urban planning.

5.4.2 Application Areas

The use of geospatial technologies for municipal, regional, and environmental planning has exploded in the last 50 years. The number of applications available is determined by the type of information a community requires, as well as purposes such as resource mapping, pollution modelling, water, traffic, and sewer management, and disaster prediction to minimize risk and build resilience. Finding land suitability for analysing urban growth (Saxena and Jat 2020) evaluating planning interventions in the existing landscape, and choosing the best site for urban facilities to cater to various needs have all been solved using GIS. GIS has also been used to discover factors that influence urban expansion and to forecast future growth (Duwal et al. 2018). Madadi et al. (2015) investigated the change in flood inundation area using GIS. Critical criteria for urban growth have been identified and analysed using a GIS-based planning support system (Long et al. 2011). To include environmental considerations into spatial planning, Carsjens and Ligtenberg (2007) employed a GIS-based support system. GIS was utilized by Baz and Nogay (2009) to assess environmental sustainability. Bansal (2011) advocated for the use of geographic information systems (GIS) in spatial decision-making. Bansal and Pal (2008) used GIS to create, evaluate, and visualize construction project schedules. Kumar and Biswas (2013) analysed possible urban development areas using GIS-based appraisal. Geospatial data has become a more significant and vital aspect in recent years, contributing both directly and indirectly to the urban development at different scales.

The geospatial technologies can also be used in the context of the sustainable development goals. Poverty is a fundamental impediment to growth and long-term sustainability. To elevate diverse communities, different techniques are required. Remotely sensed satellite data, combined with powerful machine learning approaches, has allowed researchers to map and predict poverty on a worldwide scale (Steele et al. 2017). Geospatial technology also aid in mapping and calculating agricultural yields and determining crop compatibility based on geo-climatic characteristics. Geospatial tools are mostly used at the micro level to map groundwater resources and drainage patterns (Das and Pal 2017). In general, smaller high-resolution satellites appear to offer more to small scale understanding, whereas mid-resolution satellites provide insight at a larger scale. RS in conjunction with geospatial analysis may be used to investigate possible solar and wind energy locations to identify high-priority clean energy investment prospects.

The area under consideration may be monitored for its sustainability using a combination of satellite and ground sensors combined with GIS technology (Hardin et al. 2007). High-resolution satellite or aerial imagery may be used to map urban metrics and offer a urban planning proposal. RS imagery has been used to extract and update transportation networks, buildings, and impervious areas (Yang and Li 2015), provide land use/cover data (Acharya et al. 2016), and identify urban expansion in a more integrated method. Census data has been used to enhance image categorization in urban areas which can ultimately help in estimating the population and residential densities (Acharya et al. 2018; Langford et al. 1991). These applications led to reduce the environmental harm caused by urbanization and thereby aid in the development of more sustainable and efficiently planned urban centres. Web GIS may be used to distribute this information as well (Kulawiak et al. 2010; McCord et al. 2018). RS gives real-time data on what's going on with the land surface. Forest cover is a key measure of a land surface's health. For sustainable development, it should be understood how it varies in relation to overall area change, as well as the fraction of land degraded over total area of land. These may be accomplished by employing geospatial technologies.

References

Acharya TD, Yang IT, Lee DH (2016) Land cover classification using a KOMPSAT-3A multi-spectral satellite image. Appl Sci 6:371. https://doi.org/10.3390/app6110371

Acharya TD, Yang IT, Lee DH (2018) Land cover classification of imagery from Landsat operational land imager based on optimum index factor. Sens Mater 30:1753. https://doi.org/10.18494/SAM.2018.1866

Ames L, McClelland LF (2002) Historic residential suburbs: guidelines for evaluation and documentation for the national register of historic places. Department of Interior, Washington, DC

Bansal VK (2011) Application of geographic information systems in construction safety planning. Int J Proj Manag 29(1):66–77

Bansal VK, Pal M (2008) Generating, evaluating, and visualizing construction schedule with geographic information systems. J Comput Civ Eng 22(4):233–242

Bathrellos GD, Gaki-Papanastassiou K, Skilodimou HD (2012) Potential suitability for urban planning and industry development using natural hazard maps and geological–geomorphological parameters. Environ Earth Sci 66(2):537–548

Baz AG, Nogay S (2009) Development and application of GIS-based analysis/synthesis modelling techniques for urban planning of Istanbul area. Adv Eng Softw 40(2):128–140

Beer AR, Higgins C (2000) Environmental planning for site development: a manual for sustainable local planning and design. E & FN Spon, London

Brail RK (2005) Planning support systems evolving: when the rubber hits the road. In: Portugali J (ed) Complex artificial environments. Springer, New York, pp 307–317

Brown RD, Gillespie TJ (1995) Microclimatic landscape design. Wiley, New York

Carsjens G, Ligtenberg A (2007) A GIS-based support tool for sustainable spatial planning in metropolitan areas. Landsc Urban Plan 80(1):72–83

CommunityViz (2021) CommunityViz: urban analytics for planners. www.communityviz.com. Accessed 15 May 2021

Curtis C (2008) Planning for sustainable accessibility: the implementation challenge. Transp Policy 15(2):104–112

Dai FC, Lee CF, Zhang XH (2001) GIS-based geo-environmental evaluation for urban land-use planning: a case study. Eng Geol 61(4):257–271

Das RT, Pal S (2017) Exploring geospatial changes of wetland in different hydrological paradigms using water presence frequency approach in Barind Tract of West Bengal. Spatial Inf Res 25:467. https://doi.org/10.1007/s41324-017-0114-6

Das S, Bhattacharya A, Mali S (2013) Study on urban land suitability assessment using remote sensing and GIS: a case study of Khairagarh, in Chhattisgarh. Int J Comput Appl 74(10):20–26

Dong J, Zhuang D, Xu X, Ying L (2008) Integrated evaluation of urban development suitability based on remote sensing and GIS techniques—a case study in Jingjinji Area, China. Sensors 8(9):5975–5986

Duwal S, Amer S, Kuffer M (2018) Modelling urban growth in the Kathmandu valley, Nepal. In: Van Maarseveen FAM, Martinez JA, Flacke J (eds) GIS in sustainable urban planning and management: a global perspective. CRC Press, Boca Raton, pp 205–224

Envision Tomorrow (2021) About Envision Tomorrow. http://www.frego.com/projects/documents/envisiontomorrow_info.pdf. Accessed 6 Dec 2021

Ervin S (2012) A system for geodesign. Proceedings of digital landscape architecture. Wichman, Atlanta

Ewing R, Cervero R (2010) Travel and the built environment. J Am Plan Assoc 76(3):265–294. https://doi.org/10.1080/01944361003766766

Geertman S, Stillwell J (2003) Planning support systems: an inventory of current practice. Comput Environ Urban Syst 28:291–310

Hardin PJ, Jackson MW, Otterstrom SM (2007) Mapping, measuring, and modeling urban growth, vol 141. Springer, Heidelberg. https://doi.org/10.1007/978-3-540-69417-5_8

Healey P, Khakee A, Motte A, Needham B (1997) Making strategic spatial plans: innovation in Europe. UCL Press, London

Hill J, Nelson E, Tilman D et al (2006) Environmental, economic, and energetic costs and benefits of biodiesel and ethanol biofuels. Proc Natl Acad Sci USA 103:11206–11210

I-Place (2021) I-PLACE3S. www.ruralgis.org/conference/2003proceedings/dayTwo/CommunityPlanning/morning_Session2A/place3s_McKeever.pdf. Accessed 29 July 2021

Jessel B (2006) Elements, characteristics and character–information functions of landscapes in terms of indicators. Ecol Indic 6(1):153–167

Klosterman R (1997) Planning support systems: a new perspective on computer-aided planning. J Plan Educ Res 17:45–54

Kraus M, Charvátová P (2016) Location as a determinative factor of building airtightness. Proced Eng 161:1532–1537

Kulawiak M, Prospathopoulos A, Perivoliotis L, Fuba M, Kioroglou S, Stepnowski A (2010) Interactive visualization of marine pollution monitoring and forecasting data via a web-based GIS. Comput Geosci 36:1069. https://doi.org/10.1016/j.cageo.2010.02.008

Kumar M, Biswas V (2013) Identification of potential sites for urban development using GIS multi criteria evaluation technique. J Settl Spatial Plan 4(1):45–51

LaGro JA (2007) Site analysis: a contextual approach to sustainable land planning and site design. Wiley, New York

Langford M, Maguire D, Unwin D (1991) The areal interpolation problem: estimating population using remote sensing in a GIS framework. In: Masser I, Blakemore M (eds) Handling geographical information: methodology and potential applications. Longman, London, p 55

Li K, Zhang P, Crittenden JC, Guhathakurta S, Chen Y, Fernando H, Sawhney A, McCartney P, Grimm N, Kahhat R, Joshi H, Konjevod G, Choi YJ, Fonseca E, Allenby B, Gerrity D, Torrens PM (2007) Development of a framework for quantifying the environmental impacts of urban development and construction practices. Environ Sci Technol 41(14):5130–5136

Lian X (2018) Review on advanced practice of provincial spatial planning: case of a Western, less developed province. Int Rev Spatial Plan Sustain Dev 6(3):185–202

Long Y, Shen Z, Mao Q (2011) An urban containment planning support system for Beijing. Comput Environ Urban Syst 35(4):297–307

Lynch K, Hack G (1984) Site planning. MIT Press, Cambridge

Madadi MR, Azamathulla HM, Yakhkeshi M (2015) Application of Google earth to investigate the change of flood inundation area due to flood detention dam. Earth Sci Inform 8(3):627–638

McCord P, Tonini F, Liu J (2018) The Telecoupling GeoApp: a web-GIS application to systematically analyze telecouplings and sustainable development. Appl Geogr 96:16. https://doi.org/10.1016/j.apgeog.2018.05.001

McCrea R, Walters P (2012) Impacts of urban consolidation on urban liveability: comparing an inner and outer suburb in Brisbane, Australia. Hous Theory Soc 29(2):190–206

McHarg L (1969) Design with nature. Natural History Press, New York

Ministry of Urban Development (2014) Urban and regional development plan formulation and implementation (URDPFI) guidelines. Government of India, New Delhi, p 81

Morales J, Flacke J, Martinez JA (2018) A collaborative planning framework to support sustainable development: the case of municipal housing in Guatemala City. In: Van Maarseveen FAM, Martinez JA, Flacke J (eds) GIS in sustainable urban planning and management: a global perspective. CRC Press, Boca Raton, pp 29–44

Norman J, Maclean HL, Asce M, Kennedy CA (2006) Comparing high and low residential density: life-cycle analysis of energy use and greenhouse gas emissions. J Urban Plan Dev 132:10–21

Patowary S, Sarma AK (2019) Projection of urban settlement in eco-sensitive hilly areas and its impact on peak runoff. Environ Dev Sustain 22(6):5833–5848

Paul VK, Basu C (2016) Scenario planning and risk failure mode effect and analysis (RFMEA) based management. J Constr Eng Project Manag 6(2):24–29

Persson C (2013) Deliberation or doctrine? Land use and spatial planning for sustainable development in Sweden. Land Use Policy 34:301–313

Petit S, Griffiths L, Smart SS, Smith GM, Stuart RC, Wright SM (2004) Effects of area and isolation of woodland patches on herbaceous plant species richness across Great Britain. Landsc Ecol 19: 463–471

Pickett STA, Boone CG, McGrath BP, Cadenasso ML, Childers DL, Ogden LA, Grove JM (2013) Ecological science and transformation to the sustainable city. Cities 32:S10–S20

Rubenstein HM (1996) A guide to site planning and landscape construction. Wiley, New York

Saxena A, Jat MK (2020) Land suitability and urban growth modelling: development of SLEUTH-suitability. Comput Environ Urban Syst 81:101475

Sharma BK, Chandel MK (2017) Life cycle assessment of potential municipal solid waste management strategies for Mumbai, India. Waste Manag Res 35(1):79–91

Steele JE, Sundsøy PR, Pezzulo C, Alegana VA, Bird TJ, Blumenstock J, Bjelland J, Engø-Monsen K, Montjoye YD, Iqbal AM (2017) Mapping poverty using mobile phone and satellite data. J R Soc Interface 14:20160690. https://doi.org/10.1098/rsif.2016.0690

Steinitz C, Parker P, Jordan L (1976) Hand drawn overlays: their history and prospective uses. Landsc Archit 9:444–455

Steven S, Kurt N, Jake W (2013) Slopes expressed as ratios and degrees. In: Site engineering for landscape architects, 6th edn. Wiley Publishing, New York

Tian Y, Xiao C, Wu L (2010) Slope unit-based landslide susceptibility zonation. In: Proceedings of Geoinformatics, 2010, 18th international conference, UPlan, pp 1–5. http://ice.ucdavis.edu/doc/uplan

Todes A, Karam A, Klug N, Malaza N (2010) Beyond master planning? New approaches to spatial planning in Ekurhuleni, South Africa. Habitat Int 34(4):414–420

UN-Habitat (2009) Sustainable urbanization: revisiting the role of urban planning. Global report on human settlement. UN-Habitat, Nairobi

UPlan (2021) Information center for the environment—archive. https://www.ice.ucdavis.edu//doc/uplan. Accessed 6 Jan 2021

Vonk G, Geertman S (2005) Bottlenecks blocking widespread usage of planning support systems. Environ Plan 37(5):909–924

Vonk G, Geertman S, Schot P (2007) A SWOT analysis of planning support systems. Environ Plan A 39(7):1699–1714

Watson V (2009) Seeing from the south: refocusing urban planning on the globe's central urban issues. Urban Stud 46(11):2259–2275

Wheeler SM (2013) Planning for sustainability: creating livable, equitable and ecological communities. Routledge, London

Yang J, Li P (2015) Impervious surface extraction in urban areas from high spatial resolution imagery using linear spectral unmixing. Remote Sens Appl Soc Environ 1:61. https://doi.org/10.1016/j.rsase.2015.06.005

Yang F, Zeng G, Du C, Tang L, Zhou J, Li Z (2008) Spatial analysing system for urban land-use management based on GIS and multi-criteria assessment modelling. Prog Nat Sci 18(10): 1279–1284

Yeh AGO, Wu F (1999) The transformation of the urban planning system in China from a centrally-planned to transitional economy. Prog Plan 51:165–252

Youssef AM, Pradhan B, Tarabees E (2011) Integrated evaluation of urban development suitability based on remote sensing and GIS techniques. Arab J Geosci 4:463–473

Chapter 6
Exploring Sustainable Food Systems for Urban Sustainability

Deeksha Dave

6.1 Introduction

Food is a universal discussion subject. It is not like any other commodity but is more important because it affects our overall health and well-being. Food has a multidimensional feature. It not only is essential for the survival but also relates to social fairness, gender equality, energy consumption, land use, water availability, etc. (Brand et al. 2019).

For urban middle class, living in the urban and semi urban areas, food (or eating out) is a way to rejoice and celebrate the occasions. Hotels, restaurants and hospitality industries constitute a major economic and social sector in cities. A study by Pothukuchi and Kaufman (1999) revealed the prominence and meaning of food in the city environs.

Despite the fact that food, cities and population associate in numerous ways, these regions are perceived as the recipients of the food grown in rural and semi-urban areas and neglected in urban management studies (Haysom 2021). The fact that the food transportation, food storage and food distribution involve urban policies and governance is generally overlooked. The "Food" and sustainability concerns surrounding it normally do not find place in the discourses related to urban issues because of the fact that food is considered to be more of rural/agricultural issue. The researchers also believe that since it appears to be less financially attractive as opposed to transport and housing, it is mostly neglected in urban studies, planning and development (Brand et al. 2019).

D. Dave (✉)
School of Interdisciplinary and Trans-Disciplinary Studies, Indira Gandhi National Open University, New Delhi, India
e-mail: deekshadave@ignou.ac.in

B. Pathak, R. S. Dubey (eds.), *Climate Change and Urban Environment Sustainability*, Disaster Resilience and Green Growth,
https://doi.org/10.1007/978-981-19-7618-6_6

The United Nations Sustainable Development Goals (SDGs) which are targeted to be attained by the year 2030 relate to urban food consumption directly as well as indirectly. Goal 2: Zero Hunger cannot be achieved unless the urban food sustainability is accorded due to importance and attention. Further, in one way or the other, the other sustainable development goals like pertaining to poverty, human health, education, gender equality, global partnership and responsible production and consumption are also related to food systems (Ilieva 2017).

In view of this, the current chapter throws light on the sustainable food systems and presents a viewpoint that urban sustainability is not possible without intriguing discussion and management of its food systems. The chapter begins by explaining the concept of sustainable food systems and explores food habits and preferences of the urban population. The chapter ends on the note that revitalizing the indigenous food preferences and popularizing the local cuisines can augment the food system sustainability in urban regions.

6.2 Sustainable Food Systems

Food system as a whole comprised of several processes like production through agriculture and dairy sector, processing and distribution of the food items through various channels and consumption of food at micro and macro level and the food losses at various stages (Brand et al. 2019). Agriculture in itself combines a set of actions like seeding, fertilizing, irrigating, weeding and harvesting. An ideal food system is one which aims to utilize the food, reaches to the maximum and alongside maintains its nutritive value, purity and quality. According to UN, food structures are sustainable when it means to safeguard food safety and provide nutrition to the current generation along with maintaining the same for coming generations as well. It requires balance between the interconnected social and economic determinants of good quality food and ensuring optimum food supply across the diverse population groups (Nutrition 2014). It not just makes food available but also strives for food accessibility and food affordability.

Different scholars and bodies have emphasized on different aspects while recognizing the importance of sustainable food systems. For example, HLPE, 2014 has highlighted the role of reducing food wastage but on the other hand UNEP (2016) emphasizes on the balanced nature's interaction for attaining sustainability in food systems.

The idea of environmental protection and natural resource conservation is embedded in sustainable food system. It relies on the elements of environment in its purest form (fertile soil, clean water and air), trained work force, processes and infrastructure utilizing modern technology, etc. Agriculture and allied sector does not merely influence the environmental settings but also has socio-economic consequences. The livelihoods, gender equality and migration issues form an integral part of sustainable food systems.

Because of climate change, resource depletion, widening economic gap and increase in population and changing food habits, the sustainable food systems are under threat. Today, the agriculture is industrialized, food is commercialized, human health is compromised and food sustainability is threatened. It is predicted that the population of the world will surge to over 9 billion in 2050 from 7 billion at present and it is estimated that demand for food will grow by over 60% (Searchinger et al. 2019). The demands are not just for more food but also the quality, type (vegetarian/non-vegetarian/vegan/keto) and diversity in food are demanded by the growing urban middle class. More so, it is abysmal to note that roughly 17% of total global food production is wasted (Index 2021). It includes the food losses occurring during supply chains and production. As per The Global Hunger Index Report 2021, it is projected that 47 countries of the will not be able to achieve zero hunger by 2030. The political unrest, weather related extremes, the economic and health challenges associated with the Covid-19 pandemic are the causes for driving hunger and malnutrition. For example, the directive of sustainable development goals (SDGs) to include all human beings without leaving anyone seems difficult as many of the African and Asian countries are struggling to fight over the poverty and hunger.

The General Assembly of UN declared the decade from the year 2016 to 2025 for taking stringent actions for nutrition. The objective was to initiate action and formulate plans to eradicate malnutrition at all levels and in all aspects in all the countries. This is possible only through effective implementation of policies, programmes and increased investments. The Global Hunger Index Report, 2021 highlights that the prevalence of undernourishment is increasing worldwide. It is clear that the problem of undernourishment is common across the age groups, gender in various regions and countries. The prevalence of excess body weight and obesity is increasing in almost every country of the world. It is estimated that globally more than 30% of adults are obese.

The production of ample food and its affordable consumption depends to a large extent on ecosystems and natural capital and vice versa. It is also true that the food production processes have caused irreversible changes in ecosystems, deforestation and loss of biodiversity. Food processes play a central role in ensuring sustainable development by addressing food security, alleviation of poverty, and strengthening communities for participation. In 2020, Herrero et al. (2020) raised the importance of innovations for creating an ecosystem for sustainable food system.

6.3 Urban Food Supply

Urban areas are characteristic by less accessible land for the food cultivation. With the improvements in centralized infrastructure and modern means of food preservation, the urban food supply ensures safe transformation of food from peri-urban to urban areas. A study by Boyer et al. (2019) highlighted the growing realization of cities to ensure a well-functioning food system. Urban food supply means the whole system to not just meet the demand of food but also to look into the critical aspects of agriculture, transportation of food, the distribution channels and management of food waste in cities. The food supplies in urban regions are not just limited to residential areas but also tends to include commercial establishments like restaurants, hostels, food processing industries and so on. The environmental footprint of urban food supply including the impact on water, energy, resource consumption can be mitigated by adopting suitable measures.

The present food system is known by unidirectional flow of resources from rural areas into urban areas. The Ellen MacArthur Foundation Report (Biocycles Urban 2017) reveals that the cities consuming more than 70% of natural resources like water, land, energy, etc., are responsible for generating 50% of global waste, and more than 60% of greenhouse gas emissions are released by the cities. Urban food supply is influenced by several tenacious issues like high population growth, resource scarcity, overcrowding and sanitation and hygiene. To deal with these issues, food supply in urban areas must be sustainable in its practice, as well as resilient to face the threatening weather disruptions like urban flooding, droughts, cold waves, heat and so on. Most often the mechanism of urban food supply deals with providing access to healthy and nutritious food through well connected cities, strengthened local economy, reduced food wastage and active participation of the participants in effective decision-making. Bricas (2019) has described the distancing between the food and the urban settlers. He has explained that there is a geographical distance in cities as the place where the food is consumed is far away from the place from where it is grown. There is economic distance also as lot of intermediaries are also there and political distance leaves little or no control over food. A lot of studies have demonstrated that shortening of the distance both geographical and economic is sustainable in the long run. To address income disparities and encourage livelihoods of the communities, it is envisaged that the urban food supply is efficient and comprehensive. It is possible through appropriate policies and regulatory frameworks (Byerlee and Fanzo 2019; FAO 2012).

The urban centres must also focus on the output of the food systems. These outputs relate to food distribution across the different strata of population, affordability and accessibility of food and also social importance of food in terms of its nutritional value.

With the growth of agriculture and allied sector during 2003–2004, India became the exporter of food and became self-sufficient in food production. Urbanization was the primary factor along with fossil fuel consumption that led to the increase in food production. After independence, India gained food sovereignty but we are still fighting to accomplish food security for everyone.

6.4 Diets and Food Habits of Urban Population

At times, food security and nutrition were perceived primarily as a rural subject, but with rapid urbanization and changing food system penetration in urban households, there are upcoming challenges. The demand for sustainable, safe and nutritious food is growing, and ensuring food security has become an urban issue (Tefft et al. 2017).

There was a time when diets of the people living in different geographical regions varied significantly. Their diets were influenced by the availability of the grains, seasonal vegetables and fruits and climatic conditions at a given place. However, today the food preferences and customer expenditure on food like cereals, fruits vegetables and dairy items in urban areas of India have changed dramatically. Throughout the year, all fruits and vegetables appear in the market which was at once available only during a particular season. As the consumption patterns have been altered, the convenience food, fast food, packaged food and the uniform type of food is present in the shelves of the marts and so does has the consumption patterns changed. The market pressures and commodification of food along with long working hours of the women who working out, etc. trends are designing our diets and meals. We tend to eat what is packaged properly (even though the packaging is not required), what appears to be healthy (even though it is not) and what others are eating (even if it does not match with our geographical conditions and climate). It has made food easily available and accessible but degraded the agricultural ecosystems. Moreover, consumption of similar food across the country does not suit the dietary requirements of people living in different areas of distinct climatic conditions. It causes acute as well as chronic health ailments and unhealthy food habits are recognized as a likely cause for obesity particularly in children (Ebbeling et al. 2004). It is worrisome that India is slowly entering into the pandemic of non-communicable diseases (NCDs) like obesity, diabetes and hypertension.

The study done by Eker et al. (2019) highlights that the sustainability in the urban food systems can be brought only when the traditional kitchen is revived and traditional culinary skills are celebrated. The cultural aspects relate to ethnicity, food choices and ways of cooking and choice of consumption. In the long run, all these determine the socio-economic and environmental sustainability (Carrus et al. 2018). The social and cultural elements related to food are crucial for promoting sustainable urban food style. In this context, the contribution of traditional food and importance of urban agriculture must be highlighted (Marsden and Sonnino 2012). We must restore the ancient culinary practices because slowly and slowly the traditional knowledge related to food and its ways of consumption is being lost. Revival of indigenous practices can help overcome this concern.

No matter how globalized the food systems become, food systems inherently get affected by space and time. In developed economies, food security and sustainability have been the key governance concerns. It is revealed that in spite of being the leading producer of milk, pulses, rice, wheat, sugarcane, groundnut, vegetables, there has been no improvement in the calorie uptake by the poor and nutritional poverty continued to increase in India. This categorically affects the economic weaker sections, marginalized groups and evokes gender disparities.

The purchase decisions of the millennial are based on health concerns like calorie content and origin of the edible food items. The coming generation is more aware of their actions in terms of ecological footprint. This generation demands healthy and less processed food. Therefore the urban planners must integrate these issues while framing future food policy in urban areas.

6.5 Food Wastage in Urban Areas

Food waste comprised of agricultural waste, dairy waste and also the waste generated while cooking and the leftover food. It also includes the waste generated from various sources like food related industries, food joints, restaurants, hotels, hospitality sector along with the residential areas. With the changing consumption patterns, the food waste in urban and rural areas of both the developed and developing countries differs strikingly.

Food is also wasted along the supply chain while transporting the food from rural to urban areas. FAO estimates that more than a billion tonnes of rotten vegetables, unused fruits, dairy products, poultry, meat and other edible items are wasted in various processes of supply chain like food transportation, packaging and distribution (FAO 2012). Studies have revealed that by the year 2025, the amount of food wasted in the cities of Asian countries could be more than 400 million tonnes (Melikoglu et al. 2013).

The food wastage also leads to detrimental environmental consequences. Apart from food, the wastage in the form of resources like land, energy and water, the carbon footprint is growing and greenhouse gas emissions increase the concentration of carbon dioxide in the atmosphere. Further, most of the municipal solid waste comprised of food waste which is dumped in landfills, thrown in rivers or incinerated. All these practices contribute to environmental pollution.

Shortening of the food cycle is intrinsic to reduce the wastage of food in several aspects. In urban areas, we find that food before coming to our plate passes through various stages like beginning from farmer to middlemen, transportation and packaging and distribution. This is unlike the case wherein the food grown in the kitchen gardens is utilized directly and the food chain is greatly reduced.

A number of innovative measures are taken to address the problems related to food and minimize its wastage (Dave 2021). These innovations can range from appropriate land use, emissions control, improving and diversifying production so as to improve diets and reduce wastage.

Food sharing is playing a major role in urban areas and has personal, social, economic and cultural implications (Davies and Evans 2019; Gurven and Jaeggi 2015). Food Apps and online platforms are utilizing ICT to encourage sharing among the people with the support of NGOs, business groups and consumers.

6.6 Untangling the Challenges of Sustainable Food System

The urban food sustainability faces enormous challenges in the form of increased dependence on global imports, political instability and conflicts, environmental degradation, climate change and increased resource depletion (Olsson 2018). Urban food issues remained neglected by the urban planners considering it to be mainly falling under the rural domain. However, there is a growing realization today that rural as well as urban areas converge in terms of food and eating habits of the people. Food has a multidimensional character and influences public health and nutrition, social justice, access to energy and economic development.

With the growing share of the urban population, changing social structure, globalized economy, changing environment and the fast growing cities, there is a need to acclimatize to the forthcoming challenges for sustainable urban development (Zhang et al. 2018). These adaptations must be towards more resilience and must ensure food security, better human health and well-being, social inclusion, sustained economic growth and improved environmental quality (UNEP 2016). These adaptations coupled with self-reliant food systems, reduced food footprint, increased resource efficiency and shortened food cycle can bring revolution in the urban food system (Grewal and Grewal 2012). Food safety is another major concern in the urban areas where adulteration and poisoning, growing health concerns of the population and digital transformation are leading the food businesses in the urban areas.

6.7 Conclusion

Food lies at the heart of our health, and urban food practices determine sustainability challenges (Gordon et al. 2017). Food policies must be such that interconnect human health, environmental quality and sustainability. It has a potential to create synergies to evolve efficient urban food sustainability. Further, food is the best way to know about the sustainable development. Unhealthy food consumption and intake practices are depriving the urban population especially the children with adequate nutrition and limiting appropriate physical and mental growth. This could be challenged through generating awareness, educational platforms, lifestyle changes, government policies and sound implementation.

Urban sustainability is possible through effective food systems management. With the income levels rising in the urban areas, food affordability and accessibility can also lead to better food utilization. The infrastructure in our cities coupled with smooth supply chain models can minimize food losses at various levels.

With the major share of the population going to be urban in few years, the food related issues of sustainability must be well thought of and effective policies and citizen engagement can help to achieve sustainable urban food sustainability. A sustainable food system in urban areas is only possible when factors including food production, transportation, preparation and utilization are in accordance with the varied needs and requirements of urban population.

References

Biocycles Urban (2017) Ellen MacArthur Foundation

Boyer D, Sarkar J, Ramaswami A (2019) Diets, food miles, and environmental sustainability of urban food systems: analysis of nine Indian cities. Earth's Future 7(8):911–922

Brand C, Bricas N, Conaré D, Daviron B, Debru J, Michel L, Soulard CT (2019) Designing urban food policies. In: Concepts and approaches. Springer Nature, Berlin, p 142

Bricas N (2019) Urbanization issues affecting food system sustainability. In: Designing urban food policies. Springer, Cham, pp 1–25

Byerlee D, Fanzo J (2019) The SDG of zero hunger 75 years on: turning full circle on agriculture and nutrition. Glob Food Sec 21:52–59

Carrus G, Pirchio S, Mastandrea S (2018) Social-cultural processes and urban affordances for healthy and sustainable food consumption. Front Psychol 9:2407

Dave D (2021) Solid waste management and its main aspects in the Indian context. In: Industry, innovation and infrastructure. Springer, New York, pp 1018–1026

Davies A, Evans D (2019) Urban food sharing: emerging geographies of production, consumption and exchange. Geoforum 99:154–159

Ebbeling CB, Sinclair KB, Pereira MA, Garcia-Lago E, Feldman HA, Ludwig DS (2004) Compensation for energy intake from fast food among overweight and lean adolescents. JAMA 291(23):2828–2833

Eker S, Reese G, Obersteiner M (2019) Modelling the drivers of a widespread shift to sustainable diets. Nat Sustain 2(8):725–735

FAO (2012) Towards the future we want: end hunger and make the transition to sustainable agricultural and food systems. FAO, Rome

Gordon LJ, Bignet V, Crona B, Henriksson PJ, Van Holt T, Jonell M, Queiroz C (2017) Rewiring food systems to enhance human health and biosphere stewardship. Environ Res Lett 12(10): 100201

Grewal SS, Grewal PS (2012) Can cities become self-reliant in food? Cities 29(1):1–11

Gurven M, Jaeggi AV (2015) Food sharing. In: Emerging trends in the social and behavioral sciences: an interdisciplinary, searchable, and linkable resource. Wiley, New York, pp 1–12

Haysom G (2021) Integrating food sensitive planning and urban design into urban governance actions. Urban Forum 32:289–310

Herrero M, Thornton PK, Mason-D'Croz D, Palmer J, Benton TG, Bodirsky BL, West PC (2020) Innovation can accelerate the transition towards a sustainable food system. Nat Food 1(5): 266–272

Ilieva RT (2017) Urban food systems strategies: a promising tool for implementing the SDGs in practice. Sustainability 9(10):1707

Index GH (2021) Global hunger index: hunger and food systems in conflict situations

Marsden T, Sonnino R (2012) Human health and wellbeing and the sustainability of urban–regional food systems. Curr Opin Environ Sustain 4(4):427–430

Melikoglu M, Lin CSK, Webb C (2013) Analysing global food waste problem: pinpointing the facts and estimating the energy content. Cent Eur J Eng 3(2):157–164

Nutrition HLPOEHOFSA (2014) Sustainable fisheries and aquaculture for food security and nutrition. In: A report by the High Level Panel of Experts on Food Security and Nutrition

Olsson EGA (2018) Urban food systems as vehicles for sustainability transitions. Bull Geogr Socio-Econ Ser 40:133–144

Pothukuchi K, Kaufman JL (1999) Placing the food system on the urban agenda: the role of municipal institutions in food systems planning. Agric Hum Values 16(2):213–224

Searchinger T, Waite R, Hanson C, Ranganathan J, Dumas P, Matthews E, Klirs C (2019) Creating a sustainable food future: a menu of solutions to feed nearly 10 billion people by 2050, final report. WRI, Washington, DC

Tefft J, Jonasova M, Adjao R, Morgan A (2017) Food systems for an urbanizing world. World Bank, Washington, DC

UNEP (2016) Food systems and natural resources. In: Westhoek H, Ingram J, Van Berkum S, Özay L, Hajer M (eds) A report of the working group on food systems of the international resource panel. UNEP, Nairobi

Zhang J, Chang Y, Zhang L, Li D (2018) Do technological innovations promote urban green development? A spatial econometric analysis of 105 cities in China. J Clean Prod 182:395–403

Chapter 7
Seagrasses in the Indian Ocean Region with Special Reference to Urbanization

Balaji Vedharajan, Ragavan Pandisamy, E. P. Nobi, Shaha Hashim, Akbar John, Kannan Sivakumar, Chandani Appadoo, and P. Anantharaman

7.1 Introduction

Seagrasses are a functional group of plants (Short et al. 2011), living in the marine environment (Short et al. 2016) and form the submerged aquatic vegetation (Koch et al. 2006). Like terrestrial plants, they have roots, leaves, rhizomes, seeds and they reproduce through vegetative and sexual methods. Seagrasses grow naturally between high tide and low tide areas (intertidal areas), and also the majority of them are found completely submerged underwater and are distributed in tropical and sub-tropical areas; they can grow in various substrata like sand, silt, clay and coral rubbles

B. Vedharajan (✉)
OMCAR Foundation, Velivayal, Tamil Nadu, India
e-mail: director@omcar.org

R. Pandisamy · E. P. Nobi
Ministry of Environment, Forest and Climate Change, New Delhi, India

S. Hashim
Blue Marine Foundation, Malé, Maldives

A. John
Environmental Technology Division, School of Industrial Technology, Universiti Sains Malaysia, Penang, Malaysia

Renewable Biomass Transformation Cluster, School of Industrial Technology, Universiti Sains Malaysia, Penang, Malaysia

K. Sivakumar · P. Anantharaman
Centre of Advanced Studies in Marine Biology (CASMB), Annamalai University, Parangipettai (Portnovo), India

C. Appadoo
Department of Biosciences and Ocean Studies, Faculty of Science, University of Mauritius, Le Réduit, Mauritius

B. Pathak, R. S. Dubey (eds.), *Climate Change and Urban Environment Sustainability*, Disaster Resilience and Green Growth,
https://doi.org/10.1007/978-981-19-7618-6_7

(Parthasarathy et al. 1991). Seagrasses comprise <0.02% of the angiosperm flora, represented by 72 species. They are assigned mainly to six families encompassing 14 genera and are often separated into tropical and temperate genera, with 7 genera each. Seagrasses occur in all the coastal areas of the world except the polar regions because of ice scouring. Seagrasses are one of the most widespread coastal vegetational types and highly productive ecosystems in the world; they support rich biological diversity as well as offer many ecological services to maintain the integrity of the coast and sustainability of local communities. They protect the shorelines against erosion in the middle and lower intertidal and subtidal zones with their gregarious growth and dense root systems. Seagrass ecosystem is conspicuous and often a dominant habitat in shallow coastal waters and well known for its higher primary and secondary productivity, generation of vast quantities of detritus and supporting diverse floral and faunal communities as associated organisms.

Seagrasses have developed unique morphological, ecological and physiological adaptations for a completely submerged existence, including internal gas transport, epidermal chloroplasts, submarine pollination and marine dispersal which provide with important ecological services to the marine environment and profoundly influence the physical, chemical and biological environments in the coastal waters by acting as ecological engineers. Seagrasses alter water flow, nutrient cycling and food web. They are considered to be critical aquatic plants as they maintain the water quality by effectively removing nutrients from marine waters and settling surface sediments in the coastal areas. They are an important food source for mega herbivores such as green turtles, dugongs and manatees and provide with critical habitat for many animals, including commercially and recreationally important fishery species. Seagrass leaves, rhizomes and roots are used as food source by dugongs, sea urchins, herbivorous fishes and many other invertebrates feed on the decomposed seagrasses.

Seagrasses require high light levels (25% of the incident radiation) than any of the plant groups (1 or <1%) in the world. These extreme high light requirements mean that seagrasses respond acutely to environmental changes, especially those that alter water clarity. Distribution and growth of the seagrasses are regulated by a variety of environmental characteristics such as temperature, turbidity, salinity, irradiance, nutrient availability and sediment characteristics. Nutrient characteristics and the sediment texture, which are influenced by human activities and climate change, have a control over seagrass distribution and productivity.

7.2 Seagrasses in Indian Subcontinent

India is one of the fast-developing economies in the world, due to its strong growth in its industries (United Nations Department for Economic and Social Affairs 2020). Climate change is a worldwide issue that affects all nations and their sustainable development plans (Gopalakrishnan et al. 2019) including India. Rapid urbanization without proper planning in coastal areas leads to the decline of coastal habitats (Dhiman et al. 2019).

Fig. 7.1 *Cymodocea serrulata* bed at Palk Bay, India (Photo Credit: V Balaji)

India, with a long coastline of 7517 km, is endowed with a variety of coastal habitats including seagrasses. Nonetheless, seagrasses have remained a highly ignored ecosystem for research and conservation in India (Jagtap et al. 2003); however, recently MoEF and CC have taken efforts to promote research and better conservation and management of seagrass meadows along the Indian coast (Thangaradjou and Bhatt 2018). Seagrass ecosystem in the tropics and particularly in India has always been given a low priority in studies by the scientific communities and environmentalists. But, in the late 1990s and early 2000s, the scenario has changed slowly and started to gain momentum. Indian seagrass habitats are mainly limited to mud flats and sandy regions, extending from the lower inter tidal zone to a depth of 10–15 m along the open shores and in the lagoons around the islands. Seagrass flora of India are represented by 7 genera and 15 species, out of which the Gulf of Mannar and Palk Bay (Fig. 7.1) harbour the higher number of species followed by Andaman and Nicobar (Fig. 7.2) and Lakshadweep islands (Fig. 7.3). Seagrasses serve as a habitat for endemic species (Kumaraguru et al. 2008), where the seagrass beds in the southeast coast of India serve as a biodiversity hotspot and support fisheries and livelihood of a large number of coastal communities (Jeyabaskaran et al. 2018).

In India, there are 16 species of seagrasses, as reported by Geevarghese et al. (2018), which are growing in various substrata such as sand, mud and mixed substrata and depths (Parthasarathy et al. 1991). Palk Bay and Gulf of Mannar in the Southeast coast of India are the two important and large seagrass areas, which support a variety of associated flora and fauna (Manikandan et al. 2011). In addition to this, the seagrass beds are also found in Gulf of Kachchh (Kamboj 2014) and

Fig. 7.2 Huge seagrass bed found at AN Islands, India (Photo Credit: P Ragavan)

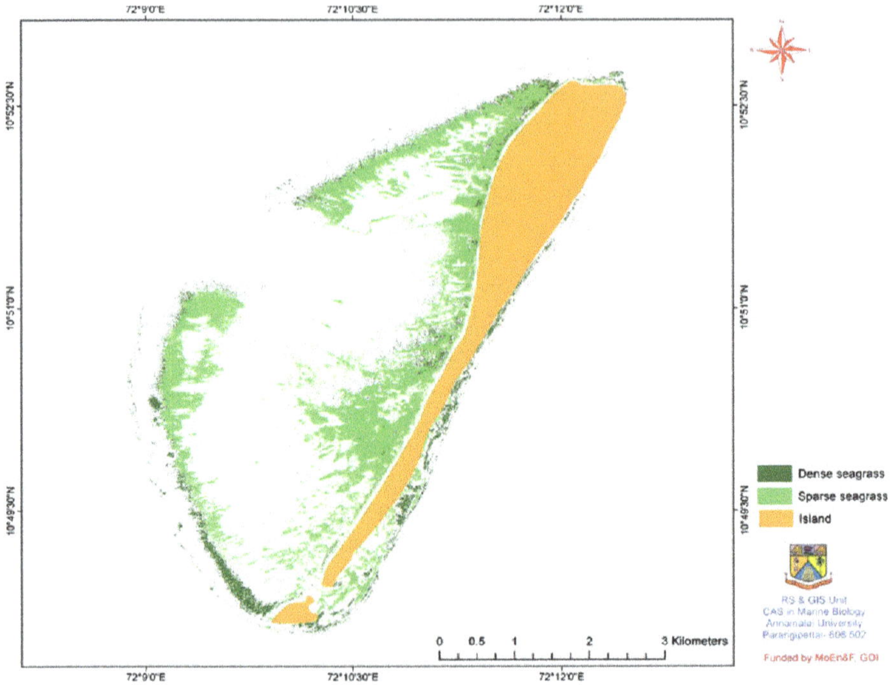

Fig. 7.3 Classified seagrass map of Agathi island, India (49 B/1) using IRS LISS IV data of 9 January 2007 (Source: Nobi et al. 2012)

Chilika lake (Geevarghese et al. 2018). In addition to this, seagrass beds have been identified in estuaries, lagoons and backwaters in most of the coastal states of India (Saravanan et al. 2013), such as Maharashtra, Kerala (Abhijith 2019), Tamil Nadu (Parthasarathy et al. 1991; Bharathi et al. 2014), Andhra Pradesh (Patro et al. 2017) and Odisha (Pati et al. 2014; Nayak 2014). *Halophila beccarii* is the most widely distributed species in India (Mishra et al. 2020).

Costanza et al. (2014) stated that the seagrass beds are highly productive ecosystems, and the estimated loss of ecological services from seagrass beds, due to land use change is at the rate of $4.3–20.2 trillion per year during the period from 1997 to 2011.

Seagrasses are used by farmers in the Palk Bay in Tamil Nadu as fertilizer for coconut trees. Similarly, they are used in other parts of the world for bedding and roofing and also for many other services (Nordlund et al. 2016).

Seagrass meadows mitigate the impacts of waves on the shorelines and erosion (Christianen et al. 2013), and the beach cast seagrasses are playing key role in protecting the shorelines from waves and erosion (Nordlund et al. 2018).

Seagrass locations in India are also the prime spots of fast-developing urbanization related activities, which threaten the existence of seagrass beds and their services. Impacts of degradation of seagrasses in Indian coastal areas directly affect their fishery productivity and livelihood of dependent fishers (Jeyabaskaran et al. 2018). In the Gulf of Kutch of north western India, the seagrass beds are facing the impacts of poor water quality due to industrial activities such as dredging, ship hot water discharge, oil spillage, port development (Kamboj 2014) and ship breaking yards (Deshpande et al. 2012). The waste waters discharged from coastal urban areas directly affect the seagrass meadows by enriching with nutrients in water and sediments (Cabaço et al. 2008). This is applicable to all the coastal urban areas of India and other parts of the world. Increased urbanization and tourism activities in the Palk Bay and Gulf of Mannar and other islands including Rameswaram would affect the seagrass beds and coral reefs in the southeast coast of India, in future.

In India, seagrass beds are not studied completely, however they are recognized as one of the sensitive ecosystems in the Coastal Regulation Zone (Patro et al. 2017). Seagrasses serve as a nursery ground (Heck et al. 1997; Jiang et al. 2020; McDevitt-Irwin et al. 2016) and breeding ground for marine animals (Bujang et al. 2006). Significant attempts have been carried out in India for the conservation and restoration of seagrass beds. Laboratory based axenic seagrass seedlings were developed by Thangaradjou and Kannan (2008). In situ seagrass restoration method was performed in the Gulf of Mannar, using PVC frames which showed higher survival rate (Patterson Edward et al. 2019). The method was further improved by using low cost and eco-friendly materials such as bamboo frames and coconut ropes (Balaji et al. 2020). Seagrass habitats in Palk Bay of Tamil Nadu are now being considered for declaring them as a "Dugong Conservation Reserve" by the State and Central Governments, recently.

Unfortunately, these seagrasses have significantly declined in their coverage and density in several parts of the country. The largest seagrass beds in India are located along the southeast coast of India, which are affected by aquaculture farms, domestic sewage from coastal towns, tourism, fishery-related disturbances and factories. Jeyabaskaran et al. (2018) mentioned that the Palk Bay seagrass beds in southeast coast of India as "biodiversity hotspots" and are disappearing faster than that of the Gulf of Mannar, which may be due to the threats to seagrass beds in Palk Bay (Kumaraguru et al. 2008). Such large-scale losses of seagrass beds have led to numerous restoration programmes worldwide. In this respect, the need to transplant in affected seagrass habitats and to collect donor materials from healthy seagrass habitats are critically important. Careful site selection is very important, and several conceptual models to optimize site selection have also been developed for seagrass restoration.

7.3 Seagrass in the Andaman and Nicobar Islands of India

As per an estimate, seagrass cover in India is 517 km^2 (Geevarghese et al. 2018), of which, 15 km^2 of seagrass beds are spread around the Andaman and Nicobar Islands (ANI) with 10 species belonging to 6 genera (Geevarghese et al. 2018; Ragavan et al. 2016). Seagrass beds in the Andaman Islands (5.79 km^2) are highly scattered, whereas in Nicobar Islands (8.81 km^2), seagrass beds are luxuriant and dense (Geevarghese et al. 2018; Ragavan et al. 2016). By surveying a total of 44 seagrass sites in the ANI, 8 sites were identified as feeding sites for dugongs through the distinct feeding trails observed at these sites (D'Souza et al. 2015). In addition, Seagrass meadows of the ANI were found to be the hot spots for blue carbon storage in Southeast Asia with the highest ecosystem carbon stock of 184.24 ± 23.84 Mg C ha^{-1}, followed by Myanmar (136.67 ± 64.77 Mg C ha^{-1}), Thailand and Vietnam (134.20 ± 73.89 and 133.16 ± 36.97 Mg C ha^{-1}, respectively), Philippines (123.49 ± 63.38 Mg ha^{-1}), Malaysia (108.63 ± 89.43 Mg ha^{-1}) and Indonesia (97.60 ± 41.49 Mg ha^{-1}) (Stankovic et al. 2021). Recently, Mishra et al. (2021) also reported relatively higher carbon stock in the seagrass beds of the Andaman Islands (128.79 ± 55.89 and 272.54 ± 164 Mg C ha^{-1}) than the other parts of India. Further, they have noted that outwelling of organic carbon from highly productive adjacent dense mangroves of the Andaman Islands would play a major role in carbon storage in the seagrass meadows of these islands. So, maintaining the ecological connectivity is imperative for the use of this valuable ecosystem for countries' climate change mitigation and adaptation measures and protecting the critically endangered dugongs.

Despite the ecological and economic benefits, seagrasses are witnessing rapid depletion around the world as well as in the ANI (Thangaradjou and Bhatt 2018). Manmade activities such as fishing, anchoring, boating and development of infrastructures in the coastal areas severe the impacts on seagrasses and their biodiversity (Ramesh and Mohanraju 2020). Recently, it has been highlighted that climate change consequences like sea-level rise and its associated soil erosion and increasing frequency of natural catastrophic events like cyclones and tsunamis would be the major factors for the degradation of coastal systems (Goldberg et al. 2020; Adame et al. 2021). Tsunami in 2004 destroyed large areas of seagrass meadows in the Nicobar Islands (D'Souza and Patankar 2011) and Lehar's cyclone damaged about 1.96 ha of seagrass beds in the Andaman Islands (Sachithanandam et al. 2014). Furthermore, the frequent occurrence of leaf reddening and leaf burning in seagrasses of the ANI indicates that seagrass meadows of the ANI are highly vulnerable to stresses like high temperatures, intense light and release of toxic sulphides (Ragavan et al. 2013; Ramesh and Mohanraju 2020). In the recent years, the seagrass research is getting more attention, which further needs to provide us with answers for conservation and management (Thangaradjou and Bhatt 2018). Since the knowledge of exact species composition and distribution is a prerequisite for better conservation and management of an ecosystem, spatial and temporal distribution of seagrass meadows in the ANI needs to be documented at the earliest irrespective of dugong represented meadows.

7.4 Seagrasses in Maldives–Lakshadweep–Chagos Archipelago

There is only limited literature about the seagrasses of the Maldives. When analysing the satellite images, seagrasses are visible in the Maldivian seas, especially around the densely populated islands or islands with more economic activities. Researches indicate that there are five species of seagrasses under four genera in the Maldives. These species are *Syringodium isoetifolium, Thalassia hemprichii, Thalassodendron ciliatum, Cymodocea rotundata* and *Cymodocea serrulata* and of these, *T. hemprichii* is the most abundant in the country. In the Maldives, seagrasses are mostly widespread in shallow reef flats and can also be observed in mangrove habitats. These are habitats of important fish species such as rabbit fishes, shrimps, sea cucumbers, sea urchins, seahorses, crabs, scallops, mussels and snails (Paz-Alberto and Sigua 2015).

The seagrass meadows around the Maldives are under threat. But the government now plans to place this valuable ecosystem under protection, as they are serving as nurseries for sea life, act as carbon sinks and protect the atolls from erosion. The Maldives welcomes tourists from around the world; many of them come here to enjoy its crystal-clear blue waters and tropical coral reefs along with seagrass meadows. Unfortunately, the meadows are not enjoying the same environmental protection as that of the better-known coral reefs. For years, tourist resorts and development projects have dredged and destroyed seagrass meadows across the country. It is thought that up to 30% of seagrass habitats have been lost over the last century.

Conservationists in the country are fighting back and have launched a campaign to help bring the spotlight to the rich biodiversity of these ecosystems and their importance for the future of the country. In 2019, Six Senses Laamu and Blue Marine Foundation (BLUE) launched the Protect Maldives Seagrass campaign which convinced more than 25% of the country's high-end resorts to protect the seagrass meadows; **37 resorts** joined the campaign and collectively pledged to protect more than **910,000m²** of seagrass beds around the resort islands across the country. The Maldives' Ministry of Tourism also supported the campaign.

Nobi et al. (2012) recorded the seagrass cover using IRS P^ LISS IV satellite data and found 2590.2 ha of seagrass meadows in the Lakshadweep islands of which 1310.8 ha of dense seagrasses and 1279.4 ha of sparse seagrasses were mapped. In Lakshadweep, seven species of seagrasses that belong to five genera were recorded and the spatial cover of the seagrasses was healthy in all the islands. *Cymodocea serrulata* was observed as the dominant species in Lakshadweep islands. Coral reefs and seagrasses were found associated, and this coexistence showed the mutual relationship between these two adjacent ecosystems, where coral reefs protect seagrasses from waves and at the same time, seagrasses prevent erosion. Diversity of seagrasses was comparatively lesser in Lakshadweep than that of mainland of India. Although seagrasses of Lakshadweep islands are healthy, increase in sea surface water temperature and anthropogenic activities cause the decrease of seagrass beds in certain islands, which are to be addressed properly for the sound management of fragile seagrass beds in order to sustain their ecosystem services. Agathi and Kadamath islands have wide seagrass beds, which serve as a feeding habitat for a large number of green turtles. The study carried out by Nobi et al indicates that increase in surface water temperature and decrease in pH are the key threats to the seagrass beds in Lakshadweep.

Kaladharan (2013) reported that there is a steep decline of standing stock of seagrass vegetation and wet biomass in four atolls such as Agatti, Chetlet, Kavaratti and Kiltan and in 5 years, which was due to the settlement of fine silt and coral rubbles possibly due to periodical dredging activities. Recently, Kaladharan et al. (2020) reported that four atolls in Lakshadweep are highly populated, which leads to the disposal of untreated domestic wastes directly into the atolls which causes increased nutrients in the lagoons that adversely alter the quality of water and algal blooming that hinder the growth of seagrasses. In the Lakshadweep Islands, the Union Territory of Lakshadweep (UTL) Administration is playing, through the Department of Environment and Forestry and Department of Science and Technology, a key role in conserving the coastal and marine ecosystems of the islands and creating new Marine Protected Areas (MPAs). The Department of Science and Technology and Department of Biotechnology, Government of India have meanwhile supported work on genomics as well as the search for novel metabolites. A Lakshadweep Action Plan on Climate Change (LAPCC) has been formulated in accordance with the principles and guidelines of the Indian National Action Plan on Climate Change (NAPCC). The LAPCC included the environmental parameters in present and future development plans in UTL along with the NAPCC guidelines and principles. Moreover, National Centre for Sustainable Coastal Management (NCSCM) of the Ministry of Environment, Forest and Climate Change, Government of India has mentioned that the integrated Island Management Programme will be implemented to promote social and ecological sustainability of the Lakshadweep.

Thalassodendron ciliatum (Forsk.) den Hartog is more extensively distributed in the Chagos Archipelago than previously reported, and *Halophila decipiens* is reported from a single locality. Extensive meadows of *Thalassodendron* are confined to the submerged banks in the north of the Great Chagos Bank, extending to the lagoon reef-flat of the Diego Garcia atoll, to the south. Vast areas of apparently suitable substrates around the other atolls and the few islands on the Chagos Bank itself are totally uncolonized by seagrasses. High water temperatures caused by coincidence of extreme low water spring tides and intense midday insolation are suggested as the main reasons for the restricted distribution of *Thalassodendron* around the atolls and islands whilst extreme geographical isolation may explain the very restricted seagrass flora (Drew 1980).

Through satellite data, Esteban et al. (2018) tracked green turtles (*Chelonia mydas*) which are known to fed on seagrasses in the remote, pristine deep-water environment in the Great Chagos Bank (Indian Ocean) which lies in the heart of one of the world's largest Marine Protected Areas (MPAs). Subsequently the study used in situ SCUBA and baited video surveys to study the day-time sites occupied by turtles and discovered extensive monospecific seagrass meadows of *Thalassodendron ciliatum*. Higher fish abundance and large predatory sharks were recorded at all sites. Given that the Great Chagos Bank extends over approximately 12,500 km^2 and many other large deep submerged banks exist across the world's oceans, it can be understood that deep-water seagrasses may be far more abundant than previously thought. Due to the remote location and lesser anthropogenic activity in the Chagos bank possibly gives space for the survival of seagrasses in the region. However, more seagrass studies are needed to know the impacts of climate change on the ecosystem of Chagos bank.

7.5 Seagrasses of Western Indian Ocean Regions

Colonial trading led to the development of many east African coastal urban centres which have been expanding with economic growth (Georgulas 1964). Majority of the coastal cities in east African coast are located adjacent to sensitive coastal habitats, which are developing faster than the predicted rate, and this urban growth leads to increasing requirement for both renewable and non-renewable resources, with negative impacts on coastal biodiversity (Celliers and Ntombela 2016).

Western Indian Ocean region includes the countries such as South Africa, Tanzania, Kenya, Somalia, Mozambique, Comoros, Reunion, Seychelles, Madagascar and Mauritius. In the western Indian Ocean region, the coastal cities such as Durban (South Africa), Maputo (Mozambique), Dar es Salaam (Tanzania), Mombasa (Kenya), Mogadisu (Somalia), Port Louis (Mauritius), Saint Denis (Reunion) and Victoria (Seychells) are developing dramatically in the twenty-first century and all these cities are vulnerable to natural disasters and climate change (Celliers and Ntombela 2016).

There are 13 seagrass species reported from Western Indian Ocean region, covering 12,000 km coastline, which are found in intertidal regions up to 40 m depths that support a variety of fishes belonging to Terapontidae, Apogonidae, Lethrinidae, Labridae, Gerridae, Centriscidae, Blenniidae, Apogonidae, Lutjanidae, Scorpaenidae, Scaridae and Monacanthidae (Gullström et al. 2002). There are only a limited number of publications about the seagrass ecosystems in Western Indian Ocean region when compared to mangroves and coral reefs (Gullström et al. 2002). Destruction of forests in the inland areas, unlawful fishing methods, discharge of untreated wastes from urban areas, sedimentation, development of unplanned infrastructures, agricultural run-off and shipping activities are the major threats to the seagrass beds in the western Indian Ocean region (Bandeira 1995).

Seagrasses serve as a source of income for many African and island countries in the Western Indian Ocean region; however, there has been an increase of pressure on seagrass beds due to expansion of urban areas and surrounding population causes manmade disturbances to the seagrass beds (Gullström et al. 2002). Urban waste disposal is mostly untreated due to the non-availability of technologically sound waste treatment infrastructure facility. So, the poor water quality, sedimentation, and salinity directly affect the health of the seagrass ecosystems in the Western Indian Ocean region countries (Ingram and Dawson 2001).

The lower part of western Indian Ocean has also experienced the impacts of urbanization. In South Africa, Durban is one of the prominent coastal cities, where Adams (2016) has reported that *Zostera capensis* is the dominant seagrass species but a huge expanse of seagrass beds of *Z. capensis* (declared as vulnerable seagrass species in the IUCN Red list) has disappeared from Durban Bay due to harbour development during the mid-decades of the twentieth century and also recently (Short et al. 2011).

Seagrass beds in urban areas of other countries in Western Indian Ocean are also facing threats in the Maputo, capital of Mozambique, due to flooding, sedimentation and pollution; however, the seagrass distribution data is not available from other urban areas of Tanzania, Zanzibar and Madagascar (Bandeira et al. 2014). In the recent years, there has been a vigorous growth of urbanization in Arabian Gulf areas due to the establishment of new industries, land reclamation projects, construction of new residential and tourism infrastructures. But there is no published reports about the degradation of seagrass beds in Arabian Gulf (Erftemeijer and Shuail 2012).

Arabian Gulf is inhibited by a large population of Dugongs (*Dugong dugon*) or sea cows that are mainly grazing seagrasses (Sheppard et al. 2010). This can be taken as evidence showing the presence of vast seagrass beds in the Arabian Gulf that includes Bahrain, Saudi Arabia, Qatar, UAE and Abu Dhabi (Green and Short 2003; Erftemeijer and Shuail 2012). Changes in the nearshore waters due to urbanization were observed about 30 years ago in Bahrain, Arabian Gulf (Zainal et al. 1992), which is a part of the Western Indian Ocean region.

7.6 Threats to Seagrasses by the Urbanization

According to Weeks (2010), urbanization is defined as "a place-based characteristic that incorporates elements of population density, social and economic organization and the transformation of a natural environment into the built environment". About 40% of the world's population is living in coastal urban areas and so the changes caused by anthropogenic factors of urban settlements in the coastal areas are altering the sediments and coastal ecosystems (Mentaschi et al. 2018).

Human activities have caused huge loss of seagrass beds all over the world (Duarte et al. 2008). Two main coastal infrastructure development activities due to urbanization are dredging and land reclamation, and it is stated that they have severe impacts on coral reefs and seagrass beds (Erftemeijer and Robin Lewis 2006). High turbidity caused by dredging, proliferation of marine algae and eutrophication in the water column by untreated sewage disposal cause the loss of seagrasses (Duarte et al. 2008). Turbidity caused by dredging activities in ports is one of the key factors limiting the seagrass distribution in urban coastal areas and has a direct impact on urban seagrass beds (Sabol et al. 2005). Removal of seabed and deposition of landfills (Newell et al. 1998) are also directly affecting the natural ecosystem in coastal urban areas.

It is true that the global distribution and abundance of seagrasses have changed over evolutionary time in response to sea-level change, physical modification of coastlines and global climate changes. But, losses of seagrasses due to anthropogenic pressures are even much more than the past. Multiple stressors including sediment and nutrient run-off, physical disturbances, invasive species, diseases, commercial fishing practices, aquaculture, overgrazing, algal blooms and global warming cause the seagrass declines at scales of square meters to hundreds of square kilometres.

Urbanization affects the function and biodiversity of an ecosystem, as it will induce the split of natural habitats and degradation (Alberti and Marzluff 2004). One of the important impacts of urbanization is the construction of barriers, landing centres and harbours in the coastal and marine environments, and it is reported that such infrastructures in the marine environment alter the species composition by favouring the growth of non-native species (Airoldi et al. 2015). So, urbanization in coastal zones threatens the coastal landscapes and ecosystems (Pasquali and Marucci 2021) including mangroves (Stiepani et al. 2021), coral reefs Eliza and significantly affect the seagrass beds (Tu et al. 2021). It is also observed that the urbanization in the marine environment affects biodiversity and leads to the overlapping of species of artificial and natural habitats (Momota and Hosokawa 2021) including coral reefs and seagrass beds.

Dredging is the major cause for seagrass degradation worldwide, which can be controlled by strict laws, periodical monitoring, assessing the impacts and taking mitigation procedures (Erftemeijer and Robin Lewis 2006). Natural recovery of seagrass beds after the manmade activities and natural disasters will take few weeks up to 5 years, based on the environmental conditions (Erftemeijer and

Robin Lewis 2006). Anthropogenic pressures exerted on the seagrass beds (a by-product of urbanization development in coastal areas) have an adverse impact on the seagrasses around the world (Unsworth et al. 2019; Erftemeijer and Robin Lewis 2006; Newell et al. 1998; Sabol et al. 2005), including India (Thangaradjou et al. 2008), Sri Lanka (Ranahewa et al. 2018), Singapore (Stéphanie and Doorn 2007), Malaysia (Freeman et al. 2008), Indonesia (Karlina et al. 2018; Riani and Purbayanto 2011), Western Australia (Fraser et al. 2017) and other southeast Asian countries (Fortes et al. 2018).

Seagrass ecosystems are among the least considered coastal habitats which frequently face cumulative pressures from coastal development, nutrient run-off and climate change. It is a well-known fact that the seagrasses of the Indian Ocean region are facing severe threats due to anthropogenic and natural pressures; even then, seagrass restorations have been undertaken only on experimental basis. Loss of seagrass habitats and its impacts on the marine ecosystems can be slowed down or even reversed by planned and effective seagrass transplantations. Healthy seagrasses provide with prospects to alleviate climate change impacts, adapt to future changes and develop resilience and the source of multiple societal benefits. This is the apt time to protect seagrasses by prioritizing timely, ambitious and coordinated programmes, relating to restoration, conservation and sustainable management.

Acknowledgements We thank Dean and Director of CAS in Marine Biology for their encouragement and Professor L. Kannan, for offering valuable comments. Authors also thank Ministry of Environment, Forest and Climate Change, Government of India.

References

Abhijith R (2019) Present status of *Halophila beccarii* seagrass bed in Kadalundi Community Reserve marine macrophytes of Indian coasts view project sustainable alternative livelihood for coastal fishing communities along Palghar district of Maharashtra: molluscan mariculture approach view project. https://www.researchgate.net/publication/343392450

Adams JB (2016) Distribution and status of *Zostera capensis* in south African estuaries—a review. S Afr J Bot 107:63–73. https://doi.org/10.1016/j.sajb.2016.07.007

Adame MF, Connolly RM, Turschwell MP, Lovelock CE, Fatoyinbo T, Lagomasino D, Goldberg LA, Holdorf J, Friess DA, Sasmito SD, Sanderman J (2021) Future carbon emissions from global mangrove forest loss. Glob Chang Biol 27(12):2856–2866

Airoldi L, Turon X, Perkol-Finkel S, Rius M (2015) Corridors for aliens but not for natives: effects of marine urban sprawl at a regional scale. Divers Distrib 21(7):755–768

Alberti M, Marzluff JM (2004) Ecological resilience in urban ecosystems: linking urban patterns to human and ecological functions. Urban Ecosyst 7(3):241–265

Balaji V, Sekar V, Murugesan G (2020) Comparison of seagrass restoration methods adopted in Palk Bay, India. J Mar Biol Assoc India 62(1):95–99. https://doi.org/10.6024/jmbai.2020.62.1.2137-13

Bandeira SO (1995) Marine botanical communities in southern Mozambique: sea grass and seaweed diversity and conservation. Ambio 24(7–8):506–509

Bandeira SO, Paula JPM, Western Indian Ocean Marine Science Association (2014) The Maputo Bay ecosystem. WIOMSA, Zanzibar

Bharathi K, Subhashini P, Ranith R, Vanitha K, Thangaradjou T (2014) Spatial variability in distribution of seagrasses along the Tamilnadu coast. Int J Curr Res 6:8897–9005. http://www.journalcra.com

Bujang JS, Zakaria MH, Arshad A (2006) Distribution and significance of seagrass ecosystems in Malaysia. Aquat Ecosyst Health Manag 9(2):203–214. https://doi.org/10.1080/14634980600705576

Cabaço S, Machás R, Vieira V, Santos R (2008) Impacts of urban wastewater discharge on seagrass meadows (*Zostera noltii*). Estuar Coast Shelf Sci 78(1):1–13. https://doi.org/10.1016/J.ECSS.2007.11.005

Celliers L, Ntombela C (2016) Urbanisation, coastal development and vulnerability, and catchments. Regional state of the coast report. United Nations, New York, pp 386–404. https://www.un-ilibrary.org/content/books/9789210601573s009-c005

Christianen MJA, van Belzen J, Herman PMJ, van Katwijk MM, Lamers LPM, van Leent PJM, Bouma TJ (2013) Low-canopy seagrass beds still provide important coastal protection services. PLoS One 8(5):e62413. https://doi.org/10.1371/journal.pone.0062413

Costanza R, de Groot R, Sutton P, van der Ploeg S, Anderson SJ, Kubiszewski I, Farber S, Turner RK (2014) Changes in the global value of ecosystem services. Glob Environ Change 26(1): 152–158. https://doi.org/10.1016/J.GLOENVCHA.2014.04.002

Deshpande PC, Tilwankar AK, Asolekar SR (2012) A novel approach to estimating potential maximum heavy metal exposure to ship recycling yard workers in Alang, India. Sci Total Environ 438:304–311. https://doi.org/10.1016/j.scitotenv.2012.08.048

Dhiman R, Kalbar P, Inamdar AB (2019) Spatial planning of coastal urban areas in India: current practice versus quantitative approach. Ocean Coast Manag 182:104929. https://doi.org/10.1016/j.ocecoaman.2019.104929

Drew EA (1980) Seagrasses in the Chagos archipelago. Aquat Bot 9:179–184

D'souza E, Patankar V (2011) Ecological studies on the Dugong dugon of the Andaman and Nicobar Islands: a step towards species conservation. Nature Conservation Foundation, Mysore, 19 pp

D'Souza E, Patankar V, Arthur R, Marbà N, Alcoverro T (2015) Seagrass herbivory levels sustain site-fidelity in a remnant dugong population. PLoS One 10(10):e0141224

Duarte CM, Borum J, Short FT, Walker DI (2008) Seagrass ecosystems: their global status and prospects. In: Aquatic ecosystems: trends and global prospects. Cambridge University Press, Cambridge, pp 281–294. https://doi.org/10.1017/CBO9780511751790.025

Erftemeijer PLA, Robin Lewis RR (2006) Environmental impacts of dredging on seagrasses: a review. Mar Pollut Bull 52(12):1553–1572. https://doi.org/10.1016/j.marpolbul.2006.09.006

Erftemeijer PLA, Shuail DA (2012) Seagrass habitats in the Arabian Gulf: distribution, tolerance thresholds and threats. Aquat Ecosyst Health Manag 15:73–83. https://doi.org/10.1080/14634988.2012.668479

Esteban N, Unsworth RKF, Gourlay JBQ, Hays GC (2018) The discovery of deep-water seagrass meadows in a pristine Indian Ocean wilderness revealed by tracking green turtles. Mar Pollut Bull 134:99–105

Fortes MD, Ooi JLS, Tan YM, Prathep A, Bujang JS, Yaakub SM (2018) Seagrass in Southeast Asia: a review of status and knowledge gaps, and a road map for conservation. Bot Mar 61(3):269–288. https://doi.org/10.1515/bot-2018-0008

Fraser MW, Short J, Kendrick G, McLean D, Keesing J, Byrne M, Caley MJ, Clarke D, Davis AR, Erftemeijer PLA, Field S, Gustin-Craig S, Huisman J, Keough M, Lavery PS, Masini R, McMahon K, Mengersen K, Rasheed M, Wu P (2017) Effects of dredging on critical ecological processes for marine invertebrates, seagrasses and macroalgae, and the potential for management with environmental windows using Western Australia as a case study. Ecol Indic 78:229–242. https://doi.org/10.1016/j.ecolind.2017.03.026

Freeman AS, Short FT, Isnain I, Razak FA, Coles RG (2008) Seagrass on the edge: land-use practices threaten coastal seagrass communities in Sabah, Malaysia. Biol Conserv 141(12): 2993–3005. https://doi.org/10.1016/j.biocon.2008.09.018

Geevarghese GA, Akhil B, Magesh G, Krishnan P, Purvaja R, Ramesh R (2018) A comprehensive geospatial assessment of seagrass distribution in India. Ocean Coast Manag 159:16–25

Georgulas N (1964) An approach to urban analysis for East African towns with particular reference to the African population. Ekistics 18(109):436–440. http://www.jstor.org/stable/43614112

Goldberg L, Lagomasino D, Thomas N, Fatoyinbo T (2020) Global declines in human-driven mangrove loss. Glob Chang Biol 26(10):5844–5855

Gopalakrishnan T, Hasan MK, Haque ATMS, Jayasinghe SL, Kumar L (2019) Sustainability of coastal agriculture under climate change. Sustainability (Switzerland) 11(24):7200. https://doi.org/10.3390/su11247200

Green EP, Short FT (2003) World atlas of seagrasses. University of California Press, Berkeley

Gullström M, de la Torre Castro M, Bandeira SO, Björk M, Dahlberg M, Kautsky N, Rönnbäck P, Öhman MC (2002) Seagrass ecosystems in the Western Indian Ocean. AMBIO J Hum Environ 31(7):588. https://doi.org/10.1639/0044-7447(2002)031[0588:seitwi]2.0.co;2

Heck KL, Nadeau DA, Thomas R (1997) The nursery role of seagrass beds. Gulf Mexico Sci 15(1): 8. https://doi.org/10.18785/goms.1501.08

Ingram JC, Dawson TP (2001) The impacts of a river effluent on the coastal seagrass habitats of Mahé, Seychelles. S Afr J Bot 67(3):483–487. https://doi.org/10.1016/S0254-6299(15)31167-4

Jagtap TG, Komarpant DS, Rodrigues RS (2003) Status of a seagrass ecosystem: an ecologically sensitive wetland habitat from India. Wetlands 23(1):161–170

Jeyabaskaran R, Jayasankar J, Ambrose TV, Vineetha Valsalan KC, Divya ND, Raji N, Vysakhan P, John S, Prema D, Kaladharan P, Kripa V (2018) Conservation of seagrass beds with special reference to associated species and fishery resources. J Mar Biol Assoc India 60(1): 62–70. https://doi.org/10.6024/jmbai.2018.60.1.2038-10

Jiang Z, Huang D, Fang Y, Cui L, Zhao C, Liu S, Wu Y, Chen Q, Ranvilage CIPM, He J, Huang X (2020) Home for marine species: seagrass leaves as vital spawning grounds and food source. Front Mar Sci 7:194. https://doi.org/10.3389/fmars.2020.00194

Kaladharan P (2013) Mariculture of seaweeds for carbon sequestration and livelihood support. In: International symposium on Towards Green Technologies in Fisheries, vol 13, p 24

Kaladharan P, Zacharia PU, Thirumalaiselvan S, Anasu Koya A, Lavanya S, Batcha SM (2020) Blue carbon stock of the seagrass meadows of Gulf of Mannar and Palk Bay off Coromandel Coast, south India. Indian J Fish 67(4):149–153

Kamboj RD (2014) Biology and status of seagrasses in Gulf of Kachchh Marine National Park and Sanctuary, India. Indian Ocean Turtle News Lett 19:8–11

Karlina I, Kurniawan F, Idris F (2018) Pressures and status of seagrass ecosystem in the coastal areas of North Bintan, Indonesia. E3S Web Conf 47:04008. https://doi.org/10.1051/e3sconf/20184704008

Koch EW, Ackerman JD, Verduin JJ, Van Keulen M (2006) Fluid dynamics in seagrass ecology - from molecules to ecosystems. In: Larkum AWD, Orth RJ, Duarte CM (eds) Seagrasses: biology, ecology and conservation. Springer, Dordrecht, pp 193–225

Kumaraguru AK, Edwin Joseph V, Rajee M, Blasubramanian T (2008) Palk Bay - information and bibliography. CAS in Marine Biology, Annamalai University/Centre for Marine and Coastal Studies, Madurai Kamaraj University, Parangipettai/Madurai, 227 pp

Manikandan S, Ganesapandian S, Singh M, Kumaraguru AK (2011) Seagrass diversity and associated flora and fauna in the coral reef ecosystem of the Gulf of Mannar, Southeast Coast of India. Res J Environ Earth Sci 3(4):321–326

McDevitt-Irwin JM, Iacarella JC, Baum JK (2016) Reassessing the nursery role of seagrass habitats from temperate to tropical regions: a meta-analysis. Mar Ecol Prog Ser 557:133–143. https://doi.org/10.3354/meps11848

Mentaschi L, Vousdoukas MI, Pekel JF, Voukouvalas E, Feyen L (2018) Global long-term observations of coastal erosion and accretion. Sci Rep 8(1):1–11

Mishra AK, Apte D, Chowk SA, Singh SB (2020) The current status of *Halophila beccarii*: an ecologically significant, yet vulnerable seagrass of India. Ocean Coast Manag 2020:0126. https://doi.org/10.20944/preprints202008.0126.v1

Mishra AK, Khadanga MK, Patro S, Apte D, Farooq SH (2021) Population structure of a newly recorded (Halodule uninervis) and native seagrass (Halophila ovalis) species from an intertidal creek ecosystem. Lakes Reserv Res Manag 26(3):e12376

Momota K, Hosokawa S (2021) Potential impacts of marine urbanization on benthic macrofaunal diversity. Sci Rep 11(1):1–12

Nayak L (2014) Studies on seagrasses in relation to some environmental variables from Chilika Lagoon, Odisha, India. Int Res J Environ Sci 3(11):92–101. www.isca.me

Newell RC, Seiderer LJ, Hitchcock DR (1998) The impact of dredging works in coastal waters: a review of the sensitivity to disturbance and subsequent recovery of biological resources on the sea bed. Oceanogr Mar Biol Annu Rev 36:127–178

Nobi EP, Dilipan E, Sivakumar K, Thangaradjou T (2012) Estimation of the aerial cover of seagrasses of Lakshadweep islands (India) using Indian Remote Sensing Satellite (IRS P6 LISS IV). J Indian Soc Remote Sens 40(3):467–481

Nordlund LM, Koch EW, Barbier EB, Creed JC (2016) Seagrass ecosystem services and their variability across genera and geographical regions. PLoS One 11(10):e0163091. https://doi.org/10.1371/journal.pone.0163091

Nordlund LM, Jackson EL, Nakaoka M, Samper-Villarreal J, Beca-Carretero P, Creed JC (2018) Seagrass ecosystem services—what's next? Mar Pollut Bull 134:145–151. https://doi.org/10.1016/j.marpolbul.2017.09.014

Parthasarathy N, Ravikumar K, Ganesan R, Ramamurthy K (1991) Distribution of seagrasses along the coast of Tamil Nadu, southern India. Aquat Bot 40(2):145–153. https://doi.org/10.1016/0304-3770(91)90092-J

Pasquali D, Marucci A (2021) The effects of urban and economic development on coastal zone management. Sustain For 13(11):6071

Pati MP, Nayak L, das Sharma S (2014) Studies on biomass of seagrass, seaweed and its associated fauna from Chilika lagoon. Int J Environ Sci 5(2):423–431. https://doi.org/10.6088/ijes.2014050100036

Patro S, Krishnan P, Deepak Samuel V, Purvaja R, Ramesh R (2017) Seagrass and salt marsh ecosystems in south Asia: an overview of diversity, distribution, threats and conservation status. In: Wetland science: perspectives from South Asia. Springer India, New Delhi, pp 87–104. https://doi.org/10.1007/978-81-322-3715-0_5

Patterson Edward JK, Diraviya Raj K, Mathews G, Dinesh Kumar P, Arasamuthu A, D'Souza N, Bilgi DS (2019) Seagrass restoration in Gulf of Mannar, Tamil Nadu, Southeast India: a viable management tool. Environ Monit Assess 191(430):1–14

Paz-Alberto AM, Sigua GC (2015) Phytoremediation: a green technology to remove environmental pollutants. Am J Clim Chang 2:71–86

Ragavan P, Saxena A, Mohan PM, Coomar TARUN, Ragavan A (2013) Leaf reddening in seagrasses of Andaman and Nicobar Islands. Trop Ecol 54(2):269–273

Ragavan P, Jayaraj RSC, Muruganantham M, Jeeva C, Ubare VV, Saxena A, Mohan PM (2016) Species composition and Distribution of Seagrasses of the Andaman and Nicobar Islands. Vegetos 29:78–87

Ramesh C, Mohanraju R (2020) Seagrass ecosystems of Andaman and Nicobar Islands: status and future perspective. Environ Earth Sci Res J 7(4):169–174

Ranahewa TH, Gunasekara AJ, Premarathna AD, Karunarathna SC, Jayamanne SC (2018) A comparative study on the diversity of seagrass species in selected areas of Puttalam Lagoon in Sri Lanka. J Oceanogr Mar Res 06(03):185. https://doi.org/10.4172/2572-3103.1000185

Riani E, Purbayanto A (2011) Challenging for seagrass management in Indonesia. Sugeng Budiharsono Ministry of Village Development of Disadvantage Region and Transmigration. https://www.researchgate.net/publication/267988083

Sabol B, Shafer D, Lord E (2005) Dredging effects on Eelgrass (*Zostera marina*) distribution in a New England small boat harbor environmental laboratory. U.S. Army Engineer Research and Development Center, Vicksburg

Sachithanandam V, Mageswaran T, Sridhar R, Purvaja R, Ramesh R (2014) Assessment of Cyclone Lehar's impact on seagrass meadows in Ross and Smith Islands, North Andaman. Nat Hazards 72(2):1253–1258

Saravanan K, Chowdhury BC, Sivakumar K (2013) Important coastal and marine biodiversity areas on East Coast of India. ENVIS Bull Wildl Prot Areas 15:292–298

Sheppard J, Marsh H, Jones R, Lawler I (2010) Dugong habitat use in relation to seagrass nutrients, tides, and diel cycles. Mar Mamm Sci 26:855–879. https://doi.org/10.1111/j.1748-7692.2010.00374.x

Short FT, Polidoro B, Livingstone SR, Carpenter KE, Bandeira S, Bujang JS, Calumpong HP, Carruthers TJB, Coles RG, Dennison WC, Erftemeijer PLA, Fortes MD, Freeman AS, Jagtap TG, Kamal AHM, Kendrick GA, Judson Kenworthy W, la Nafie YA, Nasution IM, Zieman JC (2011) Extinction risk assessment of the world's seagrass species. Biol Conserv 144(7): 1961–1971. https://doi.org/10.1016/j.biocon.2011.04.010

Short FT, Kosten S, Morgan PA, Malone S, Moore GE (2016) Impacts of climate change on submerged and emergent wetland plants. Aquat Bot 135:3–17

Stankovic M, Ambo-Rappe R, Carly F, Dangan-Galon F, Fortes MD, Hossain MS, Kiswara W, Van Luong C, Minh-Thu P, Mishra AK, Noiraksar T (2021) Quantification of blue carbon in seagrass ecosystems of Southeast Asia and their potential for climate change mitigation. Sci Total Environ 783:146858

Stéphanie M, Doorn G (2007) Environmental monitoring and management of reclamations works close to sensitive habitats. Terra et Aqua 108:3

Stiepani J, Gillis LG, Chee SY, Pfeiffer M, Nordhaus I (2021) Impacts of urbanization on mangrove forests and brachyuran crabs in Penang, Malaysia. Reg Environ Chang 21(3):1–13

Thangaradjou T, Kannan L (2008) Development of axenic seedlings of the seagrass, *Enhalus acoroides* (L.f.) Royle. Sci Trans Environ Technovation 2(2):71–77. https://doi.org/10.20894/STET.116.002.002.002

Thangaradjou T, Bhatt JR (2018) Status of seagrass ecosystems in India. Ocean Coast Manag 159:7–15

Thangaradjou T, Sridhar R, Senthilkumar S, Kannan S (2008) Seagrass resource assessment in the Mandapam coast of the Gulf of Mannar Biosphere Reserve, India. Appl Ecol Environ Res 6(1): 139–146. https://doi.org/10.15666/aeer/0601_139146

Tu NH, Uyen NT, Doc LQ, Tuan LC, Thu MA, Tin HC (2021) Impacts of urbanization and land transitions on seagrass beds in tropical lagoon in central Vietnam. Reg Stud Mar Sci 45:101860

United Nations Department for Economic and Social Affairs (2020) World economic situation and prospects 2020. United Nations, New York

Unsworth RKF, McKenzie LJ, Collier CJ, Cullen-Unsworth LC, Duarte CM, Eklöf JS, Jarvis JC, Jones BL, Nordlund LM (2019) Global challenges for seagrass conservation. Ambio 48(8): 801–815. https://doi.org/10.1007/s13280-018-1115-y

Zainal AJM, Dalby DH, Robinson IS (1992) Monitoring marine ecological changes on the East Coast of Bahrain with Landsat TM. Photogramm Eng Remote Sens 59:415–421

Chapter 8
Water Quality Index: An Important Tool to Assess Water Quality of Lake Waters for Sustainable Development

Himakshi Parmar and Prakash Samnani

8.1 Lakes

Water on Earth is huge in quantity when we consider that it covers more than two-thirds of the earth's surface. Water is woven into the fabric of all civilizations and religious groups in a variety of ways due to the strong relationship between water and life, as well as animals and plants. Surface water like rivers and lakes has been seen to become progressively polluted. It is becoming increasingly evident that the Earth is facing major environmental difficulties, with rapidly diminishing natural resources threatening the very viability of most ecosystems (Martin 2013). The next section outlines the basic definition of lake water body, its significance, classification, pollution, and need for its management.

A lake is an enclosed body of water (typically freshwater) surrounded by land and with no direct access to the sea. A lake can also be isolated, with no obvious direct water input and, on rare occasions, no direct output. Many of these isolated lakes are saline due to evaporation or groundwater inflow. A lake can occur anywhere within a river basin, depending on its origin. A headwater lake receives inflow from multiple tiny tributary streams, direct surface rainfall, and groundwater influx rather than a single river. These lakes almost always have a single river outlet. Lakes further downstream in river basins have a significant input and a major outflow, with the water balance from input to output shifting depending on other water sources (Nesje

H. Parmar
Department of Environmental Studies, Faculty of Science, The Maharaja Sayajirao University of Baroda, Vadodara, India

P. Samnani (✉)
Department of Chemistry, Faculty of Science, The Maharaja Sayajirao University of Baroda, Vadodara, India

B. Pathak, R. S. Dubey (eds.), *Climate Change and Urban Environment Sustainability*, Disaster Resilience and Green Growth,
https://doi.org/10.1007/978-981-19-7618-6_8

and Dahl 2000). A water body can be classified as a lake if it fulfils one of the following criteria:

- It should completely or partially fill a basin or a series of adjacent basins.
- With the exception of very short periods induced by wind, thick ice cover, or large inflows, it should have basically the same water level in all regions.
- Even if the water body is close to the sea coast, it does not receive frequent inflows of seawater.
- The inflow-to-volume ratio of the water body should be so low that a significant amount of the suspended silt is collected.
- At mean water level, the area of the water body should surpass a given figure, such as 1 hectare.

Lake scientific research began in the seventeenth century but was first descriptive rather than analytical. Lake measurements and observations grew increasingly focused toward the end of the nineteenth century. Alphonse Forel (1841–1912) pioneered comprehensive lake investigations on Lac Le'man (Lake Geneva) and other Swiss lakes. Forel gave data on a wide range of themes in a three-volume book (Lac Le'man: 1892, 1895, 1904), including sediments and bottom-dwelling animals, fishes and fisheries, water flow, transparency and color, temperature, and others. Thus, Forel, who coined the term "limnology" (initially the study of lakes, but later expanded to encompass other inland waters), exhibited the holistic approach to comprehending a lake as an environmental entity, but without the use of an explicit ecosystem idea (Maitland and Morgan 1997; Likens 2010).

8.1.1 Classification and Zones of Lakes

Classification of lakes is done mostly by Forel and modified by Whipple based on the temperature of the surface and bottom waters (Kar 2014) (Table 8.1).

Lakes exhibit a wide range of spatial diversity in both vertical and horizontal dimensions. Variations can be chemical, physical, or biotic, and it is critical to understanding ecosystem functions. Although some types of variation are exclusive to specific classes of lakes, others are common to most lakes and correlate to a clear spatial structure of the biota in lakes. The presence of some common patterns of spatial organization in lakes has resulted in the designation of several zones with distinct ecological characteristics as shown in Table 8.2. The zones represent some of the most critical physical and chemical parameters that influence biotically driven activities and biotic community formation. Zonation, while often a qualitative rather than a quantitative notion, reflects the accumulation of experience and measurements across a wide range of lakes. As a result, knowing the zonal borders of a lake allows for some broad predictions about the types of organisms and rates of biogeochemical processes that will occur in a given lake, as well as the spatial distribution of these animals and processes.

Table 8.1 Classification of various lakes (Kar 2014)

Classification	Types	Details	References
On the basis of their position on the globe	Polar lakes	Lakes with surface water temperatures that never exceed 4 °C	Forel (1901) and Whipple and Fair (1927)
	Temperate lakes	The surface temperature in these lakes varies above and below 4 °C	
	Tropical lakes	The surface temperature in these lakes is >4 °C	
Based on their origin	Glacial lakes	The action of glaciers leads to the formation of lakes, which are generally found in the temperate regions of the world. Glaciers deepen and widen the valleys when they break down and create piles of the moraine, which blocks valleys and dams, streams, and rivers	Hutchinson (1957)
	Ice-Scour lakes	Scours or hollow basins are generated when massive sheets of ice flow across relatively flat surfaces of hard, jointed, or fractured rocks. These hollow basins are later filled with water, resulting in lakes	
	Tectonic lakes	Tectonic lakes are lakes generated by movements of the earth's crust (tectonic plates)	
	Volcanic lakes	In cases where the volcano was large and erupted a long time ago, the floor of the massive crater is relatively flat, and water accumulate in some of the deeper zones, resulting in the development of lake	
	Riverine lakes	Lakes produced by river activity are generally referred to as riverine or "oxbow" lakes. Between two successive twists, the river cuts through the continuously degraded narrow isthmus. This may result in a river loop to one side of the new stream. This isolated area may hold enough water to produce an oxbow lake	
Based on circulation and mixing	Monomictic lakes	Only one mixing occurs in lakes	Hutchinson (1957)
	Dimictic lakes	Two pairs of mixing	
	Oligomictic lakes	These are thermally stable tropical lakes with very sluggish mixing	
	Polymictic lakes	These are high-altitude tropical lakes with almost constant circulation	
	Amictic lakes	Lakes with no or little mixing that is shielded from most of the changes at the earth's surface by ice	

(continued)

Table 8.1 (continued)

Classification	Types	Details	References
	Meromictic lakes	Meromictic means partially mixed The bottom waters of such lakes maybe devoid of free oxygen and aerobic biota	
	Holomictic lakes	Holomictic lakes are those in which the top and bottom water mix completely and on a regular basis	
On the basis of nutrition	Oligotrophic lakes	Transparent water, low nutrients, less organic matter, abundant in oxygen	Rodhe (1969) and Thienemann (1925)
	Eutrophic lakes	Less transparent water, high nutrients, more organic matter, less in oxygen	
	Mesotrophic lakes	More organic matter, increased productivity	
	Dystrophic lakes	These are claimed to be particular types of lakes with significant concentrations of humic acid in the water	

8.1.2 Significance and Threats to Lake Bodies

Since the early Vedic period, water resources like rivers and lakes are believed to be the epitome of God's creation. Primarily, the inclination for the idea of these river-waters holding the consecrated virtues led Hindus to recite the spiritual shlokas and mantras along the river banks. Such religious convictions prompted us to maintain the asserted purity of these waters in ourselves, and thus, achieve paradise. The lakes supply a diverse range of values and applications, from ecological goods and services to direct production values. These are classified as direct use values, with consumptive and non-consumptive applications such as drinking, irrigation, fishing, ecotourism, and so on. Indirect use values with beneficiaries located farther from the lake, potential future use and non-use social benefit of having a healthy water supply for future generations. The current visual portrays the pseudo-mentality of the masses where the rivers are worshipped only during the ephemeral spiritual practices while subjecting it to the blasphemous anthropogenic activities otherwise. Disposing of biodegradable as well as non-biodegradable waste, and entries of untended ETP (Effluent treatment plant) and STP (Sewage Treatment plant) effluents, in tremendous amounts, have irreversibly damaged the quality of these waters. The repercussions of these actions, however already apparent, will surface sooner rather than later and will have its long-term effects.

The political system in the world is still not capable of providing potable water at the consumer level. Hence, common masses are forced to carry out basic household activities, such as cattle bathing and cleansing at the community pond and river stream, where the threat of acute and chronic diseases lingers. Therefore, acknowledging the importance of the hydrological cycle at the residing phase, it should be mandatory to maintain/conserve the adequate standard of the universal solvent. Contamination and vitiation of these vital bodies make up the partial situation of the "tragedies of the commons".

Table 8.2 Zonation in lakes

Zonation	Temporal variability	Description
	Horizontal	
Pelagic zone	Stable	Offshore (bottom irradiance <1%)
Littoral zone		Nearshore (bottom irradiance ≥1%)

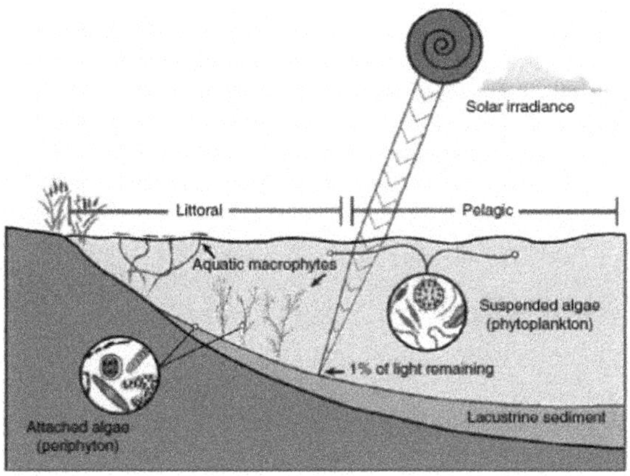

Zonation	Temporal variability	Description
	Vertical	
Water column	Stable	Water extending from the lake surface to bottom
Lacustrine sediments		Lake-generated solids below the water column
Benthic zone		The interface of the water column and lake bottom

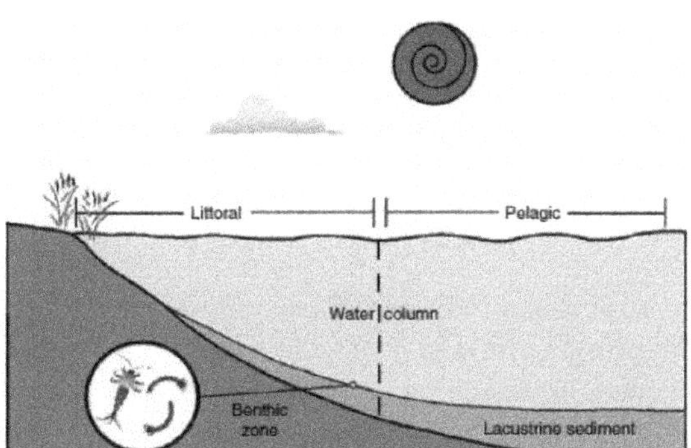

(continued)

Table 8.2 (continued)

Epilimnion (mixed layer)	Seasonal	Uppermost density layer (warm)
Metalimnion		Middle-density layer (transition)
Hypolimnion		Bottom density layer (cool

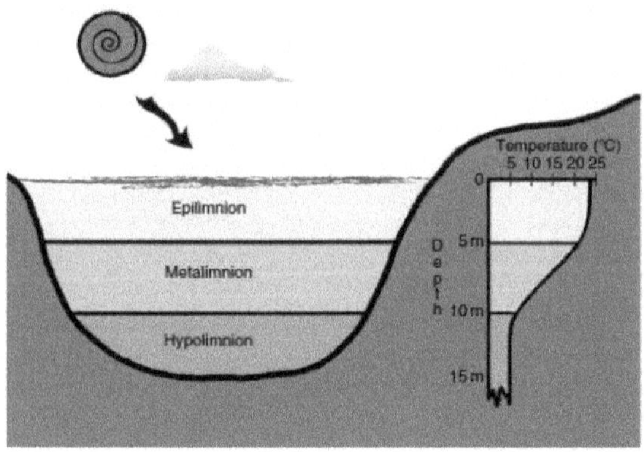

Euphotic zone	Dynamic	Portions of a lake with ≥1% light (photosynthesis)
Aphotic zone		Portions of a lake with <1% light (no photosynthesis)

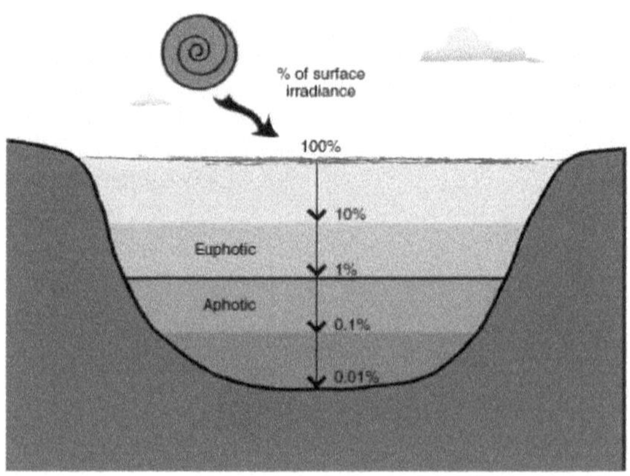

Source: (Likens 2010)

Lakes and reservoirs around India are suffering from varying degrees of environmental degradation, primarily due to encroachment, eutrophication (from domestic and industrial effluents), and siltation. Over the last century, there has been great growth in population without a corresponding increase in municipal facilities, resulting in lakes and reservoirs, particularly urban ones, becoming contaminated sinks. Causes are listed below (Smol 2008; Martin 2013; Reddy and Char 2004)

1. Pollutants emitted by fixed point sources

   Nutrients in municipal and domestic effluents; organic, inorganic, and hazardous pollutants in industrial effluents; and stormwater run-off.
2. Non-point source pollution

   Nutrients from fertilizers, harmful pesticides, and other chemicals, primarily from agricultural run-off; organic pollution from human settlements on the outskirts of lakes and reservoirs;
3. Other basin-related causes of impairment

   Lake silting as a result of increased erosion caused by urban and agricultural expansion, deforestation, road construction, and other land disturbances in the drainage basin; diversion of rivers feeding the lakes, reducing lake size; competing water uses, including drinking, irrigation, and hydropower; and untreated or inadequately treated domestic and industrial effluents from point sources located throughout the basin.

Table 8.3 lists some of the commonly observed reasons for deterioration of water quality in lakes and ponds.

In India, human settlements and public wastewater sources are the primary causes of lake deterioration, particularly in urban lakes. Anthropogenic pressures in the lake catchment produce degradation of the catchment region due to factors such as deforestation, excessive agricultural usage, and the resulting erosion and increasing silt flows, which contaminate lake water. All urban lakes have grown hypertrophic as a result of infrastructure expansion, housing pressure, and encroachment. Under this strain, many urban and rural lakes have perished. Reductions in the Osman Sagar and Himayat Sagar lakes (drinking water sources to Hyderabad in Andhra Pradesh), Udaipur lakes in Rajasthan, and Nainital lakes in Uttaranchal are a few examples of these pressures. The water quality of urban lakes has deteriorated to the point where it is causing significant disruptions to the biodiversity of lake habitats. Using bioremediation procedures alone, like with the Powai lakes in Mumbai, the

Table 8.3 Lake water quality deterioration parameters

Lake water quality deterioration parameters	
Pathogens	Occasional or regional deterioration mostly in small and shallow water bodies
Suspended solids	Not relevant
Decomposable organic matter	Occasional or regional deterioration
Eutrophication	Important deterioration
Nitrate	Rare deterioration
Salinization	Rare deterioration
Heavy metals	Important deterioration
Organic micro pollutants	Occasional or regional deterioration
Acidification	Important deterioration
Changes in hydrological regimes	Important deterioration

Kodaikanal and Ooty lakes in Tamil Nadu, and so on, has proved ineffective in achieving full lake balance. Water hyacinth growth has been prolific in many lakes, resulting in the reproduction of water vectors and, as a result, the transmission of endemic diseases. Loktak Lake, Bhopal Lakes, Ropar Lake, Sukhna Lake, Kanjli Lake, and Pong Dam Lake are notable examples. Cultural siltation, in the form of idol immersion during specific annual festivals in India, has been a significant source of heavy metal contamination in lakes. The Bhoj Wetlands and city lakes in Bombay, Hyderabad, and Bangalore are three examples. Furthermore, unchecked tourist pressure has harmed the biodiversity of lake-related flora and animals in many lakes. Tsu Moriri, Pongsho, and Dal, for example, are high-altitude lakes. Coastal lakes have also been severely impacted by salinity imbalances, which have been attributed to a lack of balance between freshwater from the lakes' inland catchment and seawater input at the estuary's mouth. Chilika Lake in Orissa, Pulicat Lake in Tamil Nadu, and Kuttanad Lake in Kerala are just a few examples. The development of satellite ports, chemical companies, and thermal power plants, such as at Pulicat Lake, has also contributed to siltation and pollution. Water shortages in lakes, which have significantly hampered water replenishment, have had a catastrophic impact on bird sanctuaries and fisheries. Keoladeo National Park (Bharatpur Lake), Nalsarovar Bird Sanctuary Lake, and Dal Lake are a few examples (Reddy and Char 2004).

8.1.3 Importance of Monitoring Lake Bodies

Water is a fundamental substance on which the dynamics of a water body, and thus the life of the biota within it, rely. The following are the primary goals of physical and chemical water testing (Bartram and Ballance 1996; Duck 2002):

- To determine the precise composition of the sample at the moment of collection.
- To categorize water based on its overall level of gaseous and mineral ingredients.
- To determine the presence or absence of components that affect various beneficial applications of water.
- To ascertain the number of organic contaminants and determine the degree of clarity and the type of the matter in suspension.

Some important parameters analyzed for assessment of lake bodies are listed below in Table 8.4.

Table 8.4 List of important parameters for assessment

Parameters	Sub parameters
Physical	Temperature
	Turbidity
	Transparency
Chemical	pH, dissolved oxygen (DO), free carbon dioxide (FCO_2), total alkalinity (TA), specific conductivity
Nutrients	Nitrates, phosphates, and chlorophyll

8.2 Water Quality Index

A water quality index seeks to assign a single value to a source's water quality based on one or more systems that translate the list of constituents and their concentrations in a sample into a single value. In ecology, indices have been used to represent species richness, evenness, diversity, and other characteristics. As a result, the Shannon Index, the Simpson Index, and so on have been developed. Indices are widely used in a variety of other fields, including medicine, sociology, and process safety. Indexes (the singular of indices is index) are composite representations of a condition or situation derived from a combination, done in specific ways, of several relevant but non-commensurate observed facts/measurements. The combination yields a single ordinal number, which aids comprehension and interpretation of the overall significance of the facts that contributed to that number (Song and Kim 2009; Abbasi and Abbasi 2012).

Water quality indices are a major component of environmental indices. Regulatory agencies use them as communication tools to describe the "quality" or "health" of a specific environmental system (e.g., air, water, soil, and sediments) and to assess the impact of regulatory policies on various environmental management practices (Pusatli et al. 2009; Sadiq et al. 2010; Abbasi and Abbasi 2012).

Once developed and implemented, the WQIs are a useful tool for examining trends, highlighting specific environmental conditions, and assisting governmental decision-makers in evaluating the effectiveness of regulatory programmes. Indices serve all of the purposes for which water quality is monitored, including assessment, utilization, treatment, resource allocation, public information, R&D, and environmental planning. Furthermore, indices greatly simplify and clarify the transfer and use of water quality data. WQIs have become a pivotal component for interpreting the environmental variation or condition for a prevailing water body (Terrado et al. 2010; Sutadian et al. 2017). Information obtained from WQIs can be used for various purposes (Sutadian et al. 2016):

1. To provide an overall status of water quality to the water authorities and the wider community (Chang et al. 2001; Ocampo-Duque et al. 2006).
2. To study the impact of regulatory policies and environmental programs on environmental quality (Swamee and Tyagi 2007).
3. To compare the water quality of different sources and sites, without making a highly technical assessment of the water quality data (Sarkar and Abbasi 2006).
4. To assist policymakers and the public to avoid subjective assessments and subsequent biased opinions (Ocampo-Duque et al. 2006; Rehana and Mujumdar 2009).

8.2.1 WQI Model

The basic form of WQI models is depicted in Fig. 8.1, revealing that the majority of WQIs consist of four major steps (Abbasi and Abbasi 2012; Sutadian et al. 2017; Lumb et al. 2011):

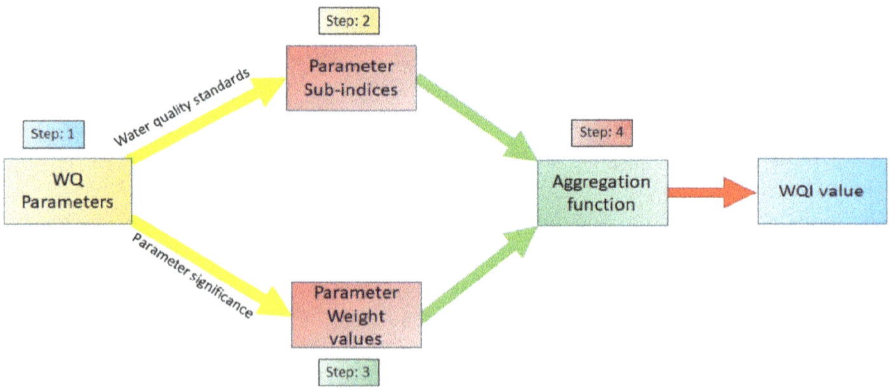

Fig. 8.1 WQI development model [Source: (Uddin et al. 2021)]

- The water quality parameters are chosen for inclusion in the assessment.
- Parameter subindices are created by converting parameter concentrations to unitless subindices.
- Weighting of parameters: weightings are assigned to parameters based on their importance to the assessment.
- The water quality index is computed using an aggregation function: individual parameter subindices are combined using weightings to yield a single overall index.

The following sections discuss in detail the four steps.

1. Parameter selection

The initial step in the WQI process is parameter selection, and there is significant variation between various models in terms of the type and number of parameters chosen, as well as the reasons for doing so. Water quality varies according to its spatiotemporal dimensions during its cycle, as well as its allocations and uses. The latter determine the water quality variables to be used, the analytical method, and the sampling period. The most commonly included parameters are temperature, turbidity, pH, suspended solids (SS), total dissolved solids (TDS), fecal coliforms (FC), dissolved oxygen (DO), biochemical oxygen demand (BOD), nitrate, ammoniacal nitrogen (NH_3-N), and phosphate. Parameter selection depends on expert opinions plus type of index being used; many models have 8–11 parameters (Bhutiani et al. 2016; Sargaonkar and Deshpande 2003) while others have 20–26 parameters also in consideration (Khamis et al. 2019; "Calculation of the Drinking Water Quality Index" Government of Newfoundland and Labrador 2001). The Delphi method can be used to select these variables, but it is still reliant on expert opinion (Grisham 2009). As a result, depending on the panels of experts consulted, the final WQI can be highly variable (Kachroud et al. 2019).

2. Subindexing

   Various parameters occur in diverse ranges, are expressed in different units, and behave differently in terms of the concentration–impact relationship. All of this must be transformed into a single scale, usually beginning at zero and ending at one, before an index can be created. The range of some index scales is 0–100.

Subindices can be classified as one of four general types (Abbasi and Abbasi 2012; Uddin et al. 2021):

- Linear
- Nonlinear
- Segmented linear
- Segmented nonlinear

3. Parameter weighting

   Weightage assignment, like parameter selection, is a matter of opinion, and thus subjective. Well-formulated opinion-gathering techniques, such as Delphi, are used in this case as well to reduce subjectivity and increase credibility. If inappropriate weightings are used, i.e., a parameter is given more importance than it deserves, it can have a negative impact on model evaluation.
4. Aggregation

   The WQI model's final step is the aggregation process. It is used to combine the parameter subindices into a single score for the water quality index. Types of functions used are:

   - Additive
   - Multiplicative
   - Combined aggregating functions
   - Square root of the harmonic mean function
   - Minimum operator function

Figure 8.2 explains the overall roadmap for obtaining and using WQIs, from sample collection to calculating WQIs and applying them, using 4 steps process mentioned above.

Different WQI models used worldwide are mentioned in Table 8.5 (Uddin et al. 2021; Sutadian et al. 2016; Akhtar et al. 2021).

Figure 8.3 shows use of various WQIs on global map.

8.2.2 *Limitations of WQIs*

Use of WQIs for monitoring water quality has some limitations, as was indicated earlier. Some aspects of the same are discussed below (Swamee and Tyagi 2007; Abbasi and Abbasi 2012; Sarkar and Abbasi 2006; Paun et al. 2016; Lumb et al. 2011; Sutadian et al. 2016).

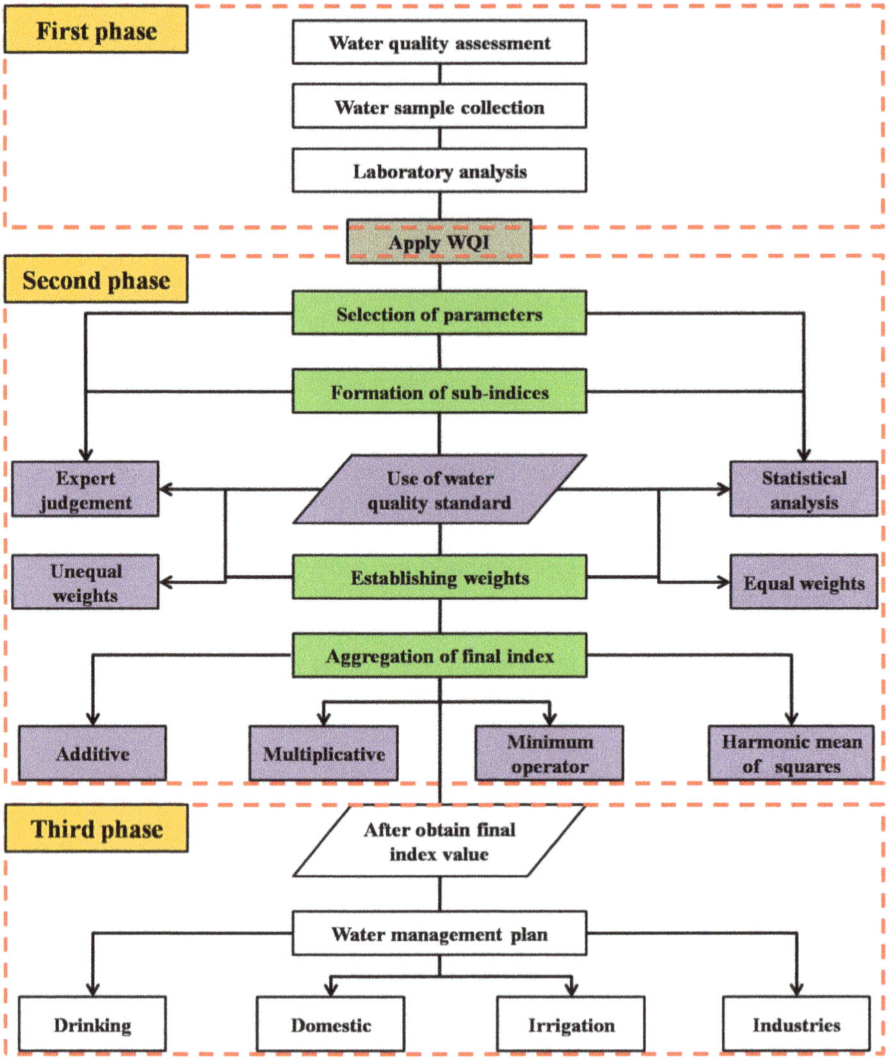

Fig. 8.2 Development of WQI in 3 phases. [Source: (Akhtar et al. 2021)]

1. Ambiguity

   When index exceeds the critical level (unacceptable value) without any of the subindices exceeding the critical level, an aggregation method becomes ambiguous. Ambiguity is defined as a situation in which the overall index suggests poorer water quality than would be expected based on all determinant subindex values. This is primarily a problem with weighted indices, depending on how the weights are assigned.

Table 8.5 Different type of WQIs and their functions

S. no.	Index model	Parameters	Subindex function	Weightage method	Aggregation method	References
1	Horton Index	8	✓	✓	Additive	Horton (1965)
2	National Sanitation Foundation (NSF) Index	11	✓	✓	Additive and multiplicative	Gradilla-Hernández et al. (2020) and Kumar and Alappat (2009)
3	Scottish Research Development Department (SRDD) Index	10	✓	✓	Additive and multiplicative	Banda and Kumarasamy (2020)
4	Dinius Index	11	✓	✓	Multiplicative	Dinius (1987)
5	Ross Index	4	✓	✓	Additive	Uddin et al. (2021)
6	Bascaron Index	26	✓	✓	Two additive functions: subjective and objective	Zemed et al. (2021) and Khamis et al. (2019)
7	Oregon Index	8	✓	✓	Weight arithmetic mean function	Borisenko and Hubler (2010) and Brown (2016)
8	EQ Index	9	✓	✓	Additive	Dash et al. (2019) and Uddin et al. (2021)
9	House Index	9	✓	✓	Used SRDD aggregation technique	House and Ellis (1987)
10	Smith Index	7	✓	✓	Minimum operator function	Carpenter et al. (1998) and Abbasi and Abbasi (2012)
11	Dojildo Index	26	▪	▪	Square root of the harmonic mean function	Dojlido et al. (1994)
12	British Colombia Index	10	▪	▪	Simple specific mathematical formula	Abbasi and Abbasi (2012) and Dash and Kalamdhad (2021)

(continued)

Table 8.5 (continued)

S. no.	Index model	Parameters	Subindex function	Weightage method	Aggregation method	References
13	Dalmatian Index modified version of SRDD index	8	✓	✓	Used automatic index formulas	Stambuk-Giljanović (2003)
14	Canadian Council of Ministers of the Environment (CCME) Water Quality Index	4	▪	▪	Used fixed mathematical functions	Regmi and Mishra (2016) and Lumb et al. (2006)
15	Liau Index	13	✓	✓	Additive and multiplicative	Liou and Lo (2004)
16	Said Index	5	✓	✓	Simple specific mathematical formula	Abbasi and Abbasi (2012)
17	Malaysian Index	6	✓	✓	Additive	Abidin et al. (2015)
18	Hanh Index	8	▪	▪	Additive and multiplicative	Abbasi and Abbasi (2012) and Uddin et al. (2021)
19	Almeida Index	10	✓	✓	Multiplicative	Abbasi and Abbasi (2012) and Uddin et al. (2021)
20	West Java Index	13	✓	✓	Non-equal geometric technique	Abbasi and Abbasi (2012) and Uddin et al. (2021) and Dash and Kalamdhad (2021)
21	Fuzzy-based indices	No guidelines	▪	▪	Fuzzy	Ocampo-Duque et al. (2006) and Chang et al. (2001)
22	Bhargava's Index	4	✓	✓	Modified multiplicative	Bhargave and Asce (1985)
23	Status and Sustainability Index	15	▪	▪	Minimum operator	Uddin et al. (2021) and Dash and Kalamdhad (2021)

(continued)

Table 8.5 (continued)

S. no.	Index model	Parameters	Subindex function	Weightage method	Aggregation method	References
24	Dalmatian Index	9	✓	✓	Additive or multiplicative	Nives (1997)
25	Ved Prakash's Index	4	✓	✓	Additive	Bhutiani et al. (2016) and Kachroud et al. (2019)
26	Water Pollution Index	15	▪	▪	Root mean square	Akhtar et al. (2021) and Sutadian et al. (2016)
27	Contact Recreation Index	8	▪	▪	Minimum operator	Akhtar et al. (2021), Uddin et al. (2021) and Sutadian et al. (2016)
28	Hallock's Index	8	▪	▪	Additive	Akhtar et al. (2021) and Uddin et al. (2021)
29	Boyacioglu's Index	12	✓	✓	Additive	Boyacioglu (2007)
30	Harkins' Index	No guidelines	▪	▪	Statistical method	Gupta et al. (2003)
31	Indian Pollutant Index	13	✓	✓	Additive	Akhtar et al. (2021) and Uddin et al. (2021)
32	Prati's Index	13	✓	✓	Additive	Prati (1971)
33	Stoner's Index	13	✓	✓	Additive	Stoner (1978)
34	Wolski and Parker's Index	11	✓	✓	Geometric mean	Abbasi and Abbasi (2012)
35	Weighted Arithmetic Index	No guidelines	✓	✓		Abbasi and Abbasi (2012)
36	Tiwari and Mishra Index	14	✓	✓	The logarithm and antilogarithm	Kachroud et al. (2019) and Abbasi and Abbasi (2012)
37	Drinking Water Quality Index	17	✓	✓	Additive	Akhtar et al. (2021) and Uddin et al. (2021)

(continued)

Table 8.5 (continued)

S. no.	Index model	Parameters	Subindex function	Weightage method	Aggregation method	References
38	Wastewater Water Quality Index	23	✓	✓	Additive	Akhtar et al. (2021) and Uddin et al. (2021)

✓: Includes all 4 steps for index formation
⍰: One or more common steps are not followed

2. Eclipsing
When index does not exceed the critical level (unacceptable value) despite one or more of the subindices exceeding the critical level, eclipsing occurs. The masking of low-value subindices in an overall high WQI value is referred to as eclipsing. Attempts to circumvent this issue include the use of alternative indices based on either weighted or unweighted multiplicative models, new index formulations, and/or novel approaches to subindex value determination.
3. Rigidity
Rigidity is defined as the inability of an index to accommodate additional or alternative water quality determinants. This is especially important when an impairment occurs in a determinant(s) not included in the index or when an index is used in an area with concerns other than those for which it was designed. For example, a regulatory agency may already have an overall index but would like to add one or more additional parameters. This situation can occur when the index indicates that the water quality is good at a particular site, but the water is negatively impacted by constituents not included in the index. Alternatively, an agency may wish to use an index developed for one region in another where weather and other environmental conditions may differ significantly.
4. Compensation
A good compensation aggregation method is one that is not biased toward extremes (i.e., the highest or lowest subindex value). However, when ambiguity-free and eclipsing-free models are desired, this property becomes a hindrance. Maximum (or minimum) operators, for example, that are free of ambiguity and eclipsing have poor compensation because they are biased toward the highest (or lowest) subindex values. As a result, the benefits of compensation must be balanced against the drawbacks of ambiguity (and eclipsing).

8.2.3 Gap in Current Situation and Future Directions

Various water quality indices have been developed to address problems on a regional scale, a highly specific problem, or a specific water body. These specific indices become highly regional, and the researchers' suggested methodologies may not be applicable on a global scale. Furthermore, the indices developed or proposed take

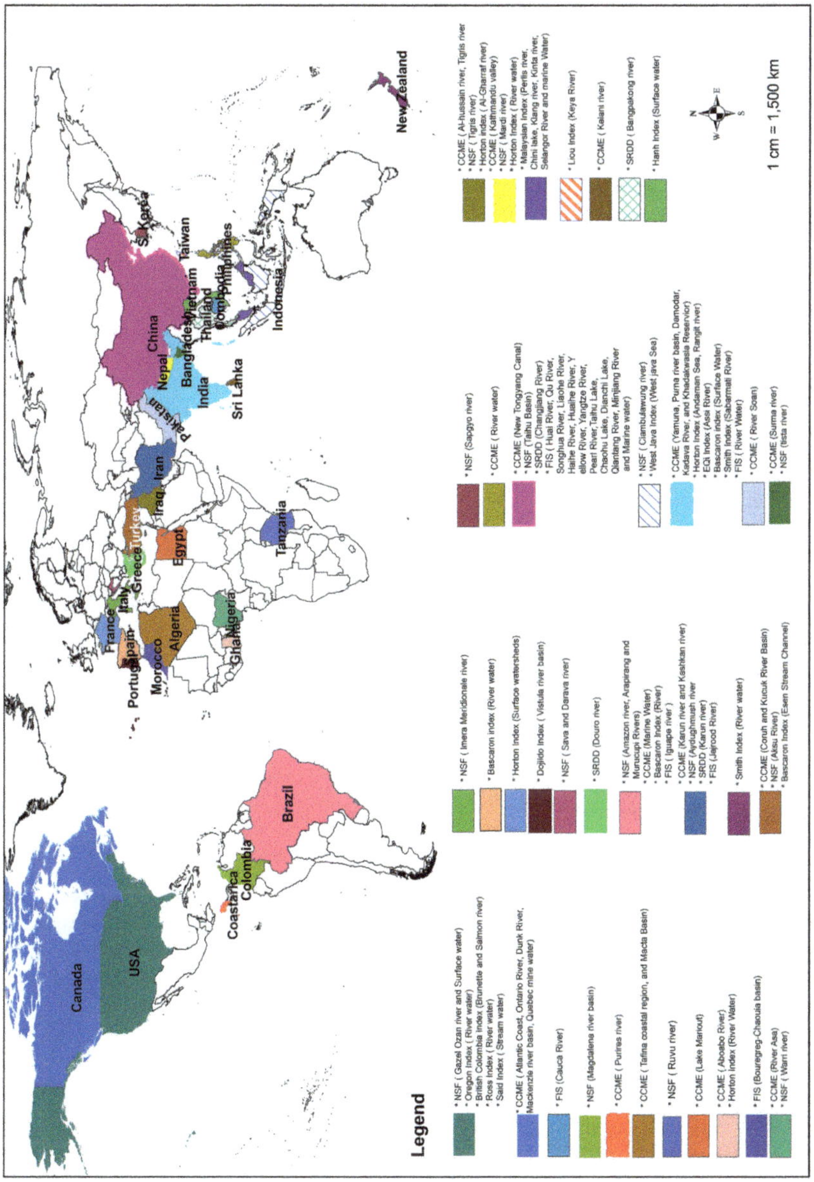

Fig. 8.3 Different types of WQIs used worldwide [Source: (Uddin et al. 2021)]

into account those water quality parameters that have been identified as problematic in those regions, even though those problems may not be applicable to other water bodies or regions under consideration. Each index developed for a specific end-use of water, such as drinking, irrigation, industries, heavy metals, and so on, which considers specific parameters, requires a comprehensive assessment when it comes to carefully selecting the water quality parameters. Listed below are some of the gaps which leads to inappropriate WQI value (Table 8.6).

Table 8.6 WQI research area and future research directions

Current Research Area	Research themes	Research Directions
1. Highly Regional 2. Consideration of specific parameters depending on the problems associated to that area.	Specific Indices	1. Benchmarking Indices which can be applied on a global scale. 2. Comprehensive assessment of parametric consideration for different indices, independent of the requirements of the area of interest.
3. Highly subjective 4. Considers ranking of parameters depending on the responses recorded by people or experts. 5. Parameters are considered based on the problems associated to a specific region of interest.	Human Intervention	1. Challenges like rigidity, eclipsing and ambiguity needs careful assessment. 2. Domain or reach of expert should be expanded. 3. Subjective methods may be followed by certain mathematical statistical based techniques
4. No specific tool to determine the efficacy of use of WQIs. 5. Very limited studies carried out on regression modelling and prediction through artificial neural network models.	Performance Assessment	1. Indices developed or proposed methodologies need performance assessment. 2. Sensitivity analysis of different parameters considered for developing various indices need to be carried out.
3. Multivariate statistics 4. Probability 5. Randomness of water quality datasets.	Emerging Technologies	1. Modification in existing mathematical models. 2. Integrating one or more mathematical techniques for a more reliable and effective WQI.

Source: (Dash and Kalamdhad 2021)

- Selection of parameters
- Comparison with standard values
- Parameters with no predefined standards
- Unit weightage
- Range selection
- Specific standards for lake/pond
- Climatic factors
- Geographical factors
- Seasonal index adoption

In our research study, we are assessing the status of lake bodies in Vadodara city of Gujarat, India, by use of WQIs and have found that when different indices are applied to a single data set of a sample body, resultant classes are different (Parmar and Samnani 2022). For example, if by applying NSF index water body falls under good water quality same water body falls under bad quality class when overall water quality index is applied. It is observed that significant differences are recorded between classes for the same lake water body as determined by different indices. These variations are the result of different summation methods, several selected parameters, range distribution, and weightage factors for selected parameters. The lakes chosen for our study are isolated ones, thought they do receive input of water and there is output as well.

From the study, it is observed that the indices are not able to predict the actual status of the water body with their original formation but when we try to modify some part of a particular index, the status of the water body changes relating it with our obtained results. Here we have modified the unit weight part that is common in both NSF and OWQ index and obtained the index value for the same data set used before. For NSF and OQW index, weightage of pH, temperature, DO, BOD, fecal coliform, total coliform, nitrate, phosphate, turbidity, and sulfate have been changed. The resultant WQI value validates the visible human interaction with the lake bodies and apparent water quality (Table 8.7; Figs. 8.4 and 8.5).

NSF INDEX—O: Original
NSF INDEX—M: Modified
OWQ INDEX—O: Original
OWQ INDEX—M: Modified

Water quality assessment in terms of WQI has emerged as a critical issue for the global sustainability index. The quality of a single lake water body is classified into three categories, ranging from acceptable to unsuitable, based on various indices that do not define the actual quality of the water body, resulting in varying interpretations. Because of the current scarcity of freshwater, decisions should be based on this understanding and any necessary changes in the computation of WQIs. Prediction of the actual status of the water body in comparison to the WQI calculated with a pre-determined range and parameter weightage is possible by modifying the indices based on parameter dominance and pollution of the water body. The pollution load received by the water body should be used to modify the universal pollution range of

Table 8.7 Original and modified weightage values for NSF and OWQ Index

Parameters	NSF Index—O	NSF Index—M	OWQ Index—O	OWQ Index—M
pH	0.11	0.08	1	1
TDS			3	3
Color			2	2
Temperature	0.1	0.17		
DO	0.17	0.2	4	4
Total solids	0.07	0.07		
5 day BOD	0.11	0.06	2	4
Sulfates			2	3
Total hardness			1	1
Chloride			1	1
Total coliform			4	3
Fecal coliform	0.16	0.08		
Turbidity	0.08	0.08	1	4
Fluoride			3	1
Nitrate	0.1	0.07	3	1
Phosphate	0.1	0.19	2	4

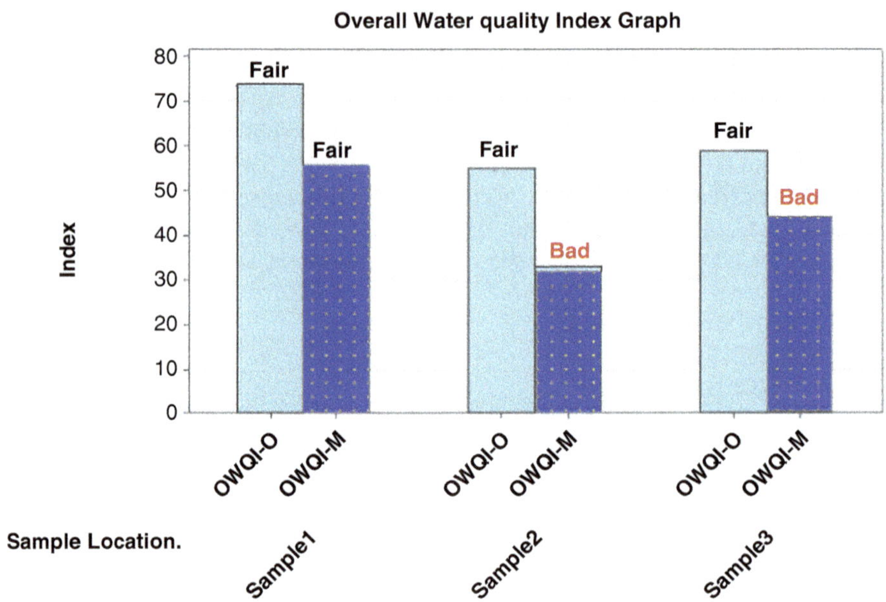

Fig. 8.4 OQWI Original Index (O) and Modified Index (M) comparison graph for 3 sample bodies

selected parameters. Parameters should be added or removed based on the nature of the water body, its geographical location, and anthropological influences. The range of weightage and subindex values should be defined differently depending on the

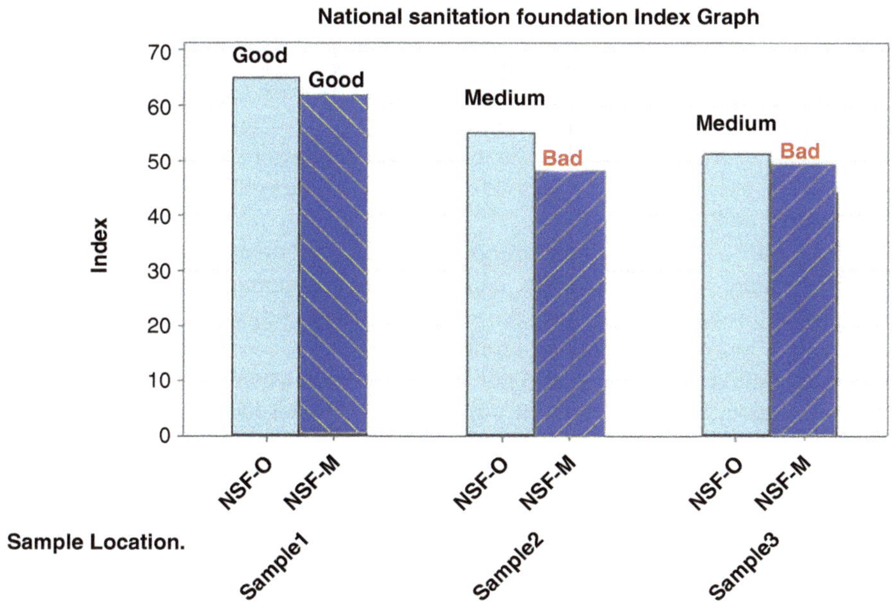

Fig. 8.5 NSF Original Index (O) and Modified Index (M) comparison graph for 3 sample bodies

pollution load received by that lake water body. Included in WQI development should be important lake parameters, their standard values, and other lake/pond characteristics that can define the actual water quality and thus treatment steps required before the water can be put to suitable use.

8.3 Summary

Lake water quality assessment is as important as that for rivers, since they are also important natural resource and their sustainable maintenance is required to fulfil the upcoming needs of future generation. Not only humans but animals and plants are fully dependent on lake bodies for water, food, habitat, etc. Thousands of flora and fauna reside in lakes adding to the biodiversity of that particular state or country. Since the concept of WQI came into existence, monitoring water bodies and communicating the status to the people have become easier as WQI is a simple tool which converts scientific data into a digit form which can be made understood the authorities and locals.

Some of the learning points from chapter

1. Lake ecosystem should be monitored on daily, monthly, and quarterly basis for knowing the health of the water body and data be stored, track with the past data for easy assessment and interpretation.
2. Assessment should not be limited to only few parameters, parameters selection, inclusion, exclusion should be done on the basis of the pollution load received by the water body.
3. People Participation Programme should be organized on certain time duration to make people value water body, understand their importance, educate them on which anthropogenic activities can pollute the water body and guide them to take measures to avoid pollution on ground level.
4. A single universal index should not be applied everywhere.
5. Modification should be done in previously developed index depending on the basis of location, past and present data, for example, parameters weightage if needs to be change should be done on basis of how that parameters values are increasing gradually so index value might get affected if weightage is low for that particular parameters. Addition or deletion of the parameters should be accepted which is not seen in case of all Indices. According to the pollution load, climatic conditions parameters should be decided for index calculation.
6. Analyzed data should be presented in a pictorial form as mathematical formulas, number, and models might not be understood to all and confuse common people.

Recently a new index concept known as Eco-Heart Index has been proposed which conveys the health of a water body in form of a heart shape. Full heart conveys excellent water quality whereas broken heart indicates bad water quality. Such development in index has engaged people as it is representing data in graphical and colorful form which is understandable to all. Not much research has been done on this index, but it draws a new line of research in index WQI field which surpasses all mathematical and statistical formula into simple drawing adding a sustainable index for the assessing a water body.

References

Abbasi T, Abbasi SA (2012) Why water-quality indices. Elsevier, Amsterdam. https://doi.org/10.1016/b978-0-444-54304-2.00001-4

Abidin IZ, Juahir H, Azid A, Mustafa AD, Azaman F (2015) Penggunaan Excel-VBA Untuk Pengiraan Indeks Kualiti Air Dan Indeks Pencemaran Udara. Malays J Anal Sci 19(5): 1056–1064

Akhtar N, Ishak MIS, Ahmad MI, Umar K, Yusuff MSM, Anees MT, Qadir A, Almanasir YKA (2021) Modification of the water quality index (WQI) process for simple calculation using the multi-criteria decision-making (MCDM) method: a review. Water (Switzerland) 13(7):905. https://doi.org/10.3390/w13070905

Banda TD, Kumarasamy MV (2020) Development of water quality indices (WQIs): a review. Pol J Environ Stud 29(3):2011–2021. https://doi.org/10.15244/pjoes/110526

Bartram J, Ballance R (1996) United Nations environment programmes world. Water quality monitoring, vol 42. CRC Press, Boca Raton

Bhargave DS, Asce F (1985) Expression for drinking water supply standards. Ill 3:304–316

Bhutiani R, Khanna DR, Kulkarni DB, Ruhela M (2016) Assessment of Ganga River ecosystem at Haridwar, Uttarakhand, India with reference to water quality indices. Appl Water Sci 6(2): 107–113. https://doi.org/10.1007/s13201-014-0206-6

Borisenko A, Hubler S (2010) Oregon water quality index (OWQI) Department of Environmental Quality, Oregon

Boyacioglu H (2007) Development of a water quality index based on a European classification scheme. Water Sa 33(1):101–106. https://doi.org/10.4314/wsa.v33i1.47882

Brown D (2016) Oregon water quality index: background, analysis and usage. Department of Environmental Quality, Eugene

Carpenter SR, Caraco NF, Correll DL, Howarth RW, Sharpley AN, Smith VH (1998) Nonpoint pollution of surface waters with phosphorus and nitrogen. Ecol Appl 8(3):559–568. https://doi.org/10.1890/1051-0761(1998)008[0559:NPOSWW]2.0.CO;2

Chang NB, Chen HW, Ning SK (2001) Identification of river water quality using the fuzzy synthetic evaluation approach. J Environ Manag 63(3):293–305. https://doi.org/10.1006/jema.2001.0483

Dash S, Kalamdhad AS (2021) Science mapping approach to critical reviewing of published literature on water quality indexing. Ecol Indic 128:107862. https://doi.org/10.1016/j.ecolind.2021.107862

Dash S, Borah SS, Kalamdhad A (2019) A modified indexing approach for assessment of heavy metal contamination in Deepor Beel, India. Ecol Indic 106:105444. https://doi.org/10.1016/j.ecolind.2019.105444

Dinius SH (1987) Design of an index of water quality. JAWRA J Am Water Resour Assoc 23(5): 833–843. https://doi.org/10.1111/j.1752-1688.1987.tb02959.x

Dojlido JAN, Raniszewski J, Woyciechowska J (1994) Water quality index applied to rivers in the Vistula river basin in Poland. Environ Monit Assess 33:33–42

Duck RW (2002) Hydrological and limnological aspects of lake monitoring. Area 34:222–223

Forel FA (1901) Handbuch Der Seenkunde: Allgemeine Limnologie. J. Engelhorn, Stuttgart

Government of Newfoundland and Labrador (2001) Drinking Water Quality Index - Environment and Climate Change Department of Environment and Conservation https://www.gov.nl.ca/ecc/waterres/quality/drinkingwater/dwqi/

Gradilla-Hernández MS, de Anda J, Garcia-Gonzalez A, Montes CY, Barrios-Piña H, Ruiz-Palomino P, Díaz-Vázquez D (2020) Assessment of the water quality of a subtropical lake using the NSF-WQI and a newly proposed ecosystem specific water quality index. Environ Monit Assess 192(5):8265. https://doi.org/10.1007/s10661-020-08265-7

Grisham T (2009) The Delphi technique: a method for testing complex and multifaceted topics. Int J Manag Proj Bus 2(1):112–130. https://doi.org/10.1108/17538370910930545

Gupta AK, Gupta SK, Patil RS (2003) A comparison of water quality indices for coastal water. J Environ Sci Health Part A Toxic Hazard Subst Environ Eng 38(11):2711–2725. https://doi.org/10.1081/ESE-120024458

Horton R (1965) An index number system for rating water quality. J Water Pollut Control Fed 37(37):300–306

House MA, Ellis JB (1987) The development of water quality indices for operational management. Water Sci Technol 19(9):145–154. https://doi.org/10.2166/wst.1987.0076

Hutchinson GE (1957) A treatise on limnology. Geography, physics and chemistry, vol 1. Wiley, New York

Kachroud M, Trolard F, Kefi M, Jebari S, Bourrié G (2019) Water quality indices: challenges and application limits in the literature. Water (Switzerland) 11(2):1–26. https://doi.org/10.3390/w11020361

Kar D (2014) Wetlands and lakes of the world, vol 52. Springer, New Delhi. https://doi.org/10.5860/choice.184969

Khamis A, Elshemy M, Rashwan I (2019) Evaluation of surface water quality for drinking in El-Beheira Governorate, Egypt, by utilizing WQI, statistical analysis approaches and an innovative software. In: Fifth international scientific research conference
Kumar D, Alappat BJ (2009) NSF-water quality index: does it represent the experts' opinion? Pract Period Hazard Toxic Radioact Waste Manag 13(1):75–79. https://doi.org/10.1061/(ASCE)1090-025X(2009)13:1(75)
Likens GE (2010) Lake ecosystem ecology: a global perspective. Academic Press, Cambridge
Liou S-M, Lo S-L (2004) A generalized water quality index in Taiwan. Environ Monit Assess 38(2):35–37. https://doi.org/10.1023/b:emas.0000031715.83752.a1
Lumb A, Halliwell D, Sharma T (2006) Application of CCME water quality index to monitor water quality: a case of the Mackenzie River basin, Canada. Environ Monit Assess 113(1–3):411–429. https://doi.org/10.1007/s10661-005-9092-6
Lumb A, Sharma TC, Bibeault J-F (2011) A review of genesis and evolution of water quality index (WQI) and some future directions. Water Qual Expo Health 3(1):11–24. https://doi.org/10.1007/s12403-011-0040-0
Maitland PS, Morgan NC (1997) Conservation management of freshwater habitats: lakes, rivers and wetlands. Chapman and Hall, London. https://doi.org/10.1007/978-94-011-5858-9
Martin JL (2013) Restoration and management of lakes and reservoirs. In: Hydro-environmental analysis. CRC Press, Boca Raton. https://doi.org/10.1201/b16108-20
Nesje A, Dahl SO (2000) Glaciers and environmental change: key issues in environmental change. Arnold, London, p 216
Nives S-G (1997) Comparison of Dalmatian water evaluation indices. Water Environ Res 75(5): 388–405
Ocampo-Duque W, Ferré-Huguet N, Domingo JL, Schuhmacher M (2006) Assessing water quality in rivers with fuzzy inference systems: a case study. Environ Int 32(6):733–742
Parmar H, Samnani P (2022) Assessment of lakes of Vadodara city in terms of water quality indices and their comparison. Int J Energy Water Resour. In press. https://doi.org/10.1007/s42108-022-00231-y
Paun I, Cruceru LV, Chiriac FL, Niculescu M, Vasile GG, Marin NM (2016) Water quality indices—methods for evaluating the quality of drinking water. Cude 2001:395–402. https://doi.org/10.21698/simi.2016.0055
Prati L, Pavanello R, Pesarin F (1971) Assessment of surface water quality by a single index of pollution. Water Resour Manag 5:741–775
Pusatli OT, Camur MZ, Yazicigil H (2009) Susceptibility indexing method for irrigation water management planning: applications to K. Menderes River basin, Turkey. J Environ Manage 90(1):341–347. https://doi.org/10.1016/j.jenvman.2007.10.002
Reddy MS, Char NVV (2004) Management of lakes in India. Lakes Reserv Res Manag 11(4): 227–237
Regmi RK, Mishra BK (2016) Use of water quality index in water quality assessment: a case study in the metro Manila. Water Urban Initiative Work 07:61920. https://doi.org/10.13140/RG.2.2.25821.61920
Rehana S, Mujumdar PP (2009) An imprecise fuzzy risk approach for water quality management of a river system. J Environ Manage 90(11):3653–3664. https://doi.org/10.1016/j.jenvman.2009.07.007
Rodhe W (1969) Sulla Productione Difi Toplankton in Laghi Transparents Di Alta Montagna. Mem 1st Ital Idrobiol 15:21–28
Sadiq R, Haji SA, Cool G, Rodriguez MJ (2010) Using penalty functions to evaluate aggregation models for environmental indices. J Environ Manage 91(3):706–716. https://doi.org/10.1016/j.jenvman.2009.09.034
Sargaonkar A, Deshpande V (2003) Development of an overall index of pollution for surface water based on a general classification scheme in Indian context. Environ Monit Assess 89(1):43–67. https://doi.org/10.1023/A:1025886025137

Sarkar C, Abbasi SA (2006) QUALIDEX—a new software for generating water quality indice. Environ Monit Assess 119(1–3):201–231. https://doi.org/10.1007/s10661-005-9023-6
Smol JP (2008) Pollution of lakes and rivers: a paleoenvironmental perspective. Bull Mar Sci 83(2): 438
Song T, Kim K (2009) Development of a water quality loading index based on water quality modeling. J Environ Manag 90(3):1534–1543. https://doi.org/10.1016/j.jenvman.2008.11.008
Stambuk-Giljanović N (2003) Comparison of Dalmatian water evaluation indices. Water Environ Res 75:388–405. https://doi.org/10.2175/106143003X141196
Stoner JD (1978) Water-quality indices for specific water uses. Geological Survey, Arlington. https://doi.org/10.3133/cir770
Sutadian AD, Muttil N, Yilmaz AG, Perera BJC (2016) Development of river water quality indices—a review. Environ Monit Assess 188(1):1–29. https://doi.org/10.1007/s10661-015-5050-0
Sutadian AD, Muttil N, Yilmaz AG, Perera BJC (2017) Using the Analytic Hierarchy Process to identify parameter weights for developing a water quality index. Ecol Indic 75:220–233. https://doi.org/10.1016/j.ecolind.2016.12.043
Swamee PK, Tyagi A (2007) Improved method for aggregation of water quality subindices. J Environ Eng 133(2):220–225. https://doi.org/10.1061/(asce)0733-9372(2007)133:2(220)
Terrado M, Barcelo D, Tauler R, Borrell E, Campos S (2010) Surface-water-quality indices for the analysis of data generated by automated sampling networks. TrAC Trends Anal Chem 29:40–52. https://doi.org/10.1016/j.trac.2009.10.001
Thienemann A (1925) Die Binnergewä Sser Mittelauropas Die Binnengewä Sser. Verlag Schweizerbart, Stutgart
Uddin MG, Nash S, Olbert AI (2021) A review of water quality index models and their use for assessing surface water quality. Ecol Indic 122:107218. https://doi.org/10.1016/j.ecolind.2020.107218
Whipple GC, Fair GM (1927) The microscopy of drinking water, 4th edn. Wiley, New York
Zemed M, Beshah M, Reddythota D (2021) Evaluation of water quality and eutrophication status of Hawassa Lake based on different water quality indices. Appl Water Sci 11:3. https://doi.org/10.1007/s13201-021-01385-6

Chapter 9
Urbanization Trends, Climate Change, and Environmental Sustainability

Asha Humbal, Neha Chaudhary, and Bhawana Pathak

9.1 Introduction

Urbanization is a long-term process of social migration from traditional rural area to modern urban regions (Haumann et al. 2020; Kumar and Rai 2014). Urban transformation of the world may be seen as the lasting legacy of the twentieth century (Madgin 2020). World's half of the population already lives in urban settlements (Liu and Li 2017). There has been a tendency to generalize about urbanization in the "Third world" (Hardoy and Satterthwaite 2019). The majority of the world's advanced countries are undergoing a phase of over urbanization, whereas India, like other developing countries of the third world, has started experiencing rapid urbanization only in the recent past (Jiang and O'Neill 2017). According to the data of year 2001, 75% of developed countries, 40.9% underdeveloped countries, and 45.3% of least developing countries are urbanized. It is estimated that approximately 64% population of developing word and 86% of developed world will be urbanized by 2050 (Gu 2019). As per data of (United Nations 2007) shows that % of urban population to total nations in developing nations was 17.61% which is around 1.64 times increase in 11 decades while in case of developed countries is more than 50% which is increase in 3.73 times (United Nations 2007). At present, about more than 50% of the world population lives in urban areas (Kourtit et al. 2013).

Moreover urbanization also symbolized as the industrial development and social-economic development. After the Industrial Revolution, urbanization increases faster (Pan et al. 2021). Geographically India is the seventh largest country in world. India occupies 2.4% of the total world's area and has a 16.7% world

A. Humbal · N. Chaudhary · B. Pathak (✉)
School of Environment and Sustainable Development, Central University of Gujarat, Gandhinagar, Gujarat, India
e-mail: bhawnapathak@cug.ac.in

B. Pathak, R. S. Dubey (eds.), *Climate Change and Urban Environment Sustainability*, Disaster Resilience and Green Growth,
https://doi.org/10.1007/978-981-19-7618-6_9

population with a 2% growth rate per year. As per the 2001 census, in India 27.8% population lives in urban area whereas according to recent data of 2011 census suggested that 31.67% population lives in India. Censes data indicate the high growth of urbanization in India (Kumar 2011). Another significant characteristic of India's urbanization has been the regional variation in the distribution of urban population concentration (Seto et al., 2012). Indian cities are growing, during first 5 year plan in 1951 there were just five metropolitan cities, their numbers increased to 12 in 1981, 35 in 2001 (Bhagat and Mohanty 2009). The share of urban population increased from 18.9%, 27.7 to 37.8%, respectively (Kumar and Rai 2014). The urban population has concentrated in these few cities and states like Gujarat, Maharashtra, Karnataka, Tamil Nadu, West Bengal, and Punjab. These states account for about half of the country's urban population. "It is Industrialization process during 1960s and 1970s which led to accelerated urban growth in these states-except Punjab" (Kumar et al. 2010). Many factors influence urbanization's growth, such as Employment opportunity, Educational facility, Industrial development, Modernization, Social factor. However, moving to a metropolis doesn't guarantee growth if it doesn't provide absolute quality of opportunity. This century has witnessed rapid change in mobilization of population, the reasons are diverse ranging from search of employment to impact of globalization (Hou et al. 2019). Rapid increase in the global population is one of the main reasons of the urbanization. Growth in the population rate leads to the scarcity of the resources such as water resources, food resources as well as the enhanced unemployment or business purposes which force the people to migrate from rural to urban area (Rashid et al. 2018). As the population is increased, industrialization and buildings in urban areas indirectly affect the energy consumption for the people because as population increases energy consumption increases and more demand for building material and food security indirectly causes the pollution and land-use changes occur (de Amorim et al. 2019).

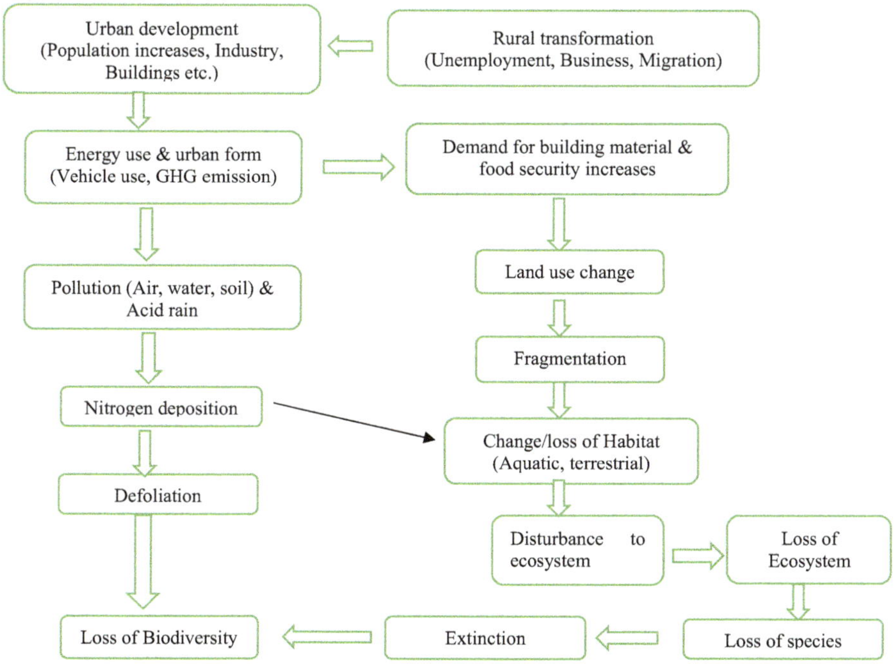

9.2 Factors That Encourage Urbanization

- Employment opportunity
- Educational facility
- Industrial growth
- Modernization
- Social factor

9.2.1 Employment Opportunity

Employment opportunity is one of the main reasons for moving people from rural areas to urban areas (Ulrich-Schad and Duncan 2018). In rural areas, people mainly depend on agriculture for their food and income source (Alamgir et al. 2021). But agriculture system depends on monsoon, so in drought seasons and natural climatic condition, people can't get any source of income which for the people migrate in urban areas (Dumenu and Obeng 2016). Whereas in metropolitan area due to industrialization and development, people get enough opportunity for the employment (Pawan and Niyazi 2016).

9.2.2 Industrial Growth

Industrial growth is one of the most significant reasons of development of urbanization. Industrial growth mainly occurs at near water bodies because industry needed a very high amount of water to manufacture products (Gebre and Gebremedhin 2019). Industrial growth also develops the transportation system for supply of raw material and transport of products. Industrial development required many labors, giving employment opportunities to the people and developing the educational system for their need of educated employees (Akan et al. 2017). Employment, educational facility, and transportation system also attract the people who live in rural areas migrating in urban areas (Lin 2019). There are numerous instances where the nation's industrialization has deteriorated the water, air, and soil texture and this became a strong reason that environmental degradation is driving industrialization (Huang et al. 2021).

9.2.3 Social Factor

Many social factors are responsible for the migration of people from rural to urban areas in which social status is one of the main reasons (Knickmeyer 2020). People in our society thought that people living in metro cities are a higher class society (Scudder and Colson 2019). So the people are moving toward the metro cities due to the thinking of our communities. Moreover, lighting of cities, better lifestyle, better educational and health care system attract the people to migrating in urban areas (Fay and Opal 2000).

9.2.4 Modernization

Urban areas having the development and better infrastructure and the lifestyle of people their attitude, food habits, transportation system, communication system, and medical facility attract the people living in rural areas to move toward the urban areas for take a benefit of modernization (Knickmeyer 2020).

9.2.5 Educational Facility

Educational facility in urban areas attracts the people to move toward urbanization. In urban areas, educational facilities are more compared to rural areas (Mlambo 2018). Urban development provides advanced education systems, such as intelligent class and online system (Babar et al. 2019). In urban areas, people get enough opportunities to use modern technology. They also get better internet facility which is very much needed current education system. Urban development provides all types of course and coaching classes, whereas in rural areas, limited courses are found (Mishra et al. 2020b). Other than this people of urban areas get a better transportation system and communication facilities. So, the better education facility of the metropolitan regions attracts people living in rural areas to migrate toward urban development (Awumbila 2017).

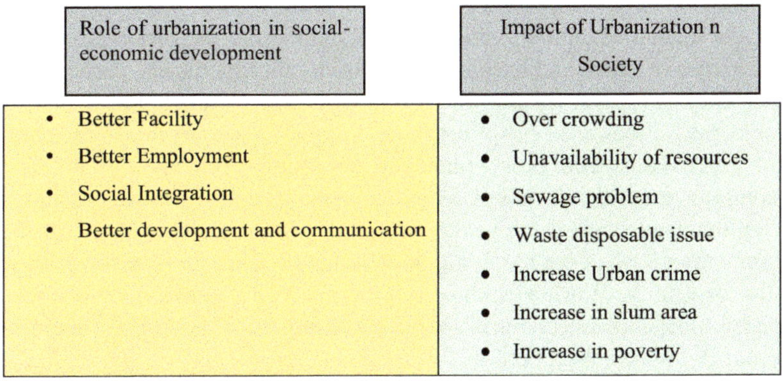

9.3 Environmental Impact of Urbanization

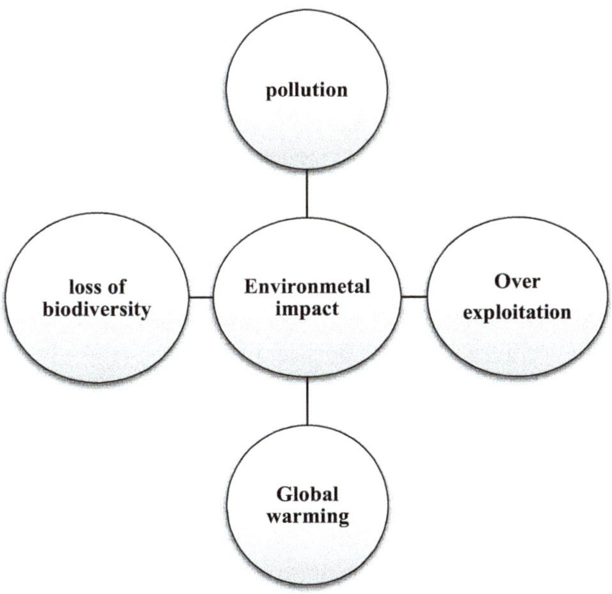

9.3.1 Pollution

9.3.1.1 Air Pollution

Urban air pollution is considered one of the most visible impacts of urbanization on the environment (Fang et al. 2019). Due to various anthropogenic activities in urban areas, the air of the metropolitan regions gets polluted, i.e., large number of automobiles and industries emitted various air pollutants such as carbon dioxide, hydrocarbons, oxides of sulfur and nitrogen, vapor of organic chemicals, particulate matter and toxic chemicals, etc. (Haque and Talukder 2021). Photochemical smog in Los Angeles is one famous example of urban air pollution, it occurs due to the reaction of sunlight with exhaust from power plant and automobiles (Goodsite et al. 2021). The main component of these smokes includes ozone, oxides of nitrogen, and VOCs. These pollutants also affect the health of humans (Manisalidis et al. 2020). Furthermore, more than 80% of people living in these urban areas are exposed to air quality levels that exceed the World Health Organization (WHO) guideline limits, where the low- and middle-income countries were suffering from the highest exposures of outdoor air pollution (WHO 2021).

9.3.1.2 Noise Pollution

Unwanted and excessive sound in any area is called as noise pollution. In urban areas, noise pollution sources include vehicles, industries, social function, and automobiles (Bello et al. 2019). Among these sources, vehicle traffic in urban areas is considered the primary source of noise pollution. An ancient city of India, Kolhapur, is now rapidly industrialized and urbanized, and it is suffering from severe noise pollution problems. A study has been carried out in Kolhapur to assess the daytime noise quality in five critical zones: (1) Education, (2) Industrial-cum-residential, (3) Commercial-cum-residential, (4) Recreational, and (5) Silence zone. For all zones, study computed L10, L50, and L90 noise pollution indices, noise climate (NC), equivalent continuous noise level (Leq), noise pollution level (Lnp), and noise exposure index (NEI). Results of the study indicate that the highest Leq of 72.25 dBA was found in industrial-cum-residential zone followed by 64.47 dBA in commercial-cum-residential zone, in educational zone observed was 63.71 dBA, 53.66 dBA in recreational zone and silence zone found 42.84 dBA. Leq observed that educational zones were above the statutory limits, while others were marginally below. According to the noise assessment study, Kolhapur is experiencing alarming levels of noise pollution (Hunashal and Patil 2012). Furthermore, noise pollution in urban areas causes significant risk to human health and affects the life of other organisms, i.e., wildlife organisms mainly depend on sound for their communication and behavior, but excessive noise breaks their contact and affects their behavior (Yuan et al. 2019).

9.3.1.3 Water Pollution

There will be an 80% increase in urban water demand by 2050. In India and China, nearly half of the 4 billion people facing severe water scarcity do so for at least 6 months out of the year. The Chennai (India) drinking water system ran dry earlier this year (Mekonnen and Hoekstra 2016; Vanham et al. 2018; Flörke et al. 2018) and Cape Town (Africa) ran out of drinking water in 2018 (Mekonnen and Hoekstra 2016; Vanham et al. 2018). Worldwide, over 80% of wastewater is not treated appropriately before being released into the environment (Connor et al. 2017). Due to rapid industrialization and urbanization, many waste disposal occurs in the river and lakes. Besides this, the discharge of untreated sewage waste from mega cities and towns into the water bodies causes water pollution (Jayaswal et al. 2018). As per the CPCB data, about 38,254 million L/day sewage is generated from class I and class II cities of the country. Treatment capacity found only for 11,787 million L/day sewage (Olds et al. 2018). So, the untreated sewage dump into water bodies increases the river's pollution load and other water bodies (Xu et al. 2019). Water pollution caused by urbanization also affects the aquatic ecosystem. Unplanned urbanization also leads to the scarcity of fresh water by destructing the path of natural water flow and also causes the hydrological changes (Mishra et al. 2020a). The dumping of solid waste on land without treatment causes leaching of harmful material to the river water and groundwater which cause water pollution (Nai et al. 2021).

A case study of city Hyderabad, one of India's fastest growing megacities with a significant impact on the environment. A city's uncontrolled growth and densification severely affect its surface and groundwater resources. A better understanding of urban growth's impact on the environment is necessary to prevent future deterioration of these essential resources and recover them. A study found that rapid urbanization of Hyderabad has led to an increase in the need for fresh water for domestic and industrial use. The use of groundwater was increased as a result of this increase in demand.

Furthermore, a quality analysis found that the groundwater samples did not meet the BIS and WHO drinking water standards. In addition, heavy metals like zinc, lead, chromium, iron, etc. were found in the water samples, indicating that industrial effluent had percolated into the aquifer. This study also proven that the sewage water and residential area had also contributed in recharging grown water. The study concluded that the urban growth had made the groundwater resources more vulnerable and in many cases it had been affecting it very severely (Wakode et al. 2014).

9.3.2 Over Exploitation of Natural Resources

Humans depend on ecosystems for a wide range of services and resources. Human intrusions are affecting soil, water, and biodiversity, which are fundamental to the ecological balance. Rapid urbanization and industrialization make the high pressure on the natural resources. Due to the urbanization, the land for agriculture, water bodies, and forest area are shrinking while in the other hand, the demand for resources is increasing rapidly (Wassie 2020). Moreover, the changing in the landing pattern directly affects the overall productivity of the land (Borrelli et al. 2017). According to the data of FAO, 2007 in 2050, the urban population will be increased around 54% and demand of domestic population will be 90 km^3 (Boretti and Rosa 2019). Due to the rise in the rate of the urban population and their lifestyle, they create the overexploitation of natural resources such as groundwater, energy, and food (Tayal and Singh 2021). Industries and other urban sectors required more water which cause the scarcity of potable and drinking water (Livingston 2021).

9.3.3 Global Warming

Result of rapid urbanization leads to the rise in global temperature. As the industries release various air pollutants, these pollutants having multiple greenhouse gases such as methane, carbon dioxide, carbon monoxide, etc. (Arsalan et al. 2020), addition of this pollutant into the atmosphere increases the concentration of greenhouse gases. Now these gases absorb reflected solar radiation and cause the rise the global temperature and ultimately global warming (Abdollahbeigi 2020). Moreover, urbanization also cause the urban heat island effect in which the temperature of the metropolitan area increases compared to the rural area (Zhou and Chen 2018). As in the urban areas, the use of air conditioners increases, releasing different ODs (ozone depletion substances). These ODs directly affect the ozone layer, which results in thinning the ozone layer. Thinning of the ozone layer leads to the more harmful radiations coming to the earth and warming the earth more and more. This increase in temperature or warming of earth affects the whole life on the earth surface (Bartholy and Pongrácz 2018). Climate change caused by unplanned construction of large buildings in urban areas absorbs solar radiation and emits heat radiations in the afternoon, increasing the atmospheric pressure. Therefore, this year was the warmest since 2010. As a result of dust condensing into rain droplets, cities receive more rain than the surrounding countryside.

9.3.4 Solid Waste Generation

Solid waste management is a severe issue in urban areas. In urban areas due to various industries like chemical and pharmacological as well as the technological development use the large quantity of material and simultaneously generate the high amount of solid waste materials (Yukalang et al. 2018). Solid waste generated from urban areas contains municipal waste, hospital waste, industrial waste, hazardous waste, etc. This waste includes bad smell and poisonous gases responsible for the generation of vector of different diseases (Haque and Talukder 2021). The dumping of solid waste on land without treatment cause leaching of harmful material to cause water pollution as well as burning of solid waste causes the air pollution the famous example of air pollution is burning on Parana landfill site (Przydatek and Kanownik 2019). Contamination of water with organic chemicals, metals, metalloids, and bacteria from untreated or poorly treated wastewater and seepage. As a result, the soil becomes salinized and solidified (Muyen et al. 2011; Jamwal and Lele 2017).

9.3.5 Loss of Biodiversity

Biodiversity is typically defined as the diversity of animals and plants native to a particular habitat or ecosystem. In an ecosystem, every species serves a specific function through the food web. When species diversity changes in a system, the biogeochemical cycles change and the system performs differently. Consequently, ecosystems stability, functionality, and sustainability depend on biodiversity (Tilman et al. 2012).

Uncontrolled urbanization causes the negative impact on the biodiversity (Tang et al. 2020). Due to increased amount of urbanization, the area of forest and wildlife habitat converted into the urban area which result into habitat fragmentation and loss (Cincotta et al. 2000). Habitat fragmentation and loss directly affect the biodiversity (Peng et al. 2019). Loss of habitat creates force on migration of wildlife animal and it decline the rate of biodiversity (Almond et al. 2020). Due to the habitat loss, some animals cannot survive, leading to the decline of the area's biodiversity. Moreover, urbanization creates a selective force that changes the wildlife communities (Ouyang et al. 2018).

Urbanization also creates the geographic isolation of species which disturb the whole life of the species as well as it also affect their dispersal and connectivity of native species (Johnson and Munshi-South 2017). Change in the habitat configuration results in the introduction of the alien or invasive species, affecting the native or endemic species. The indirect effect of expanding urban areas includes soil quality and physical transformation. Change in land use pattern also changes the soil component, which affects soil nutrient availability. Moreover, urbanization causes the abiotic stress by increasing air pollution, water pollution, invasive species and enhances the predation rates (Elmqvist et al. 2019). Therefore, understanding the impact of urbanization on biodiversity becomes imperative from the point of view of conservation and planning sustainable cities. This chapter delivers to understand the effect of urbanization on biodiversity. All animals interact with their environment for energy gain and loss, protection, and for various other reasons. Yet, in recent years, the world's human population has grown so rapidly that natural resources have been depleted, the environment degraded, and animal habitats have been destroyed. Land use, topography, and population change are the main factors that cause environmental degradation. Consequently, there will be a decline or even extinction of wildlife species used for food and business purposes. There can be no saving of wildlife that is a crucial part of biodiversity unless we address and resolve the issues that are propelling their extinction.

Gujarat State covers an area of 6.1% of the country's total geographical location and accounts for 4.9% of the total population. Situated on the Northwestern sea coast, it has the most extensive coastline of 1600 km, concealed in two of the three gulfs of the country (Bajpai 2004; Faizi and Ravichandran 2017). It also shares an international boundary with Pakistan in the North. The maritime location and long coastline have considerably influenced its people's climate, life, and activities. The mainland of Gujarat on account of its geographical advantages and historical developments has favored the concentration of urban population to 67%. Among the seven micro-geographical regions based on topography, climate, and soils, the Eastern Hilly Region covers an elongated hilly, wooded territory of the Dangs, Eastern parts of Valsad, Surat, Bharuch, Vadodara, Panchmahal and Banaskantha district.

Great Indian Bustard Population Decrease: In grassland ecosystems, fragmentation and habitat loss, change in land use/land cover pattern, desertification, harmful-thought plantation of invasive and exotic species, and neglect of state institutions due to classification of "grasslands" as "wastelands", conversion of grasslands to agriculture lands due to increasing irrigation potential and decline of nature/GIB-friendly farming practices, are all commonly and correctly responsible for the continual decline in India's GIB population. Many other birds also die due to the collision/electrocution with these transmission lines at the rate of 10 birds per km per month totaling nearly 1 lakh bird deaths annually in 4200 sq. km.

Listing out other threats to the Great Indian Bustard, farming technology and irrigation, wind turbines, mining, and connected infrastructure development have caused severe habitat degradation to birds. However, other factor is the noise pollution, which is also adversely affecting the prospects of the Great Indian Bustard's revival (The Hindu news 2019). The WII report said the GIB population has been reduced by 75% in the last 30 years. As a result of their poor vision and inability to see power lines from a distance, adult GIBs are extremely vulnerable to collisions with power lines that cross their flight paths (The Hindu news 2019).

As people interfere more with areas where birds nest and colonize, the population of these birds is declining. The results of studies done by Bombay Natural History Society (BNHS) and Wildlife Institute of India (WII) of factors most responsible for the decline of several bird species reveal that grasslands and forests, like wetlands most other habitats such as grasslands and forests, also face severe threat due to developmental pressures. Over the past decades, the drastic loss of grassland habitat has severely threatened species such as the Great Indian Bustard, Siberian Cranes, Bengal Florican, and *Jerdon's Courser* McKee et al. (2004) and Flörke et al. (2018) have predicted that maximum urban growth is expected to occur in regions around the world's biodiversity hotspots. They have also analyze the critical demographic variable and reported that about 20% of the world's population resides in biodiversity hotspot regions. Therefore, predicting patterns of urban developmental in areas of rich biodiversity are essential for conservation.

9.3.6 Urbanization Induce Climate Change

Global urbanization is accelerating, leading to cities becoming increasingly important in combating climate change. Seventy percent of greenhouse gas emissions globally are attributed to cities, according to UN Habitat. Even though it occupies only 2% of the earth's surface. Consequently, urban areas contribute significantly to climate change. Urbanization leads to the warming of the local region and establishes a direct correlation between temperature and urban population. It is studied over China that since 1978 when innovation takes place a sharp change in the air temperature has been noticed and a strong UHI effect has also been noticed in the autumn and winter season (Hua et al. 2008). In the most urbanized sites of Washington DC, a maximum and most frequent temperature surge of >7 °C has been observed due to urbanization led climate change (Nelson and Palmer 2007). It is observed that UHI and population are possessing a logarithmic relationship even with the population of a thousand. The extreme rainfall events and changes in the hydrological cycles also has seen due to the UHI effect (Huong and Pathirana 2013). It is simulated that the future urbanization will have an effect on the minimum temperature due to increased heat capacity and reduced evaporation of the city region (Argüeso et al. 2014). Moreover, due to the increased built-up in the urbanized region, the surface air temperature could be increased significantly and the humidity decreased by 10% which results in Urban Heat Island (UHI) effect. Taiwan is having one of the largest heat island effects which have a significant impact on mesoscale circulation and increases the surface convergence which has an impact on the precipitation (increased by 28%) and could lead to heavy rainfall, thus leading to urban flooding (Shepherd et al. 2002). It is also found that urbanization enhanced the rainfall intensity not only over the city but also in the nearby regions as well that have been implicated in the form of urban flooding (Liu and Niyogi 2019).

9.4 Urbanization and Sustainability

The term sustainable has been actively used in the past 15–20 years to delineate a world in which both human and natural systems can continue to exist long into the future. The concept of "Sustainable Development" has been retained to denote alternatives to traditional patterns of physical, social, economic development in both industrialized nations and what is often called the Third world's alternatives that can avoid the problems overexploitation of natural resources, pollution, overpopulation, and loss of biodiversity, degradation of ecosystems, habitat destruction, and the degradation of human living condition. The changes in cultivated land and rural residential land, forest land, and their interrelation are the most direct indication of land use transformation in rural areas, which is an important root to carry out rural land coalesce scientifically. Globally, towns and cities are rapidly increasing in area and in population; the urban area is projected to triple by 2030 (Batty 2008; Seto et al. 2012). According to Fernández-Götz (2018), the world's communities require data to compensate for and adapt to the current growth while planning for future change and the implications on infrastructure and the environment. Urbanization and climate change are among the pressing issues facing the world today because they affect the environment and population across the globe. For a sustainable future, cities must act to consume less carbon and reduce their ecological footprints on surrounding hinterland cities must act to consume less carbon and reduce their ecological footprints on surrounding hinterland cities have to act to transform themselves for a greener future, consuming less carbon and reducing their ecological footprints on surrounding hinterland. Thus, cities with a low or zero carbon footprint are in high demand. There were hints of this vision in 1992 at the Rio Earth Summit in Local Agenda 21, which called for local sustainable living agendas to be developed alongside commitments to bringing about the Millennium Development Goals (MDGs) (Niyonzima 2020). Green building is a sustainable approach that conserves energy and water resources, minimizes waste, and maximizes reuse. The concept of green construction focuses on improving the health and wealth of society as well as connecting us with nature. As an eco-friendly component, the green building is based on the REDUCE, REUSE, and RECYCLE principle (Zuo and Zhao 2014). When green building is applied, there is a 40% reduction in water usage, a 50% reduction in energy usage, a 70% reduction in solid waste, and a 35% reduction in carbon emissions (Bombugala and Atputharajah 2013). Moreover, urban gardeners can maximize their space by expanding their rooftop gardens. It is the process of cultivating a garden on the terrace of a building, whether it is a vegetable garden or decorative. The advantages of rooftop gardens, also called living roofs or green roofs, include their ability to utilize space efficiently, convert CO_2 emissions into oxygen, as well as their ability to reduce heat of buildings and energy costs. They are also beautiful to look at and they improve air quality. In addition to providing recreational opportunities, the majority of rooftop gardens also provide food for the family (Chitra 2021). In the private vehicle market, electric vehicles reduce emissions in the urban environment as they emit less pollution. In particular, this consideration does not hold true for carbon dioxide (CO_2), which is a greenhouse gas. It is worth considering that a large proportion of electric energy is generated through the use of fossil fuels and that environmental impact must be considered at a global scale, the possible reduction of CO_2 emissions should be estimated. Therefore, the concept of green building, roof top gardening, use of solar energy and electronic vehicle help the development of sustainable urban cities (Perujo et al. 2011).

9.5 Conclusion

The biggest challenge humanity faces today is environmental sustainability. In the modern world, nearly all the necessities of life—such as food, water, shelter, clean energy, and pollution-free air—are threatened by human activities. Increasing demand for energy, infrastructure, water, food, etc., causes ecological stress, contributing to emissions, resource depletion, and economic and environmental system distortion. The chapter focuses on the effects of urbanization on environmental components, namely climate, the biosphere, land use, and water availability. As the world is urbanizing. Urban decision-makers and citizens need to integrate nature into their everyday lives to ensure a global sustainable future underpinned by nature-based solutions and ecosystem-based adaptation.

Additionally, a new valuation technique that incorporates a resilience and inclusive wealth perspective is crucial to effectively capture the value of biodiversity and ecosystems in reducing urban vulnerability to shocks and disturbances. Population growth leads to increased population density, an increase in the number of people living below the poverty line, and a pressure on natural resources, which leads to environmental degradation through overexploitation. Natural resources and the environment must be given top priority if humans are to survive on Earth. Moreover, government should not be responsible for protecting the environment alone; local people and leaders should be encouraged to make dedicated efforts to prevent pollution and environmental impact due to urbanization.

References

Abdollahbeigi M (2020) Non-climatic factors causing climate change. J Chem Rev 2(4):292–308
Akan MÖA, Dhavale DG, Sarkis J (2017) Greenhouse gas emissions in the construction industry: an analysis and evaluation of a concrete supply chain. J Clean Prod 167:1195–1207
Alamgir M, Furuya J, Kobayashi S, Mostafiz RB, Ahmed M (2021) Farm income, inequality, and poverty among farm families of a flood-prone area in Bangladesh: climate change vulnerability assessment. GeoJournal 86(6):2861–2885
Almond RE, Grooten M, Peterson T (2020) Living planet report 2020-bending the curve of biodiversity loss. World Wildlife Fund, Washington DC
Argüeso D, Evans JP, Fita L, Bormann KJ (2014) Temperature response to future urbanization and climate change. Climate Dynam 42(7):2183–2199
Arsalan S, Imran A, Khawar A, Imran F, Faiz Z, Razzaq SE, Razzaq SS (2020) Greenhouse gasses emission in the highly urbanized city using GIS-based analysis. Int J Hum Capital Urban Manag 5(4):353–360
Awumbila M (2017) Drivers of migration and urbanization in Africa: key trends and issues. Int Migr 7(8):9
Babar M, Arif F, Jan MA, Tan Z, Khan F (2019) Urban data management system: towards big data analytics for internet of things based smart urban environment using customized Hadoop. Futur Gener Comput Syst 96:398–409
Bajpai N (2004) Regional economic policies, geography, and growth episodes in China's coastal growth provinces: lessons for the state of Gujarat. Academia, San Francisco
Bartholy J, Pongrácz R (2018) A brief review of health-related issues occurring in urban areas related to global warming of 1.5 °C. Curr Opin Environ Sustain 30:123–132
Batty M (2008) The size, scale, and shape of cities. Science 319(5864):769–771
Bello JP, Silva C, Nov O, Dubois RL, Arora A, Salamon J, Doraiswamy H (2019) Sonyc: A system for monitoring, analyzing, and mitigating urban noise pollution. Commun ACM 62(2):68–77
Bhagat RB, Mohanty S (2009) Emerging pattern of urbanization and the contribution of migration in urban growth in India. Asian Popul Stud 5(1):5–20
Bombugala BAWP, Atputharajah A (2013) Sustainable development through green building concept in Sri Lanka. Academia, San Francisco

Boretti A, Rosa L (2019) Reassessing the projections of the world water development report. NPJ Clean Water 2(1):1–6
Borrelli P, Robinson DA, Fleischer LR, Lugato E, Ballabio C, Alewell C, Panagos P (2017) An assessment of the global impact of 21st century land use change on soil erosion. Nat Commun 8(1):1–13
Chitra K (2021) Establishment of organic terrace garden and cultivation of nutritious vegetables. J Univ Shanghai Sci Technol 23:464–468
Cincotta RP, Wisnewski J, Engelman R (2000) Human population in the biodiversity hotspots. Nature 404(6781):990–992
Connor R, Renata A, Ortigara C, Koncagül E, Uhlenbrook S, Lamizana-Diallo BM, Brdjanovic D (2017) The United Nations world water development report 2017. Wastewater: the untapped resource. The United Nations World Water Development Report, Washington DC
de Amorim WS, Deggau AB, do Livramento GG, da Silva Neiva S, Prasath AR, de Andrade JBSO (2019) Urban challenges and opportunities to promote sustainable food security through smart cities and the 4th industrial revolution. Land Use Policy 87:104065
Dumenu WK, Obeng EA (2016) Climate change and rural communities in Ghana: social vulnerability, impacts, adaptations and policy implications. Environ Sci Policy 55:208–217
Elmqvist T, Andersson E, Frantzeskaki N, McPhearson T, Olsson P, Gaffney O, Folke C (2019) Sustainability and resilience for transformation in the urban century. Nat Sustain 2(4):267–273
Faizi S, Ravichandran MG (2017) India's biodiversity: a study of the management regime. UNEP, Nairobi
Fang C, Cui X, Li G, Bao C, Wang Z, Ma H, Ren Y (2019) Modeling regional sustainable development scenarios using the urbanization and eco-environment coupler: case study of Beijing–Tianjin–Hebei urban agglomeration, China. Sci Total Environ 689:820–830
FAO (2007) The state of food and agriculture (Report)
Fay M, Opal C (2000) Urbanization without growth: a not so uncommon phenomenon, vol 2412. World Bank Publications, Washington DC
Fernández-Götz M (2018) Urbanization in iron age Europe: trajectories, patterns, and social dynamics. J Archaeol Res 26(2):117–162
Flörke M, Schneider C, McDonald RI (2018) Water competition between cities and agriculture driven by climate change and urban growth. Nat Sustain 1(1):51–58
Gebre T, Gebremedhin B (2019) The mutual benefits of promoting rural-urban interdependence through linked ecosystem services. Glob Ecol Conserv 20:e00707
Goodsite ME, Hertel O, Johnson MS, Jørgensen NR (2021) Urban air quality: sources and concentrations. In: Air pollution sources, statistics and health effects. Springer, New York, pp 193–214
Gu C (2019) Urbanization: processes and driving forces. Sci China Earth Sci 62(9):1351–1360
Haque M, Talukder B (2021) Urbanization and quality of environment: a case study of Cooch Behar District, West Bengal, India. Academia, San Francisco
Hardoy JE, Satterthwaite D (2019) Urban change in the third world: are recent trends a useful pointer to the urban future? vol 1. Routledge, Abingdon, pp 75–110
Haumann S, Knoll M, Mares D (eds) (2020) Concepts of urban-environmental history, vol 1. transcript Verlag, Bielefeld
Hou H, Wang R, Murayama Y (2019) Scenario-based modelling for urban sustainability focusing on changes in cropland under rapid urbanization: a case study of Hangzhou from 1990 to 2035. Sci Total Environ 661:422–431
Hua LJ, Ma ZG, Guo WD (2008) The impact of urbanization on air temperature across China. Theor Appl Climatol 93(3):179–194
Huang SZ, Sadiq M, Chien F (2021) The impact of natural resource rent, financial development, and urbanization on carbon emission. Environ Sci Pollut Res 2021:1–13
Hunashal RB, Patil YB (2012) Assessment of noise pollution indices in the city of Kolhapur, India. Proced Soc Behav Sci 37:448–457
Huong HTL, Pathirana A (2013) Urbanization and climate change impacts on future urban flooding in Can Tho city, Vietnam. Hydrol Earth Syst Sci 17(1):379–394

Jamwal P, Lele S (2017) Addressing pollution in urban rivers: lessons from the Vrishabhavathy river in Bengaluru. In: Hiremath A, Rai ND, Siddhartha A (eds) Transcending boundaries: reflecting on twenty years of action research at ATREE. ATREE, Bengaluru, pp 105–113
Jayaswal K, Sahu V, Gurjar BR (2018) Water pollution, human health and remediation. In: Water remediation. Springer, Singapore, pp 11–27
Jiang L, O'Neill BC (2017) Global urbanization projections for the shared socioeconomic pathways. Glob Environ Chang 42:193–199
Johnson MT, Munshi-South J (2017) Evolution of life in urban environments. Science 358(6363): eaam8327
Knickmeyer D (2020) Social factors influencing household waste separation: a literature review on good practices to improve the recycling performance of urban areas. J Clean Prod 245:118605
Kourtit K, Nijkamp P, Partridge MD (2013) The new urban world. Eur Plan Stud 21(3):285–290
Kumar SS (2011) Land accounting in India: issues and concerns. Central Statistics Office, Ministry of Statistics and Programme Implementation, New Delhi
Kumar D, Rai AK (2014) Urbanization process, trend, pattern and its consequences in India. Neo Geogr 3:24
Kumar M, Mukherjee N, Sharma GP, Raghubanshi AS (2010) Land use patterns and urbanization in the holy city of Varanasi, India: a scenario. Environ Monit Assess 167(1):417–422
Lin Y (2019) E-urbanism: E-commerce, migration, and the transformation of Taobao villages in urban China. Cities 91:202–212
Liu Y, Li Y (2017) Revitalize the world's countryside. Nature 548(7667):275–277
Liu J, Niyogi D (2019) Meta-analysis of urbanization impact on rainfall modification. Sci Rep 9(1): 1–14
Livingston J (2021) Water scarcity and health in urban Africa. Daedalus 150(4):85–102
Madgin R (2020) Urban heritage and urban development. Concepts Urban-Environ Hist 1:235
Manisalidis I, Stavropoulou E, Stavropoulos A, Bezirtzoglou E (2020) Environmental and health impacts of air pollution: a review. Front Public Health 8:14
McKee JK, Sciulli PW, Fooce CD, Waite TA (2004) Forecasting global biodiversity threats associated with human population growth. Biol Conserv 115(1):161–164
Mekonnen MM, Hoekstra AY (2016) Four billion people facing severe water scarcity. Sci Adv 2(2):e1500323
Mishra BK, Chakraborty S, Kumar P, Saraswat C (2020a) Sustainable solutions for urban water security, vol 93. Springer International Publishing, New York
Mishra L, Gupta T, Shree A (2020b) Online teaching-learning in higher education during lockdown period of COVID-19 pandemic. Int J Educ Res Open 1:100012
Mlambo V (2018) An overview of rural–urban migration in South Africa: its causes and implications. Arch Bus Res 6(4):4407
Muyen Z, Moore GA, Wrigley RJ (2011) Soil salinity and sodicity effects of wastewater irrigation in South East Australia. Agric Water Manag 99(1):33–41
Nai C, Tang M, Liu Y, Xu Y, Dong L, Liu J, Huang Q (2021) Potentially contamination and health risk to shallow groundwater caused by closed industrial solid waste landfills: site reclamation evaluation strategies. J Clean Prod 286:125402
Nelson KC, Palmer MA (2007) Stream temperature surges under urbanization and climate change: data, models, and responses. JAWRA J Am Water Resour Assoc 43(2):440–452
Niyonzima T (2020) Urbanization trends and green development policies in Rwanda. Int J Curr Res 12(07):12533–12540
Olds HT, Corsi SR, Dila DK, Halmo KM, Bootsma MJ, McLellan SL (2018) High levels of sewage contamination released from urban areas after storm events: a quantitative survey with sewage specific bacterial indicators. PLoS Med 15(7):e1002614
Ouyang JQ, Isaksson C, Schmidt C, Hutton P, Bonier F, Dominoni D (2018) A new framework for urban ecology: an integration of proximate and ultimate responses to anthropogenic change. Integr Comp Biol 58(5):915–928
Pan Y, Ma L, Tang H, Wu Y, Yang Z (2021) Land use transitions under rapid urbanization in Chengdu–Chongqing region: a perspective of coupling water and land resources. Land 10(8):812

Pawan S, Niyazi A (2016) From Mahalla to Xiaoqu: transformations of the urban living space in Kashgar. Inner Asia 18(1):121–134
Peng Y, Mi K, Wang H, Liu Z, Lin Y, Sang W, Cui Q (2019) Most suitable landscape patterns to preserve indigenous plant diversity affected by increasing urbanization: a case study of Shunyi District of Beijing, China. Urban For Urban Green 38:33–41
Perujo A, Thiel C, Nemry F (2011) Electric vehicles in an urban context: environmental benefits and techno-economic barriers. IntechOpen, London
Przydatek G, Kanownik W (2019) Impact of small municipal solid waste landfill on groundwater quality. Environ Monit Assess 191(3):1–14
Rashid H, Manzoor MM, Mukhtar S (2018) Urbanization and its effects on water resources: an exploratory analysis. Asian J Water Environ Pollut 15(1):67–74
Scudder T, Colson E (2019) From welfare to development: a conceptual framework for the analysis of dislocated people. In: Involuntary migration and resettlement. Routledge, Abington, pp 267–287
Seto KC, Güneralp B, Hutyra LR (2012) Global forecasts of urban expansion to 2030 and direct impacts on biodiversity and carbon pools. Proc Natl Acad Sci 109(40):16083–16088
Shepherd JM, Pierce H, Negri AJ (2002) Rainfall modification by major urban areas: observations from spaceborne rain radar on the TRMM satellite. J Appl Meteorol 41(7):689–701
Tang M, Li Z, Hu F, Wu B (2020) How does land urbanization promote urban eco-efficiency? The mediating effect of industrial structure advancement. J Clean Prod 272:122798
Tayal S, Singh S (2021) Covid-19 and opportunity for integrated management of water–energy–food resources for urban consumption. In: Environmental resilience and transformation in times of COVID-19. Elsevier, Amsterdam, pp 135–142
The Hindu news (2019) Great Indian bustard nearing extinction due to high voltage power lines
Tilman D, Reich PB, Isbell F (2012) Biodiversity impacts ecosystem productivity as much as resources, disturbance, or herbivory. Proc Natl Acad Sci 109(26):10394–10397
Ulrich-Schad JD, Duncan CM (2018) People and places left behind: work, culture and politics in the rural United States. J Peasant Stud 45(1):59–79
United Nations (2007) World urbanization prospects: the 2007 revision population database, vol 2. United Nations, New York
Vanham D, Hoekstra AY, Wada Y, Bouraoui F, De Roo A, Mekonnen MM, Bidoglio G (2018) Physical water scarcity metrics for monitoring progress towards SDG target 6.4: an evaluation of indicator 6.4. 2 "Level of water stress". Sci Total Environ 613:218–232
Wakode HB, Baier K, Jha R, Ahmed S, Azzam R (2014) Assessment of impact of urbanization on groundwater resources using GIS techniques-case study of Hyderabad, India. Int J Environ Res 8(4):1145–1158
Wassie SB (2020) Natural resource degradation tendencies in Ethiopia: a review. Environ Syst Res 9(1):1–29
World Health Organization (2021) World health statistics 2021: monitoring health for the sustainable development goals. WHO, Geneva
Xu SC, Miao YM, Gao C, Long RY, Chen H, Zhao B, Wang SX (2019) Regional differences in impacts of economic growth and urbanization on air pollutants in China based on provincial panel estimation. J Clean Prod 208:340–352
Yuan M, Yin C, Sun Y, Chen W (2019) Examining the associations between urban built environment and noise pollution in high-density high-rise urban areas: a case study in Wuhan, China. Sustain Cities Soc 50:101678
Yukalang N, Clarke B, Ross K (2018) Solid waste management solutions for a rapidly urbanizing area in Thailand: recommendations based on stakeholder input. Int J Environ Res Public Health 15(7):1302
Zhou X, Chen H (2018) Impact of urbanization-related land use land cover changes and urban morphology changes on the urban heat island phenomenon. Sci Total Environ 635:1467–1476
Zuo J, Zhao ZY (2014) Green building research—current status and future agenda: a review. Renew Sustain Energy Rev 30:271–281

Chapter 10
Novel Technologies and Eco-Friendly Lifestyle for Sustainable Cities

Arun Arya

10.1 Introduction

Climate change and environmental degradation is a growing problem both locally and internationally and will continue to be an issue in the near future (Joldersma 2017). Environmental degradation has been preceded by a long-standing erosion of environmental values from the human value system. This means that it can be managed by changing relevant behaviour and we should adopt sustainable solutions. The idea of creating green urban cities for a sustainable future for the planet was conceived. But not practised in a right way to produce real green ecosystem for inhabitants. Now we will have to modify our cities and focus on new and innovative ways to tackle the problems in many upcoming new towns. It is argued that in an increasingly urbanized world, the cities must exercise concrete actions to reduce the release of greenhouse gas emissions (Corfee-Morlot et al. 2009). Because of the pollution and climate change, destruction of our eco-environment, genetic resources, and the massive killing of wild animals, the planet earth is experiencing the sixth mass extinctions (Hutson et al. 1985). Estimates according to Intergovernmental Panel on Climate Change (IPCC) speculate for a rise of 1.1–6.4 °C during twenty-first century. The rise in sea level may affect the city of Charleston, South Carolina, Boston, Philadelphia, New York, Baltimore, and Miami. Best tourist beaches and places like Maldives and Marshal Islands would disappear from globe. About 205 of most populated areas in India and Bangladesh will be washed away (Arya 2009). According to Farooqui (2007), the climatic changes in the past have served as a basis for present changes in environment.

A. Arya (✉)
Department of Environmental Studies, Faculty of Science, The Maharaja Sayajirao University of Baroda, Vadodara, India

B. Pathak, R. S. Dubey (eds.), *Climate Change and Urban Environment Sustainability*, Disaster Resilience and Green Growth,
https://doi.org/10.1007/978-981-19-7618-6_10

Urbanization is an important demographic shift worldwide. According to NFHS-3 report, India's 285.4 million population live in urban areas (Anonymous 2008a). Out of these 80.8 million people live below the poverty line and this creates slums in megacities. Such urban poor have various health and environmental issues (Jadav and Suryawanshi 2013). Urbanization has been accompanied by unstable ecosystems including reduction in surface water bodies, construction of high-rise concrete towers and declining ground water table. Green cover is drastically disturbed (Gupta et al. 2015). Average annual growth rate of urban India was 3%, against 5% growth recorded in slums during 1991–2001. Local authorities will have to work for their social uplift and ensure better health services to each citizen. Problem of air pollution causes serious health hazards. Pollution emitted by large number of vehicles, dust and pollutant from thermal power plants and cement industries (Mehta and Dubey 2003). Pollution caused by crackers during festival of *Deepawali* caused severe pollution and SMOG leading to asthma and respiratory problems (Arya and Bhatt 2020).

Ideally, humankind should work to achieve the growth above Human Development Index (HDI) and reduce the ecological footprint per capita. Living above the minimum HDI would guarantee that human needs such as education or health are satisfied. An ecological footprint may be defined as a limit of consumption per person according to Earth ecological capacity. Living below it wouldn't compromise the future generations, as the planet would be able to regenerate itself. But the fact is that every year the Earth overshoot day comes earlier. This day represents the date when humankind gets in debt with the planet. This is because our demand for ecological resources in a given year exceeds, what the planet can regenerate in that same year. We are using more ecological resources than the planet can handle to lose. Over tendency to consume more is against the good development.

Asian Philosophies related to different civilizations have generated the interest in students ask and answer the big questions of life, which despite significant cultural differences are basically the same everywhere. Studying the great philosophical traditions of Asia, makes it possible for us to understand these traditions' carefully considered answers to the most important questions of life, answers that are supported by profound insights and good reasons. Because these answers have guided the thought and action of the peoples of Asia over the centuries, they provide the basic clues to the guiding ideas and values of Asian societies today. And in today's world, where the very future of humankind depends upon understanding and cooperation among people with diverse values and ideas, it is imperative that these values and ideas be understood (Koller 2018).

Indus Valley Civilization (IVC) showed technological advanced and developments of first urban or organized settlements. We have seen how Babylonian hanging gardens were developed and a civilization called Mesopotamian developed in modern Iraq. The knowledge of city planning can be visualized their approach for cleanliness and good health.

The Wari Empire was present in high Andes of America, extending north nearly thousands of km. from the southern border now known as Peru. It had a rival Tiwanaku, near the shores of Lake Titicaca in Bolivia. Both had a ruling elite that oversaw the construction of grand cities, temples, and palaces. They also shared certain religious symbols and beliefs. Without any knowledge of writing system, 40,000 people of Wari (A.D. 600–1000) were used to be governed systematically. People were skilled as road builders and potters. The occurrence of drought probably destroyed these civilizations (Morell 2002).

The word sustainability is derived from the Latin sustinere (tenere, to hold; sus, up). That is to "maintain", "support", or "endure". Since the 1980s sustainability has been used in the sense of human sustainability on planet Earth and this has resulted in the most widely quoted definition of sustainability and sustainable development, that of the Brundtland Commission of the United Nations on March 20, 1987: "sustainable development is development that meets the needs of the present without compromising the ability of future generations to meet their own needs". At the 2005 World Summit, it was noted that this requires the reconciliation of environmental, social, and economic demands—the "three pillars" of sustainability. The simple definition "sustainability is improving the quality of human life while living within the carrying capacity of supporting eco-systems".

10.2 Sustainable City Goals

Humans have been using land and its resources for centuries in a pursuit of their better lives. It is a well-known fact that city is not a problem, it is a solution for more facilities in a limited space. But at the same time, uncontrolled urbanization may lead to deterioration of environment and reduction in quality of life (Gupta et al. 2015). United Nations Environment Programme (UNEP) suggested that 17 goals and SDG 11th goal are devoted to sustainable cities and human settlements with specific focus on access to housing and basic services including transport, energy efficient sustainable buildings, resource efficiency, and preparedness for disaster risks. Cities generate over 80% of gross domestic product in many countries in Asia and the Pacific and are engines of economic growth that have lifted millions from poverty. Today, approximately 700 million people live in urban slum. Urban migration promotes the economic growth. Environmental footprints of cities can threaten the natural resources. The main areas for the current work on sustainable cities are:

1. Sustainable consumption and production roadmap for cities covering all the sectors
2. Upstream interventions through policy, technology, and financing to reduce and manage pollution and waste

Private sector and other civil society groups along with government can help to achieve resource efficient and environmentally sound green technologies (Fig. 10.1). Proper planning of the urbanization is necessary for a successful society for future. A good environmental sense has been of the fundamental features of Indian philosophy. The Earth was symbolized as mother earth and rivers were worshipped as mothers (Gupta et al. 2015). Climate change and deteriorating environmental effects

Fig. 10.1 Role of different stakeholder in sustainable cities

can be reduced by adopting newer and safer technologies to boost the socioeconomic standards, live in harmony with nature in Indian cities, and conserve the valuable resources for sustainable use. Nature is to be preserved from pollution.

10.3 Societies in Ancient Times

10.3.1 Indus Valley Civilization

Urban planning was seen in Harappa, Mohenjo Daro, and the recently partially excavated Rakhigarhi. The total population of the civilization is thought to have been upward of five million, and its territory stretched over 1500 km along the banks of the Indus River and then in all directions outward. Indus Valley Civilization sites have been found near the border of Nepal, in Afghanistan, on the coasts of India, and around Delhi, to name only a few locations (Mark 2020). This urban plan included the world's first known urban sanitation systems. Each house got water from wells present nearby. There was a room for bathing, and waste water was collected through covered drains. Houses were having inner courtyards and connected to small lanes. The massive walls of Indus cities most likely protected the Harappans from floods and may have dissuaded military conflicts (Morris 1994). Some theories proposed are:

- There was a single state, given the similarity in artefacts, the evidence for planned settlements, the standardized size of bricks, and the establishment of settlements near sources of raw material.

- There was no single ruler but several cities like Mohenjo Daro had a separate ruler, Harappa another, and so forth.
- Harappan society had no rulers, and everybody enjoyed equal status. However, as in other cultures, actual weights were not uniform throughout the area.

Both Mohenjo Daro and Harappan civilizations were characterized by having "differentiated living quarters, flat-roofed brick houses, and fortified administrative or religious centres" (Possehl 2002a). The discovery of the cotton fragments (*Gossypium arboreum*) was made at Mohenjo Daro in Sind in an expedition led by Sir John Marshall, Director General of the Archaeological Survey of India from 1902 to 1928. Sir Marshall in his book on Mohenjo Daro and the Indus Civilization speaks of how some fragments of cloth came wrapped around a silver perfume jar and a salt cellar (Balasubramaniam 2016). Cotton was woven and dyed for clothing; wheat, rice, and a variety of vegetables and fruits were cultivated; and a number of animals, including the humped bull, was domesticated, as well as " fowl for fighting" (Possehl 2002b). In Harappans, age trade routes were developed along the Indus River that went as far as the Persian Gulf, Mesopotamia, and Egypt. Some of the most valuable things traded were carnelian and lapis lazuli (Possehl 2002a).

It is clear that Harappan society was not entirely peaceful, with the human skeletal remains demonstrating some of the highest rates of injury (15.5%) found in South Asian prehistory (Singh 2008). Palaeopathological analysis demonstrated that leprosy and tuberculosis were present at Harappa, with the highest prevalence of both disease and trauma present in the skeletons from Area G (an ossuary located south-east of the city walls) (Coningham and Young 2015). The Harappans also made various toys and games, among them cubical dice which were found in sites like Mohenjo Daro (Lal 2002). The Indus Civilization's economy appears to have depended significantly on trade, which was facilitated by major advances in transport technology. The IVC may have been the first civilization to use wheeled transport (Bellis 2021). Archaeologists have discovered a 200 m by 35 m dock at the coastal city of Lothal, an excavation of which revealed stone anchors and marine shells believed to have originated in the Persian Gulf (UNESCO World Heritage 2022). An extensive canal network used for irrigation was discovered by H.-P. Francfort (Singh 2008).

During the early Harappan period (about 3200–2600 BCE), similarities in pottery, seals, figurines, ornaments, etc. document intensive caravan trade with Central Asia and the Iranian plateau (Parpola 2005). Judging from the dispersal of Indus Civilization artefacts, the trade networks economically integrated a huge area, including portions of Afghanistan, the coastal regions of Persia, northern and western India, and Mesopotamia, leading to the development of Indus-Mesopotamia relations contacts extended to Crete and possibly to Egypt (Doniger 2010).

According to Gangal et al. (2014), there is strong archaeological and geographical evidence that Neolithic farming spread from the Near East into north-west India, but there is also "good evidence for the local domestication of barley and the zebu cattle at Mehrgarh". The authors speculated the food-balls to be of a ritualistic significance, given the founds of bull figurines, adze, and a seal in immediate vicinity (Tewari 2021).

10.3.2 Egyptian Society: Lifestyle

In 1994, Belgian archaeologists found the earliest remains of human child who lived approximately 55,000 years old. This finding near Luxor is significant, because it's on a possible dispersion route of modern human from Africa into Asia and Europe between 50,000 and 100,000 years ago (Fletcher 2016). Egypt was a very fertile land, and under normal circumstances no one went hungry. Food could be home grown, earned in the form of rations (there was no money), hunted, fished, or bartered at market. The roof over the kitchen was made from matting that would allow smoke and cooking smells to escape (Roveri 2008).

It is well known that from ancient Egypt, the main centres of civilization came the greatest achievements in the writing language which began as drawings with its four developments, the geometry where the pyramids were designed and built according to the greatest constant 3.14 or π this constant could be considered the crown of Egyptian geometrical achievements, no equation up till now in space area, in atomic age, in wireless communication could be correct without it. This constant is the relation between the circumference and the diameter of the circle which is the heart of the universal geometry Also in the Egypt, great achievements had been accomplished like astronomy, medicine, chemistry, poetry, etc. Now our curve goes through north India not through its south, according to Gustavo Lo Bon in his book about the civilization in India there were no literary effects in north India, and the stony buildings found there do not exceed the fifth century A.D. In China the curve goes through its south part in ancient times, according to Linton (1949). To the left hand from the main centre of civilization in its olden times, the curve goes through Atlantic Ocean to Mexico in middle America with the Maya people, the similarity between Maya civilization and Egyptian is still one of the miracles in the history in the style of hieroglyphic writing (Eid 2021).

The process of preserving the dead body called embalming or mummification was developed. Body organs were removed and preserved in canopic jars, placed near a mummy. The heart, in contrast, was left in place and the corpse packed with natron salt. It was left for up to 40 days, until entirely dry. Finally, the desiccated body was washed, oiled, and bandaged. Not everyone could afford this treatment, however. The vast majority of the population were buried unmummified, in simple desert graves. What kind of an afterlife was expected by the Egyptians is not well understood (Lane 2011).

10.4 Cities During Mughal Period

Mughal emperors were ruling from the cities like Agra, Awadh, Khajuha near Fatehpur, Delhi, Fatehpur Sikri, and Lahore. Although these were capitals only in a formal sense, yet actually they were moving in a camp with their emperor. There were winter and summer capitals. To get away from the sweltering heat, the emperor may visit a suitable place like Kashmir.

To establish a city was fun for them, when King Akbar was blessed with a son with the blessings of Sheikh Salim Chisti a new city called Fatehpur Sikri was built 36 km away from Agra in 1571. It was named after the village of Sikri. The famous Buland Darwaza was built to honour the victory in Gujarat. Then the city came to be known as *Fatehpur Sikri*—"The City of Victory". But the royal family lived there for a short time. In Agra fort as well as Sikri one can see the places to keep animals like elephant and horses (Catherine and Blanshard 1992). There were places for stay to pigeons to be used for communication in Sikri. The reason for its abandonment in 1610 is usually given as the failure of the water supply, though Akbar's loss of interest may also have been the reason since it was built solely on his whim (Asher 1992). Fitch (Srivastava 1973) described, "Agra and Fatehpur Sikri are two very great cities, either of them much greater than London, and very populous. Akbar visited the city only once in 1601 after abandoning it. William Finch, visiting it 4–5 years after Akbar's death, stated". It is all ruinate, " and lying like a waste desert" (Eraly 2000). Jahangir stayed here for 3 months in 1619 during epidemic of bubonic plague from 1616 to 1624 (Fig. 10.1).

10.5 Concept of Smart Cities

The concept of smart cities began as far back as the 1960s and 1970s when the US Community Analysis Bureau began using databases, aerial photography, and cluster analysis to collect data, direct resources, and issue reports in order to direct services, mitigate against disasters, and reduce poverty. This led to the creation of the first generation of smart cities. The first generation of smart city was delivered by technology providers to understand the implications of technology on daily life. Use of renewable energy, novel technologies, and eco-friendly lifestyle hold the key for sustainable cities in future. Combing automation, machine learning and functions like digital payment are effectively tried. Another example would be smart traffic management to monitor traffic flows and optimize traffic lights to reduce congestion, whilst ride-sharing services can also be managed by a smart city infrastructure. Smart city features can also include energy conservation and environmental efficiencies, such as streetlights that dim when the roads are empty.

Undoubtedly today, the conception and the digital logistics of a "Cyborg Networking Computing City", described at length at the beginning of this chapter, are the dominant model for progress towards a Smart City. This dominance is the result of the soft power of the global economy and its hi-tech industries that have managed to convince most of the decision-makers around the world that technology is a paternalistic intelligence, capable of making megacities liveable. Critics of the Cyborg City are still in the minority but are gaining ground. If urban life is only analysed through big data statistics, if each inhabitant is subjected to the permanent surveillance by an inquisitive Big Brother, if diagnoses are made and solutions automatically chosen by algorithms, the city becomes a cybernetic machine where the psychology and sociology of the residents no longer count, the decision-makers of the cities no longer have any power and the citizens are nothing more than puppets (Cathelat 2019).

10.6 Use of Novel Technologies

The process of urbanization includes dense settlements having changes in land use pattern. Whilst the antecedents of urbanization are long, contemporary urbanization is now predominantly a developing-country phenomenon, centred largely in Asia. Urbanization in Asia involves around 44 million people being added to the population of cities every year. To put this in perspective, each day a further 120,000 people are added to the populations of Asian cities, requiring the construction of more than 20,000 new dwellings, 250 km of new roads, and additional infrastructure to supply more than six mega-litres of potable water (Roberts and Kanaley 2006). Various newer technologies can help to solve the problems of energy and water supply and help in treatment of the solid waste and sewage into useful products. Everitt et al. (2007) used Quick Bird false colour satellite image of the area in Texas in 2003 and were able to locate spiny aster (*Leucosyris spinosa*), a noxious weed. This technique can help to identify the occurrence of trees. Remote measurement techniques from satellites be exploited to measure regional and global distributions of gaseous pollutants in the troposphere. Use of drones and balloon platform can also be done with satellite sensors (NASA 1971). Production of industrially important materials from animal waste (Shah 2020), fibre from banana leaves and pseudo stem (Francies 2020), and use of solar power and hydrogen and biofuel for vehicles are tried successfully. And there is a need to be exploited for making the cities sustainable. Some of these are listed below.

10.6.1 Eco-Friendly Green Buildings

Green design increases the economics, social equity, and environmental impact of affordable housing projects. According to the National Centre for Healthy Housing (NCHH) 5.7 million of American families live in substandard housing (Anonymous 2011). Existing low-cost units tend to be older than the housing stock in general. For example, in 1995, 68% of the affordable rental housing stock was constructed before 1970 and another 21% were built before 1980. The city buildings should use recycled construction material and try to reduce costly wood. Use of waste plastic into plastic boards for door windows and furniture can be made. With conservation of water and energy utilization smart buildings are planned. Efficient LED lighting can reduce the power budget of inhabitants. Efforts are done to put reflective tiles or reflective white paint on terrace is usually tried to reduce the heating in buildings in hot summer.

10.6.2 Energy Uses for House-Related Activities

The impact of energy consumption on GHG emissions depends not just on the amount consumed, but also on the GHG emissions generated by the energy source, which in turn depend on the mode of energy production. It has been found that in second largest city of South Africa, Cape Town has comparatively low per capital electricity consumption than Geneva in Switzerland, but its consumption has a higher GHG emissions factor, due to South Africa's use of coal for 92% of its electricity generation whilst Geneva uses hydropower. Technology used is very important. When urban areas depend on wasteful energy sources, they contribute more GHG emissions.

As reported in International Energy Outlook 2017, world energy consumption is going to increase by 28% by 2040, with India and China being a few of the largest consumers because of their strong economic growth. So, it has become almost certain that the world now needs plenty of renewable energy resources to satisfy the growing need for energy (https://yearbook.enerdata.net).

Energy expenditures for low-income households have risen over the past decades, and the energy burden (energy expenditures as share of household income) has increased since 1997. Mean residential energy expenditures increased by 27% from 2001 to 2005, to $1522, and by almost 20% to $1822 by fiscal year 2009 (Anonymous 2011). Solar water heaters and solar electricity panels can be placed at the roof tops, canal, and bridges. Methods should be developed to use non-renewable.

Electricity can be generated by microbial fuel cells (MFC). Air cathode in which oxygen is fed by self-diffusion instead of aeration to take part in the oxygen reduction reaction was used by Wu et al. (2018). The non-limiting availability of oxygen to the air cathode ultimately makes it possible to achieve higher current densities. However, the use of air cathode in small, especially miniaturized MFC devices is limited by the inevitable water loss due to evaporation through the air cathode of MFC, which leads to unstable power generation. Under the synergistic actions of both forces, water is pumped in *Sequoia sempervirens*, a tree with height over 115.72 m. Inspired by this concept of auto-feeding MFC (AF-MFC) similar mechanism was used in MFCs without any external equipment. First, the hydrogel performance was studied and optimized for the water evaporation rate and feeding capacity, and then the performance of the AF-MFC was evaluated for power generation. The MFCs generated bioelectricity from lignocellulosic biomass using two different microbial communities (*Acetobacter aceti* and *Gluconobacter roseus*) (Krishnaraj et al. 2015).

10.6.3 Smart Traffic System

Baroda had good roads during the time of the Maharaja Sayajirao Gaekwad III. Water was sprinkled on roads in busy markets, every evening to settle the dust and reduce pollution. The citizens were using gardens regularly for leading a healthy life. According to an estimate, the average traffic has increased by 14% on all roads between 1995 and 2002 in Delhi (Arya 2020). The speed was slow due to heavy traffic congestion in 13% of the network on at least half of the days of the year (NAO 2004). The traffic can move faster if normal traffic rules are strictly followed by users. The traffic consisting of Ambulance, fire brigade, and VIP vehicles should be provided with electronic/sensitive tagging and this will help to monitor the traffic lights for clearing the path, and movement of these vehicles could be uninterrupted. Multilane express traffic roads are built to provide speed to the moving vehicles. The designing of motor vehicles should consider laws of aerodynamics and efficient fuel utilization. The rout for transport movement is determined by traffic congestion as reported by use of GPS and remote sensing. The trend will soon start with driverless cars running with high speed having remote control.

10.6.4 Waste Management Strategy

On my visit to Torino, Italy in 2008, I was introduced with the waste collection plan of cities there. In 2011, the city of Milan started implementing an ambitious scheme to separately collect bio-waste and recycle it. With 1.4 million inhabitants and an extremely densely populated area, this wasn't an easy task as bio-waste collection schemes are more difficult to set up in big cities. However, after 10 years, Milan is now one of the leading examples with 95 kg of bio-waste collected per inhabitants and a 62% waste collection rate. Among the different success factors behind the strategy, two of them have played an essential role for Milan:

- The thorough planning of the household collection. As household bio-waste collection is the most complex to tackle, Milan's roll-out phase only started after a 2 years pilot project and was then progressively adopted in the four quadrants of the city.
- The awareness-raising campaign. The campaign was made all along the roll-out phase, not only focusing on general communication but also on targeted communication with workshops in areas during the roll-out phase. Among the many tools used, Milan also set up a website with all the information available in 10 languages.

This allowed Milan to currently collect approximately 130,000 tonnes of bio-waste per year thus avoiding around 9000 tonnes of CO_2 emissions. But it is not only about quantity as the quality is also extremely high with an average contamination rate below 5%. As bio-waste is the first step of any successful waste management strategy, this also has an impact on the average collection rate which is now around 63% and the economic savings with an estimation of 30 euros saved per ton of waste properly recycled.

10.6.5 Clean Fuel for Vehicles

In India oil exploration started in 1867, whilst first oil well was started in 1859 in America and current oil production is less than 100 million barrel which will soon become 111 million barrel per day (Khamesra 2015). The exhaust emission control regulations were introduced in 1970. The European Union passed directives for clean fuels for passenger and light commercial vehicles beginning in 2000 (Euro III) and 2005 (Euro IV). Achieving the full benefit of SULEV emission control technologies requires use of low-sulphur fuel, which is required in California but not yet available nationwide (Patel 2005). The fuel is lead and sulphur free to reduce the pollution levels. In Vadodara, there is a need of Metro for pollution free movement. The number of two wheelers increased in the city from 85,000 in 1985 to 8 lakhs in 2005 and from 10,300 cars to 1 lakh during same period. The mega cities are having more pollution due to increasing number of vehicles (Bhatt and Singh 2005).

The biofuel and alternative fuels like CNG, electric and chemical or fuel cells can be used as vehicular fuel solely or in hybrid mode. The blending of ethanol with petrol is suggested. In Brazil, the use of 100% ethanol is allowed for car fuels, whilst in India only 10% blending with petrol is done. Use of hydrogen as cleanest fuel is suggested but technologies are not yet standardized to be applied at commercial scale. All such fuels are suggested to reduce vehicular pollution and reduce transportation cost.

10.7 Certain Eco-Friendly Model Cities

It is interesting to note that more than one billion people in the developing world are currently unable to access clean water. This trend is more prevalent among the rural communities in the developing world (WHO-UNICEF 2018). Water is undoubtedly essential for life and no living thing can survive without it. Access to safe and reliable drinking water is a concern throughout the developing world (Galiani et al. 2005). More innovations in construction material and adopted technologies are needed. Eco-friendly green and smart buildings with all living amenities can provide sustainable lifestyle living and better happy index.

10.7.1 Vadodara

A town Ankottaka took roots some 2200 years ago on west banks of Vishwamitri River, which is known as Baroda or Vadodara, the third-largest city of Gujarat (Figs. 10.2 and 10.3). Vadodara or Vatodar is derived from presence of vad of *Ficus* in the belly. A large number of *Ficus* trees are found here (John 2012). Originating from Pavagadh hills, Vishwamitri is a seasonal river.

Fig. 10.2 Picture of traffic in old city of Vadodara (**a**) 1885 (**b**) 2021

Fig. 10.3 One of the lovely public parks in Portland, Oregon

The city is located on the fertile plains of Mahi and Narmada Rivers. Due to sudden and heavy rainfall, city faces floods. The municipal corporation gets water from Sayaji Sarovar constructed near Azwa in early twentieth century by Maharaja Sayajirao Gaekwad III.

To clean up river, a workable solution is needed (Patel 2020). More than 50% of the total waste of the city is collected by door-to-door garbage collection system. A study conducted by Saraswat and Veerkumar (2020) showed that in Vadodara only one-third of respondents separate the waste to get more income, whilst others fail to do so due to time consuming. Unless separation is proper, the recycling waste is a problem. Two percent of the total waste generated is handled as biomedical waste. Water from Sayaji sarovar is treated at Nimeta plant run by VMC and then supplied. An average of 240 million litres of water per day or 190 L/person/day is provided to the citizens. A large number of individual houses also generate power on their roof tops, which is uploaded to grid. Vadodara is a cosmopolitan city with many historical places like Lukshmi Vilas palace, Kirti Mandir, EME temple, Hazira a mausoleum of Qutb-ud-din Muhammad Khan, a teacher of Salim. It is famous for The Maharaja Sayajirao University. Robert Chisholm designed the buildings like Baroda College, Museum, and Khanderao, whilst another architect and city planner Patrick Geddes who worked to build Jerusalem helped to develop Mandvi, Sursagar, and Kothi (John 2012). The city is clean with magnificent gardens and a large number of trees on roadsides. However public transport system is poor in the city.

Pollution due to particulate matter $PM_{2.5}$ is more than permissible limits in ambient air. The problem of underground water pollution is reported from certain industrial areas in outskirts of the city.

10.7.2 Copenhagen

Copenhagen, the capital of Denmark is reputed to be happiest city in the world. Copenhagen is known for its canals, excellent food, Tivoli gardens. Less than 2% of the city's waste goes to landfill sight in Denmark; we do this by promoting greener choices and by broadening tourism both in terms of geography and seasonality, etc. Just as we have a dedicated focus on sharing Copenhagen's sustainable solutions with the rest of the world. The branding is carried out on Western and Chinese SoMe channels and by relevant media. The efforts are based on Wonderful Copenhagen's knowledge of the destination and reflects our core stories: Art and Design, Monarchy-History, and Modern Architecture (www.visitcopenhagen.com).

There were several reasons when Gehl, a Danish urban planner, first started exploring the "people issue" in cities. One was marrying his wife Ingrid, a psychologist who engaged him in countless conversations about the human side of architecture. They wondered how and what we build influences our lives and what we do. Another was to work for a home that was "good for people" (Beacom 2012).

10.7.3 Portland

The city of Portland had a population of 652,503, making it the 25th big city in the United States, the sixth-most populous on the West Coast, and the second-most populous in the Pacific Northwest, after Seattle (Danver 2013). It has smart city planning, plentiful public parks, free libraries, zero waste vegan restaurants. Legacy Health, a non-profit healthcare system in Portland, operates multiple facilities in the city and surrounding suburbs (Anonymous 2013). The city-owned Portland streetcar serves two routes in the Central City—downtown and adjacent districts. The first line, which opened in 2001 and was extended in 2005–2007, operates from the South Waterfront District through Portland State University and north through the West End of downtown, to shopping areas and dense residential districts north and northwest of downtown. The second line has tracks on the east side of the Willamette River and across the Broadway Bridge to a connection with the original line (Rose 2009).

10.7.4 Singapore

Singapore is a densely populated most sustainable and green city in the Asia. It has world-famous indoor garden called Gardens by the Bay (Fig. 10.4). Other important ones are Jewel Changi and Marina. By 2030, the 80% of the city's architecture will be sustainable and eco-friendly. Climate is suitable to inhabitants and Singapore

Fig. 10.4 Singapore's gardens by the Bay aims to be a model for sustainable development and conservation

National gardens help in conservation of rare plant species. Citizens try to implement zero waste strategy.

10.7.5 London

London is an ancient name, already attested in the first century AD, usually in the Latinised form Londinium for example, handwritten Roman tablets recovered in the city originating from AD 65/70–80 include the word *Londinio* ('in London'). The river Thames flow through London. The city came into pollution hazards when great smog of London caused loss of 10,000 persons due to respiratory diseases in December, 1952 (Stone 2002). Citywide administration is coordinated by the Greater London Authority (GLA), whilst local administration is carried out by 33 smaller authorities (Anonymous 2021). Further urban expansion is now prevented by the Metropolitan Green Belt, although the built-up area extends beyond the boundary in places, producing a separately defined Greater London Urban Area. Beyond this is the vast London commuter belt (Anonymous 2008b).

10.7.6 Yokohama

Yokohama means "horizontal beach". The current area surrounded by Maita Park, the Ōoka River and the Nakamura River have been a gulf divided by a sandbar from the open sea. This sandbar was the original Yokohama fishing village. Since the sandbar protruded perpendicularly from the land, or horizontally when viewed from the sea, it was called a "horizontal beach". It is the second largest in Japan and capital city of Kanagawa, with a population of 3.8 million (in the year 2020). A military garrison was established here in 1862 and Yokohama quickly became the base of foreign trade in Japan (Fig. 10.4). Yokohama was rebuilt, only to be destroyed again by U.S. air raids during World War II. The city has good bus, taxi, and train transportation facility. A neat city with China market in the centre. Good plantation can be seen in public parks and on rod side. To protect from frost, trees are covered by strips of jute or other ribbons. Clean drinking water is available. Recycling of waste material is done (Fig. 10.5).

10.7.7 Marrakesch

One possible origin of the name Marrakesch in Morocco is from the Berber (Amazigh) words *amur (n) akush*, which means "Land of God". According to historian Susan Searight, however, the town's name was first documented in an eleventh-century manuscript in the Qarawiyyin library in Fez, where its meaning

Fig. 10.5 Port city of Japan-Yokohama

was given as "country of the sons of Kush". The city is located in the Tensift River valley, with the Tensift River passing along the northern edge of the city. The Ourika River valley is about 30 km south of Marrakesch (Searight 1999). The "silvery valley of the Ourika River curving north towards Marrakesch" and the "red heights of Jebel Yagour still capped with snow" to the south are sights in this area (Rogerson and Lavington 2004). The city is known for beautiful gardens. The Agdal gardens were established during the reign of Abu Ya'qub Yusuf (r. 1163–1184) and extend over a larger area today. National and Islamic festivals are celebrated in Marrakesh, and one of the main cultural festivals celebrated in Marrakesh is the National Folklore Festival.

The economy of Marrakesch is governed by unique culture of Morocco and growing tourism industry. However, the majority of the citizens are still not having good source of water and are poor. Trade and crafts are extremely important to the local tourism-fuelled economy. There are 18 *souks* (market) in Marrakesch, employing over 40,000 people in pottery, copperware, leather, and other crafts. The UNESCO world heritage site since 1985, Jemaa el-Fnaa is one of the best-known squares in Africa (Fig. 10.6). It has been described as a "world-famous square", a metaphorical urban icon, a bridge between the past and the present. It is one of the most interesting souks. An adventure to immerse yourself in. It is a genuine labyrinth filled with numerous alleys, stairways, and dead-end street. It is

Fig. 10.6 UNESCO World Heritage Centre Jemaa El-Fnna in Marrakesch

difficult to imagine until we visit the place having sounds, colours, and exotic aromas blended with unique location (Kich 2008). One can find old Moroccan traditions and articles of interest here.

10.8 Expo 2020 Dubai: Digitally Connected

Architecture enthusiasts might easily overlook one of the most exciting technological innovations. The entire constructed Expo site is being digitally connected by Siemens, the official Infrastructure Digitalization Partner for Expo 2020 Dubai—from air-conditioning and elevators to access control. More than 130 buildings on the 4.38-square-kilometre site will be interconnected via the Internet of Things (IoT). Siemens is working closely with the organizers of Expo 2020 Dubai to develop a smart city app that will be used for the first time at the mega event. The app conveys the idea of the potential of a pervasive digital infrastructure. For instance, it can supply detailed environmental data such as air quality and weather conditions and can monitor and control irrigation systems in the area which will be monitored through App governed by artificial intelligence. "Our solution in Dubai will demonstrate how smart infrastructure can contribute to more liveable, more sustainable cities around the world", says Oliver Kraft, Siemens' Head of Project and Account Management for Expo 2020 Dubai. "Ultimately, our work is not just about ingenious technology, but about creating added value for society and making our cities better places to live".

10.9 Eco-Friendly Lifestyle Needed

Way of life is a word that shows how we bear our life, how we act in this life, how we connect with others and it likewise implies the manner in which we consider each part in our lives. We do not utilize bicycle extensively as a mode of conveyance, but we use bike and other vehicles. Our way of life is transforming from the crude one quite a whilst back to a cutting edge way of life in this globalization period. We have realized that with industrialization our way of life has changed. We simply utilize our mobiles, tablets, PC, or mini-computer to do accounting and communication with each other. Really with increase in population and massive industrialization, there is increase in pollution. The anthropogenic pollution from vehicles, industries, and agricultural activities has resulted into loss of many human lives. An eco-friendly way of life along with integrated traffic and organic urban farming strategy will lead a healthy and sustainable society. People should prefer local fruits and vegetables and try to utilize their leisure time in nearby picnic parks and gardens. Reading habits should be promoted in younger generations than playing video games. People should learn to live in harmony with nature. Habits of gardening and bird watching should be promoted. Schemes should be launched for the interaction of the children with nearby villagers to acquaint them with the adaptations and traditions they practice for healthy living. There is a need for promotion of urban farming and tree plantation. People should follow good living habits. Vegetarian diets, practicing yoga, and passing morning hours in sunlight are recommended to enhance immunity against the diseases.

10.9.1 Eco-Friendly Housing

More than one half of the population in developing countries still resides in huts or earthen facilities. To build shelter in these developing countries with traditional industrial building materials, i.e. brick, concrete, and steel has proven impossible to fulfil the requirements for providing affordable houses in developing countries (Minke 2009). Use of wood and fly ash bricks should be preferred. The proper light and ventilation through cross windows should be ensured. Water recharge facility and solar panels should be used to cut the energy costs.

Green affordable houses lower the lifetime cost of units and improve the quality of life. Newer technologies and the requirement of especially skilled force are felt for reducing the cost of green buildings. Many new types of eco-friendly or recycled products can be used in affordable housing. The construction of houses should be near the work place.

In Vadodara, there is a society named Green Planet in Atladra area. There are 40 houses with fly ash brick walls, solar and wind power arrangements. The society has excellent rain water harvesting facility. Each house has an underground storage tank, rain water is stored and is used in kitchen, the water used in washing vegetables is used for watering the trees, and even water used for washing a car is not wasted but it is used in garden. The kitchen waste is converted into biogas and then into electricity. Grey water from kitchen and bathroom is purified and used in lawns of the society. A common garden is also developed to organize public functions. The road side plants are attached to a drip irrigation and foot path is made of paver blocks so the water can percolate. Persons occupying the houses are environment conscious and follow the principle of reduce, recycle, and reuse.

10.9.2 Different Alternatives for Waste Management

We should work to reduce the pollution and develop cities with zero waste. Europe's best performing systems in this field are from Wales, Italy, and Spain. Different cities recognize the value of cleanliness and benefit of adopting a zero waste approach. Bioconversion of wastes can also improve the environmental quality by helping easy disposal of the solid organic wastes. Disposal of solid wastes is already a major problem in the big cities the world over. For example, in Delhi alone over 2000 tonnes of garbage is being produced daily (Singh et al. 1992). The average collection efficiency for MSW in Indian cities is about 70% (Nema 2004; Rathi 2006). Manual collection is 50% in class I cities rest is collected by trucks (CPCB 2000). Necessary steps are taken to reduce and collect the municipal solid waste door to door in Vadodara. Management of kitchen and municipal and agro waste is suggested by Dabke et al. (2022). Stubble burning in certain northern states in India has created an alarming air pollution condition, which can be taken care by converting it into compost or bio pellets for further fuel substitute (Yelda and Kansal 2003).

Zero waste European network works together with municipality and waste company officials who recognize the urgent need for change, and the benefits that can bring their local community. The industrial waste should be treated separately. For individual houses, waste segregation should be done and bio-waste generated by individual houses should be converted into compost at local level in houses or in societies. This will cut the transportation cost. The people should work to separate glass and e-waste separately and these should be recycled at home or by ward wise facilities developed by Corporation. The reduction in organic bio-waste will help to reduce the methane and CO_2 emissions causing increase in global warming. At different places, waste is used in different ways like production of bio-bricks, organic fertilizer or compost and get energy through production of biogas. Plastic waste can be converted to energy fuel, building material or in road making. In Vadodara, a tree garden with beautiful walking track is built by Municipal Corporation in Vadsar dumping site (Dabke et al. 2022).

10.9.3 Safe Transport Facility

Transportation in cities is heterogeneous in India, the number of two wheelers are many times more than four wheelers. The people should adopt public transport. In different cities, the BRTS bus system is implemented. Metro trains should be available in big cities. The people should use vehicles with biofuel. Regular tuning and PUC check should be done to reduce pollution. In small cities or towns, non-motorized vehicles like horses and bullocks can be used for transportation of material. Now electric vehicles are in trend. Govt. has suggested use of biofuel, and hydrogen as future alternatives. People should prefer cycling for better health and to reduce the pollution from transportation, as is common in certain Chinese cities. At different places, footpaths and separate cycling path should be available for pedestrians.

In Spanish city, Vitoria non-motorized transport is promoted, which increased from 52% in 2006 to 65% in 2016, and boosted by the city council's Sustainable Mobility Plan. Seville stands out for its promotion of cycling, with the highest modal share of cyclists among the cities analysed. A Zem2All project (project with 200 electric cars and 243 charging points) was initiated in Malaga.

10.9.4 Energy Uses in Future

No doubt the cities are vital to the world's economies but they also consume energy and are responsible for 70% of greenhouse gas (GHG) emissions. Public authorities can change the energy model and suggest measures for developing cities as had been seen in case of Spain. In terms of municipal support for sustainability, for example, Madrid publishes a regular energy consumption and emissions inventory, which is put together according to European Environment Agency guidelines. The Barcelona municipal bus depot is one of the most sustainable in Europe and has electric buses. In terms of buildings, Zaragoza has made significant progress. It has earmarked funds to refurbish more than 3500 buildings that are over 40 years old with a focus on energy sustainability. It has also constructed 10,000 new bioclimatic homes in the Valdespartera Ecocity, with energy savings of up to 90%. Steven Chu former US Secretary of Energy and Nobel Physicist advocated for the use of greener energy and reduce the dependency on carbon-based fuels.

10.9.5 Education and Entertainment

Environmental education is desirable to understand the climate change, its mitigating and adaptations. A sense of self-efficacy, or belief in one's ability to achieve meaningful goals, is also associated with health (Bandura 1997). Therefore, it is not surprising that one of the most frequent recommendations to support students facing daunting environmental challenges is to enable them to do their own investigations, select action steps that they find personally meaningful, and experience agency (Kelsey and Armstrong 2012). Environmental education is thus the key catalyst (Kaur 2020), for the needed change, and a primary means of achieving environmental protection (Sharma 2016). It is among the most effective strategies that does not require a large enough cost to change people's perspectives and attitudes towards the environment (Habibie 2020).

Education about clean environment is included in syllabus for every graduate degree according to Supreme Court directive proposed in 1991. The school students should be aware about pollution, global warming, and climate change. The Govt. is working to increase the forest cover and save the biodiversity. Hands on training should be imparted to students from childhood and with the entertainment they should develop the hobby of caring and planting the trees (Biswal 2016). This will help them to become more knowledgeable and physically healthy. Participation in outdoor activities will ensure better utilization of their time away from their virtual world spend on mobiles.

10.9.6 Yoga and Healthy Life

Yoga is a 5000-year-old discipline practised in ashrams or places of rishis in India. All yoga styles help us in balancing the body, mind, and spirit, but they are achieved in different ways. Some yoga practices are intense and vigorous. Others are cooling, relaxing, and meditative. Each yogic pose targets certain specific muscles. This helps any one to increase the flexibility and thus reduce the risk for injury. Yoga can lower stress levels thus increase our concentration.

Hatha yoga is a form of yoga most popular in the U.S. It is known as the yoga of force (Hofmann et al. 2016). It emphasizes strengthening and purifying the body. One has to strengthen physically by different postures called asanas and performing deep breathing termed pranayama. The style of yoga which focuses on alignment is known as **Iyengar yoga**. It is fluid and dancelike. It uses props like wooden blocks, straps, chairs, bolsters, and blankets to help you achieve and hold postures you otherwise couldn't hold.

Ashtanga yoga is designed to build your endurance and strength. You do a series of postures in continuous, flowing movement. These motions affect the breathing patterns. Challenging yourself physically is the basic exercise. At this "edge", the focus is on your breath whilst your mind is accepting and calm. Weintraub, author of *Yoga for Depression*, makes yogic tools accessible for use in a variety of settings through this innovative card deck. These 52 cards present what the author considers to be the most therapeutic yoga practices for managing your mood, along with basic instructions and illustrations to follow. She also provides the benefits of each technique so you can choose the right practice for your temperament. The deck's colour coding guides users to help lift depression (red), calm anxiety (blue), or ground for balance (green). As a yoga practitioner, I found that the colour-coded scheme deepened my understanding of the benefits offered by each practice. "The purpose of yoga is to build strength, awareness and harmony in both the mind and body". It include sessions of breathing and physical asana or posture to relaxation. Yoga is a great tool for staying healthy because it is based on principles of meditation.

Anxiety symptoms can be cured by yoga. Whilst Hatha yoga treatment in nearly all of the included studies contained postures and breathing, some studies involved additional treatment components, e.g. meditation, mindfulness, and thus further research is needed (Hofmann et al. 2016). The observations are consistent with a time-limited effect of the yoga intervention on thalamic GABA levels. The lack of tonic changes in this study and the lack of baseline differences between the experienced yoga practitioners and controls in the prior study suggest that tonic GABA levels are stable in subjects screened to exclude low GABA states (Streeter et al. 2010).

10.10 Conclusion

Most of the civilizations were developed near water bodies particularly rivers. The towns today are facing massive congestion from near and far off people, who come in search of jobs. The developing cities have failed to provide even minimum facilities like clean drinking water and sanitation. The people living in slums are poor and posing health issues. Recent Corona crisis has shown the way that people can work from home. Urbanization needs to be monitored as shown in certain major cities around the globe. Paper describes salient features of these cities. Government plans to develop sustainable and smart cities are discussed. Novel technologies should be applied to get solutions of problems like efficient use of energy and waste management. With good education, people can learn to care for the natural world. A single inspiring teacher can leave a lasting influence. But the scale of the risks ahead calls for a fundamental reorientation of our education. Emphasis should be more on environmental education. New education policy should prepare the students to lead their communities and country through the worldwide changes that are already underway.

References

Anonymous (2008a) Ministry of Health, and Family Welfare, Govt. of India 2005–2006. National Family Health Survey (NFHS-3). www.nfhsindia.org

Anonymous (2008b) London in its Regional Setting. London Assembly. Archived from the original (PDF) on 27 May 2008

Anonymous (2011) National Centre for Healthy Housing. "State of Healthy Housing". Archived from the original on 21 November 2011

Anonymous (2013) Human Rights Campaign. Healthcare Equality Index 2013. HRC
Anonymous (2021) Who runs London—find out who runs London and how. London Councils. Retrieved 28 March 2021
Arya A (2009) Global warming and climate changes- challenges ahead. SOCLEEN 43:7–10
Arya A (2020) Intelligent traffic and fast transport system to reduce vehicular pollution in certain Indian cities. In: Arya A (ed) Environment at crossroads. Scientific Publication, Jodhpur, pp 140–146
Arya A, Bhatt A (2020) Effect of cement dust pollution on the colonization of leaf surface microorganisms. In: Arya A (ed) Environment at crossroads. Scientific Publication, Jodhpur, pp 40–46
Asher CEB (1992) Age of Akbar. Architecture of Mughal India, (Part 1). Cambridge University Press, Cambridge
Balasubramaniam C (2016) The case of the missing cloth: in the quest of a piece of cotton unearthed from Mohenjo Daro. The Hindu December 2, 2016
Bandura A (1997) Self-efficacy. W. H. Freeman, New York
Beacom E (2012) Copenhagenizing the world, one city at a time. The Copenhagen Post November 10, 2012
Bellis M (2021) The invention of the wheel. Thought Co, Feb. 11, 2021. thoughtco.com/the-invention-of-the-wheel-1992669
Bhatt B, Singh J (2005) Air pollution and its impact; some reflections. In: Arya A, Bedi SJ, Jasrai YT, Patel VS (eds) Urban pollution issues and solutions. Nidhi Book Centre, Delhi, pp 115–134
Biswal BN (2016) Bio heritage education: a pragmatic innovation of Sri Sai Vidya Niketan to mitigate the impact of climate change. In: Arya A, Basu S (eds) Anthropogenic pollution- causes and concern. The Readers Paradise, New Delhi, pp 132–139
Cathelat B (2019) Smart cities: shaping the society of 2030. UNESCO, Paris, p 351
Catherine E, Blanshard A (1992) Architecture of Mughal India, part 1, volume 4. Cambridge University Press, Cambridge, pp 51–53
Coningham R, Young R (2015) The archaeology of South Asia: from the Indus to Asoka, c. 6500 BCE—200 CE. Cambridge University Press, Cambridge. https://en.wikipedia.org/wiki/Robin_Coningham
Corfee-Morlot J, Kamal-Chaoui L, Donovan MG, Cochran I, Robert A, Teasdale PJ (2009) Cities, climate change and multilevel governance, OECD environmental working papers N° 14, 2009. OECD Publishing, Paris
CPCB (2000) Status of municipal solid waste generation, collection, treatment and disposal in class I cities series: ADSORB/31/1999-2000
Dabke S, Sejpal CA, Arya A (2022) Fungal flora associated with kitchen waste and management of municipal and agro waste: turing garbage into gold. In: Arya A, Patel VS, Agrawal M, Murthy CN, Bhatt B, Padate G (eds) Environmental pollution, climate change and altered lifestyle during Covid. Astral International Pvt. Ltd., New Delhi, pp 35–62
Danver SL (2013) Encyclopaedia of politics of the American West. Sage Publication, Thousand Oaks, pp 533–534
Doniger W (2010) The Hindus: an alternative history. Oxford University Press.p, Oxford, p 67
Eid S (2021) Moving curve of civilization. Adv Soc Sci Res J 8(5):500–503. https://doi.org/10.14738/assrj.85.10140
Eraly A (2000) Emperors of the peacock throne: the Saga of the great Mughals. Penguin Books, New Delhi, p 179
Everitt JH, Yang C, Drawe DL (2007) Mapping spiny aster infestations with QuickBird imagery. Geocarto Int 22(4):273–284
Farooqui A (2007) Can climate change in the past serve as a basis for future indications? Environ Forensic 13(3):9–10
Fletcher J (2016) The big questions of ancient Egypt. Knowl Mag India 6(2):40–43
Francies H (2020) Go, banana, wear banana-vegan leather from wild banana leaves. In: International conference on Ecohealth and environmental sustainability. Navrachana University, Vadodara, p 89

Galiani S, Gertler P, Schargrodsky E (2005) Water for life: the impact of the privatization of water services on child mortality. J Polit Econ 113(1):83–120

Gangal K, Sarson GR, Shukurov A (2014) The near-eastern roots of the Neolithic in South Asia. PLoS One 9(5):e95714. https://doi.org/10.1371/journal.pone.00957141

Gupta R, Gupta N, Gupta N (2015) Understanding urbanization vs. sustainable development. Science 49(6):363–367

Habibie A (2020) The effect of personality over student insights on the environment. Int J Multicult Multi-relig Underst 7(2):74–81

Hofmann SG, Andreoli G, Carpenter JK, Curtiss J (2016) Effect of Hatha Yoga on anxiety: a meta-analysis. J Evid Based Med 9(3):116–124. https://doi.org/10.1111/jebm.12204

Hutson V, Law R, Lewis D (1985) Dynamics of ecologically obligate mutualisms—effects of spatial diffusion on resilience of the interacting species. Am Nat 126:445–449

Jadav S, Suryawanshi DS (2013) Status of environmental factors of urban health in Jalgaon, India: a geographical review. J Environ Res Dev 8(1):159–168

John P (2012) Baroda- know your roots. Bennett, Coleman and Co. Ltd., New Delhi, p 84

Joldersma C (2017) Education, change, and development, educational theories and philosophies. Oxford research encyclopaedia of education. Oxford University Press USA, New York

Kaur M (2020) Impact of environmental education on environment awareness among school students: a case study of district Amritsar (Punjab). Studies in Indian place names (UGC care journal) ISSN: 2394-3114

Kelsey E, Armstrong C (2012) Finding hope in a world of environmental catastrophe. In: Wals A, Corcoran P (eds) Learning for sustainability in times of accelerating change. Wageningen Academic Publishers, Wageningen, pp 187–200

Khamesra NS (2015) Tel Aevam gas utpadan Ka Bhartiya Paridrashya (in Hindi). Bhuvigyan 5:13–18

Kich A (2008) All Marrakech. Escudo de Oro, Sabadell, p 96

Koller JM (2018) Asian philosophies, 7th edn. Routledge, Abington, p 428

Krishnaraj RN, Berchmans S, Pal P (2015) The three-compartment microbial fuel cell: a new sustainable approach to bioelectricity generation from lignocellulosic biomass. Cellulose 22(1): 655–662

Lal BB (2002) The Sarasvatī flows on: the continuity of Indian culture. Aryan Books International, New Delhi

Lane A (2011) Myths and legends of ancient Egypt and Tutankhamun's curse: the developing history of an Egyptian king

Linton R (1949) The tree of culture. Trans N Y Acad Sci 11:171–175

Mark JJ (2020) Indus Valley civilization. World history Encyclopedia

Mehta N, Dubey PS (2003) Air quality in three different cement producing areas. In: Jasrai YT, Arya A (eds) Air pollution development at what cost? Daya Publishing House, Delhi, p 231

Minke G (2009) Building with earth: design and technology of a sustainable architecture. Birkhäuser, Basel

Morell V (2002) Empires across the Andes. Natl Geogr 201(6):106–129

Morris AEJ (1994) History of urban form: before the industrial revolutions, 3rd edn. Routledge, New York

NAO (2004) Tackling congestion by making better use of England's motor ways and trunk roads. Report by comptroller auditor General HC 15 session 2004–2005

NASA (1971) Remote measurement of pollution. Scientific and technical information. NASA, Washington, DC, p 253

Nema AK (2004) Collection and transport of municipal solid waste. In: Training program on solid waste management. Springer, Delhi

Parpola A (2005) Study of the Indus script. Archived from the original on 6 March 2006. (50th ICES Tokyo session 19th May 2005)

Patel VS (2005) Beyond the exhaust emission limits. In: Arya A, Bedi SJ, Jasrai YT, Patel VS (eds) Urban pollution; issues and solution. Nidhi Book Centre, Delhi, pp 10–16

Patel VS (2020) Clean up strategy for the river Vishwamitri. In: Arya A (ed) Environment at crossroads. Scientific Publication, Jodhpur, pp 178–190
Possehl GL (2002a) Harappans and hunters: economic interaction and specialization in prehistoric India. In: Morrison KD, Junker LL (eds) Forager-traders in south and Southeast Asia: long-term histories. Cambridge University Press, Cambridge, pp 62–76
Possehl GL (2002b) The Indus civilization: a contemporary perspective. Rowman Altamira, Lanham
Rathi S (2006) Alternative approaches for better municipal solid waste management in Mumbai, India. J Waste Manag 26(10):1192–1200
Roberts BH, Kanaley T (2006) Urbanization and sustainability in Asia: good practice approaches in urban region development. In: Project: the new urban agenda needs to recognize a future of city-to city networks and trade. Asian Development Bank, Mandaluyong
Rogerson B, Lavington S (2004) Marrakech: the Red City. Sickle Moon
Rose J (2009) TriMet's open source heaven: the 5 best transit-rider apps. The Oregonian
Roveri AMD (2008) Egyptian museum. Barisone Editore, Giaveno, p 80
Saraswat S, Veerkumar V (2020) Waste disposal practices of homemakers and satisfaction with door to door waste collection service of municipal Corporation of Vadodara. In: International conference on Ecohealth and environmental sustainability. Navrachana University, Vadodara, p 59
Searight S (1999) Maverick guide to Morocco. Pelican Publishing, Gretna, p 403
Shah ND (2020) Production of industrially important materials from animal wastes—a potent alternative for waste management. In: International conference on Ecohealth and environmental sustainability. Navrachana University, Vadodara, p 89
Sharma PK (2016) Assessment of environmental literacy of school students. Calorx Teachers' University, Ahmedabad
Singh U (2008) A history of ancient and early Medieval India: from the stone age to the 12th century. Pearson Education India, New Delhi
Singh RN, Balakrishna K, Varma AK (1992) Refuse for reuse. Sci Rep 29(10):19–22
Srivastava AL (1973) Akbar the great, Vol. III: society and culture in 16th century India. Shiva Lal Agarwala, Agra, p 10
Stone R (2002) Counting the cost of London's killer smog. Science 298(5601):2106–2107. https://doi.org/10.1126/science.298.5601.2106b
Streeter CC, Whitfield TH, Owen L, Rein T, Karri SK, Yakhkind A, Perlmutter R, Prescot A, Renshaw PF, Ciraulo DA, Jensen JE (2010) Effects of yoga versus walking on mood, anxiety, and brain GABA levels: a randomized controlled MRS study. J Altern Complement Med 16(11):1145–1152. https://doi.org/10.1089/acm.2010.0007
Tewari M (2021) Harappan people ate multigrain, high-protein 'laddoos': Study—Times of India
UNESCO World Heritage (2022) Archaeological remains of a Harappa port-town, Lothal. UNESCO World Heritage Centre, Quebec
WHO-UNICEF (2018) Joint monitoring programme for water and sanitation
Wu S, Patil SA, Chena S (2018) Auto-feeding microbial fuel cell inspired by transpiration of plants. Appl Energy 225:934–939
Yelda S, Kansal S (2003) Economic insight into MSWM in Mumbai: a critical analysis. Int J Environ Pollut 19(5):516–527

Chapter 11
Spatial Analysis of Precipitation Climatology Over India: Using Satellite Remote Sensor

Sandeep Kalyan, Aditya Sharma, and Saurabh Choubey

11.1 Introduction

Precipitation is one of the most valuable and important components in a climate structure to regulate the hydrological cycle. Besides that, the data quality is also a matter of concern as it is the driving parameter for forecasting and hydrological simulation that depends on rainfall (Tarek et al. 2017). However, it is a tough task for exact simulation and model interpretation of precipitation because of its spatial and temporal irregularity and heterogeneity nature of data (Li et al. 2020). With the help of climatology, we can predict future scenarios based on historical precipitation data but to do so we should also understand all the aspects of the driving parameters (Seneviratne et al. 2012). Changing rainfall patterns and associated effects on water resources is one of the possible consequences of climate change. Therefore, adequate information about precipitation and associated changes is necessary on different spatio-temporal resolutions (Fleming et al. 2011). In India, there is a great variation in rainfall, not only from place to place but also from season to season (Sharma et al. 2018; Pradhan et al. 2019). Precipitation in India is very significant as more than 70% of the total Indian rural population depends greatly on monsoon rains for agricultural activities (FAO 2022) as agriculture has a very big share in the Indian Economy (almost equal to 18.1% of Indian GDP) (Bhatt and Nakamura 2005; Arjun 2013). Agriculture being primary activity is directly dependent on precipitation because of the lack of proper sources of irrigation for most of the available agricultural lands. India is fundamentally an agrarian economy, and the deficiency of

S. Kalyan (✉) · S. Choubey
School of Environment and Sustainable Development, Central University of Gujarat, Gandhinagar, India

A. Sharma
Department of Atmospheric Science, Central University of Rajasthan, Ajmer, India

B. Pathak, R. S. Dubey (eds.), *Climate Change and Urban Environment Sustainability*, Disaster Resilience and Green Growth,
https://doi.org/10.1007/978-981-19-7618-6_11

monsoon rains tends to disturb the lives of many people, involved in different sectors of the Indian economy (Mohapatra et al. 2018). The entire burden due to lack of proper rainfall falls on the farmers who eventually have to bear the entire cost of crop failure trapping them in a vicious cycle of rural indebtedness. It is evident that the changing pattern of rainfall and precipitation causes disasters like floods, land degradation and drought, etc. which could lead to huge devastation of life and property (Seneviratne et al. 2012; Kalyan et al. 2021).

The agriculture sector in India is mainly dependent on monsoon rainfall, in which almost 68% of the cultivated land of India is primarily dependent on monsoon rainfall that supports 60% and 40% of the livestock and human population, respectively (Kumar et al. 2010). Hence, the climatological study of the precipitation or rainfall will surely help to understand its variability and spatial pattern and that could be useful for sustainable water resource management (Loucks and van Beek 2017). The report of Guhathakurta and Rajeevan (2008) shows that the monsoon rainfall is having a decreasing trend over Eastern Region Plains while an increase in the rainfall has been observed over the middle part of India and some part of the Southern (Goa) region. Wagholikar et al. (2014) observed that between 1951 and 2011 over most of the middle part and some of the Eastern part of the Indian region shows a negative trend while some parts of the Eastern and Western regions are having a positive trend. The mean annual rainfall of July and August is contributed 24.2% and 21.2%, respectively, of total rainfall from 1901 to 2003, while approximately 11% of the rainfall has received in each pre- and post-monsoon season, and no significant trend of rainfall has observed in the period over the region (Guhathakurta and Rajeevan 2008). According to India Meteorological Department (IMD), India receives maximum rainfall in the months of June, July, August, and September (monsoon season as designated by IMD) while months from December to May receive minimum rainfall when south-west monsoon is inactive. Also, Pai et al. (2014) have reported that from 1901 to 2014, India receives highest rainfall during the southwest monsoon and minimum rainfall during the winter months, respectively. The variation in the spatial pattern of extreme rain events has also been observed during the monsoon season for 1901–2004 along the west coast, North-East, and Central India (Rajeevan et al. 2008).

The rains in India did not follow a particular pattern and therefore it is very difficult to predict the amount and intensity of precipitation in the upcoming seasons. Rainfall is the primary and important source of water in many areas, and the availability of water for several purposes depends significantly on the amount of rainfall they get in that area (Endris et al. 2021). Now it is possible with the invention and development of technologies that study and monitor the precipitation patterns for a long period will provide a better prediction of precipitation for upcoming years (Ohba 2021). Satellite-based measurement of precipitation complements ground-based measurements to provide a more efficient and clear picture of rainfall or hydrological system. The TRMM (Tropical Rainfall Measuring Mission) precipitation product provides the impetus for precipitation algorithm improvement, for global analysis of precipitation and development as the satellite data have demonstrated more promise for studies of the tropical convective system and monsoon

(Mitra et al. 2009; Tarek et al. 2017). This study analyses and investigates the trend of TRMM precipitation data from 1998 to 2014. The understanding of precipitation trends provides a practical reference for water resources management, agricultural planning, and for a forecast of precipitation in different regions of India which may give a better climate change indication. Therefore, it seems essential to understand the past trend using high-resolution satellite datasets. Hence, this chapter exemplifies the investigation and importance of satellite data for precipitation climatology and trend analysis.

11.2 Study Area

India is a subcontinent country in the South-Asia possessing a megadiverse characteristic not limited to the cultural aspect but climatic and geographical variation is also having equal importance. India is the seventh-largest country in the world and has a total area of 32,87,263 km^2 with 8°4′N to 36°6′N and 68°7′E to 97°25′E latitude and longitude, respectively (Fig. 11.1). Climatic conditions in the region are varied place by place as it has desert on the west side, snow and glaciers on the northern side, coasts on the southern side, and dense forest cover in the eastern part. It also has 15 different climatic zones as classified by the Koppen–Geiger (Peel et al. 2007). The temperature of the region varies between −4 and 48 °C but most of the region doesn't even experience the below 10 °C, in some areas it could go as low as −10 °C and as high as 50 °C (Changnon 1971; Blasco et al. 1996). The monsoon season alone carries almost 80% of the total yearly rainfall (Bagla 2006).

11.3 Data and Methodology

For the study of the contribution of precipitation over weather and climate research Tropical Rainfall Measurement Mission (TRMM), a joint mission of the National Aeronautics and Space Administration (NASA) and the Japan Aerospace Exploration (JAXA) Agency was carried out (Li et al. 2020). TRMM carried 5 instruments: a 3-sensor rainfall suite (Precipitation Radar, TRMM Microwave Imager, Visible Infrared Radiometer Scanner) and 2 related instruments (Lighting Imaging Sensor, Cloud, and Earth Radiant Energy Sensor) (Heiblum et al. 2012). TRMM covers 180° W–180°E, 50°N–50°S in a non-sun-synchronous orbit, providing spatial and temporal datasets for the tropics and sub-tropical regions (Hashemi et al. 2017). An hourly precipitation data was collected at level 3 product 3B43: Multi-satellite precipitation, Version 7 and a spatial resolution of 0.25° × 0.25° for 1998–2014 (Cao et al. 2018) (Table 11.1). The NetCDF format of the data was assessed using MATLAB and monthly, and seasonal data were prepared (Fig. 11.2).

The seasonal categories (Guha et al. 2021) were made and described as winter, pre-monsoon, monsoon, and post-monsoon. After seasonal and monthly data

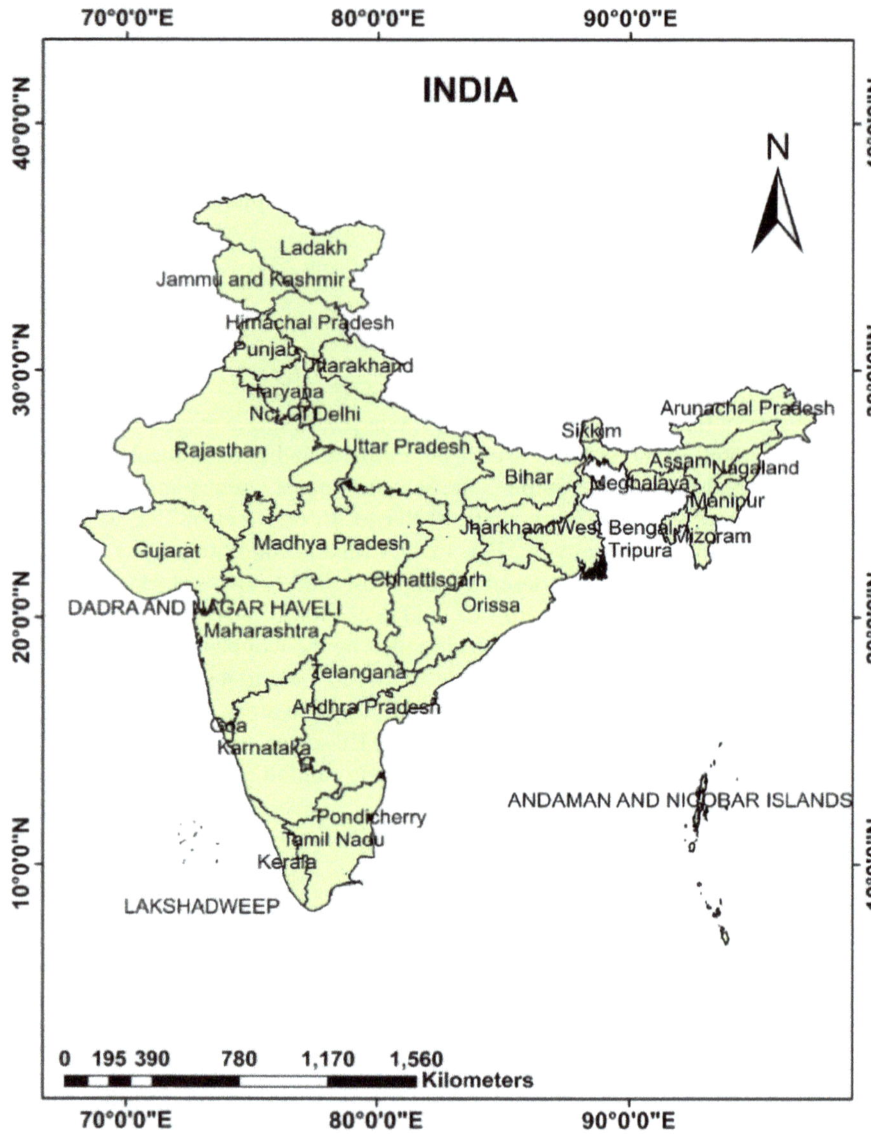

Fig. 11.1 Study area map

Table 11.1 Data and source

Satellite	Product	Data type	Resolution	Source
TRMM	3B43: Multi-satellite precipitation	Precipitation (1998–2014)	0.25° × 0.25°	TRMM, 2011. https://doi.org/10.5067/TRMM/TMPA/MONTH/7

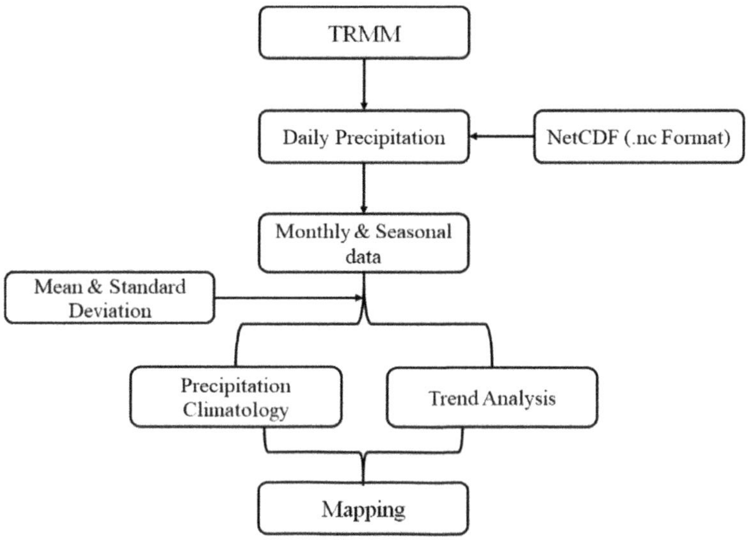

Fig. 11.2 Methodology flowchart

preparation, simple arithmetic mean and standard deviation of the data were calculated by using the following statistical formula:

$$\overline{X} = \frac{\sum_{i=1}^{n} \mu_i}{N}, \tag{11.1}$$

where $\overline{X}$ = total mean of all the observations, $\sum_{i=1}^{n} \mu_i$ = sum of the total number of observations, N = total number of observations.

Standard deviation is the deviation from a mean value of total observations which was calculated by using the following statistical formula:

$$\sigma = \sqrt{\frac{\sum_{i=1}^{n} \left(X_i^2 - \mu^2\right)}{N}}, \tag{11.2}$$

where σ = standard deviation, $\sum_{i=1}^{n} \left(X_i^2 - \mu^2\right)$ = sum of total observation minus mean of that observations, N = total number of observations.

A trend analysis was also performed for all the data set, and a color map was prepared for all mean, standard deviation, and trend data for all seasons and months. Simple Linear Trend explains "how to intercept value changes with slope." Color maps of trend analysis show how monthly and seasonal rainfall patterns have changed over the Indian subcontinent from 1998 to 2014. Color maps for monthly and seasonal precipitation data for mean, standard deviation, and linear trend are discussed in the next section.

11.4 Results

11.4.1 Seasonal Variation and Trend in Precipitation

Figure 11.3 shows that in pre-monsoon (March, April, and May) and post-monsoon (October, November, and December) season, there is the same pattern of rainfall over eastern (Assam, Manipur, Nagaland, etc.) and Bay of Bengal including some parts of Tamil Nadu, Orissa, Andhra Pradesh, etc. with less than 0.3 mm/h while in the winter season (January, February) there is less than 0.3 mm/h rainfall over most of the region.

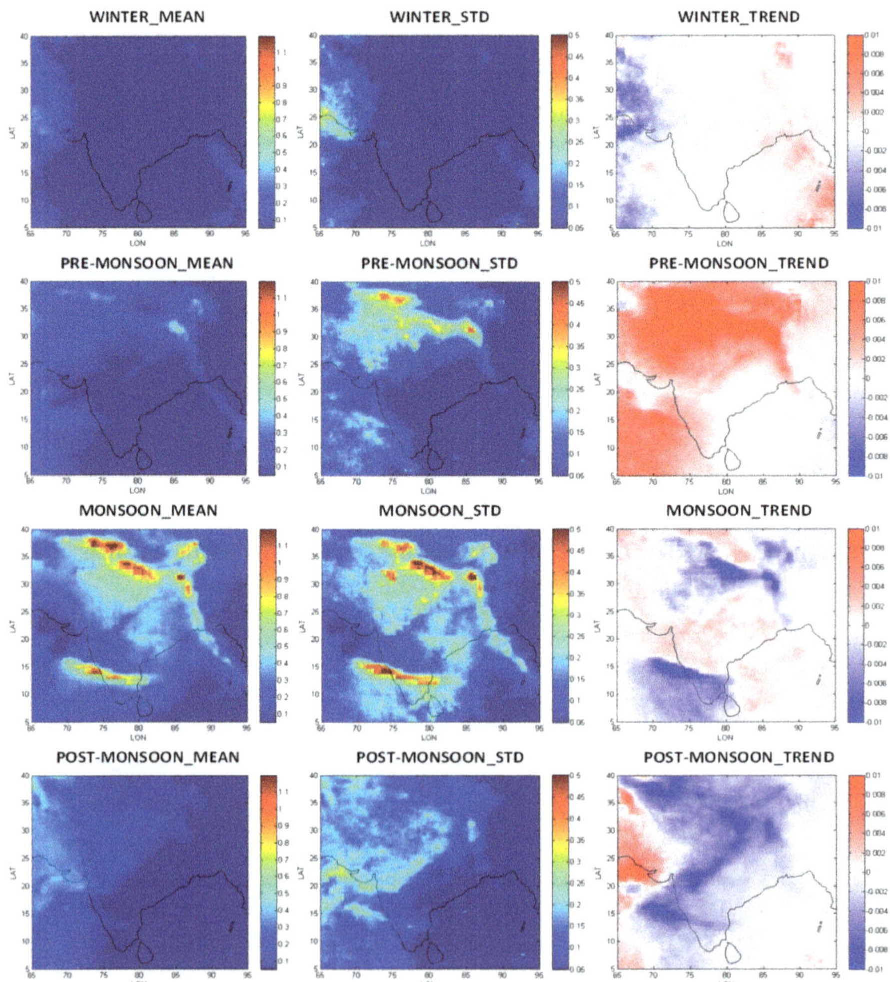

Fig. 11.3 Spatial variation of seasonal mean precipitation, associated standard deviation, and trend over the Indian region

In the monsoon season (June, July, August, and September), maximum precipitation is observed over the Northern region and some parts of Karnataka and the Arabian sea with more than 1.1 mm/h of intensity. Maximum seasonal variation in precipitation is also observed over where maximum precipitation was experienced in monsoon season while in the post-monsoon season the variation from mean precipitation has been observed over the part of middle to Eastern region (Rajasthan and Gujarat) with a negative trend. In the monsoon season, a negative trend over the highest precipitation region while a positive trend was found over the areas with minimum precipitation. In pre-monsoon, the trend is positive over the whole region and it is decreasing toward the post-monsoon through monsoon season and depicts an interesting scenario of precipitation.

11.4.2 Monthly Variation and Trend in Precipitation

Though the variation in precipitation over the period is not significant, there is a noticeable variation in trend in January, February, March, and April (Fig. 11.4).

As it can be seen from the climatology maps that minimum precipitation has been received over most of the region from January to April with 0.1 mm/h as these months are cold and considered in the pre-monsoon season which would be a reason for the low precipitation over the region. Some part of the Rajasthan and Gujarat region also shows high precipitation with an intensity of >0.5 mm/h. It is also interesting to see that the western part of India has received more precipitation as compared to the rest of the region. A variation in the trend maps could be seen over all the months. The red and blue colors in the maps are representing a positive and negative trend, respectively. In March and April month, a slightly positive trend has been observed in the Southern to the middle region.

From May onwards an increased intensity of the precipitation has been observed through August as June, July, and August months are considered as the Indian monsoon time and the maximum precipitation has also been received in the same period. A positive trend of the precipitation has been observed in July and August month (Fig. 11.5). Over Northern and Eastern region with more than 1.1 mm/h of intense precipitation has observed, and interestingly the deviation of >0.5 mm/h from the mean precipitation value has also been observed in the same region from May to August. A dominated region with a negative trend has been observed in May, only some part of middle India has experienced a positive trend over the years.

September has been considered as monsoon month, so it gets the precipitation throughout the month as it is also revealed by trend map as the almost whole region is showing a positive trend (Fig. 11.6). Though the precipitation is lower than 0.5 mm/h in all the months (Fig. 11.6) except September still the trend in September–December is dominated as positive especially in the western part of the Indian subcontinent. In December only over the Gujarat and Rajasthan state precipitation is around 0.5–0.6 mm/h while the deviation is >0.5 mm/h over the coast below Gujarat has observed.

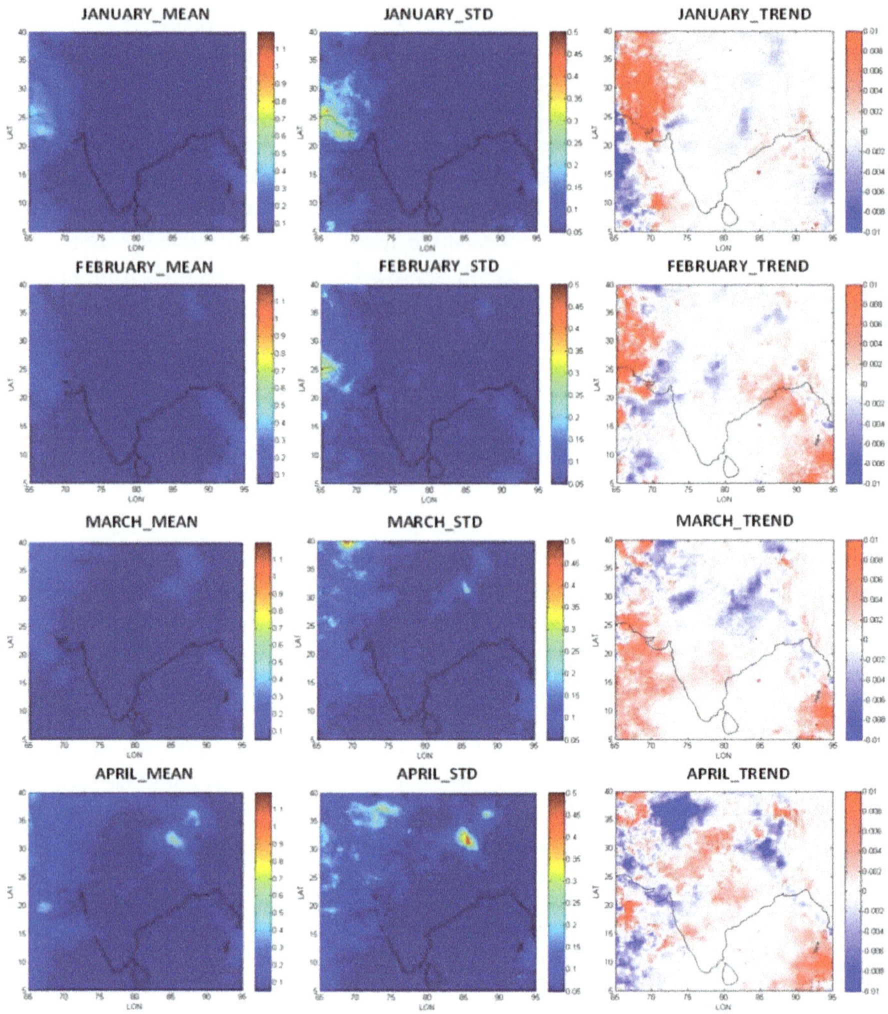

Fig. 11.4 Spatial variation of monthly (January–April) mean precipitation, associated standard deviation, and trend

11.5 Discussion

Precipitation data of TRMM was analyzed and interpreted over the Indian region for precipitation climatology and trend analysis. The monthly and seasonal data was prepared for the Indian region at 0.25° × 0.25° spatial resolution for 1998–2014. The intercomparison of the seasonal map shows that the maximum precipitation is observed in monsoon than pre/post-monsoon season, the same results were also found in Parthasarathy et al. (1993) study, and the maximum deviation is also observed in monsoon season where maximum precipitation was received. A

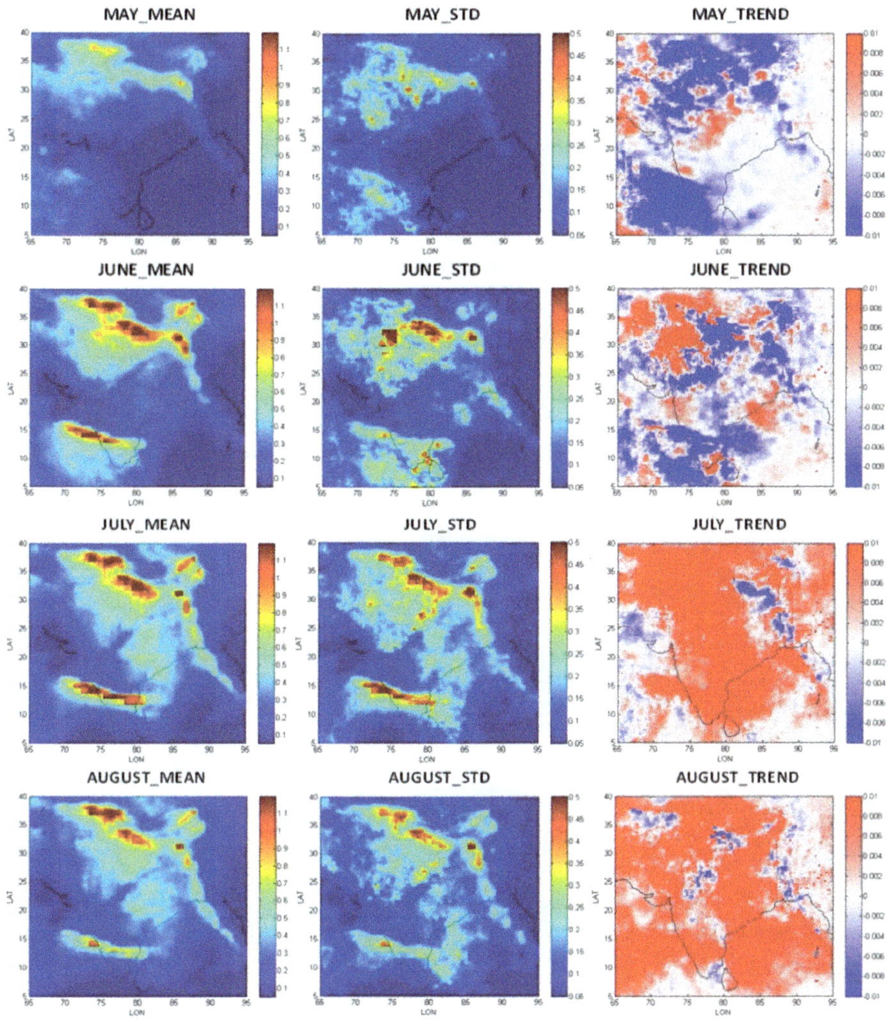

Fig. 11.5 Spatial variation of monthly (May–August) mean precipitation, associated standard deviation, and trend

negative trend of precipitation over the southern part particularly over Goa was observed and interestingly a significant decreasing trend was also observed by Guhathakurta and Rajeevan (2008) in their study.

There is a positive trend in the pre-monsoon season while most of the region is dominated by a negative trend in post-monsoon and monsoon season. It is observed that a positive trend in July, August, September, October, November, and December months same as Ramanathan et al. (2005) found while in April, May, and June month a clear negative trend has been observed. In monsoon season, the mean

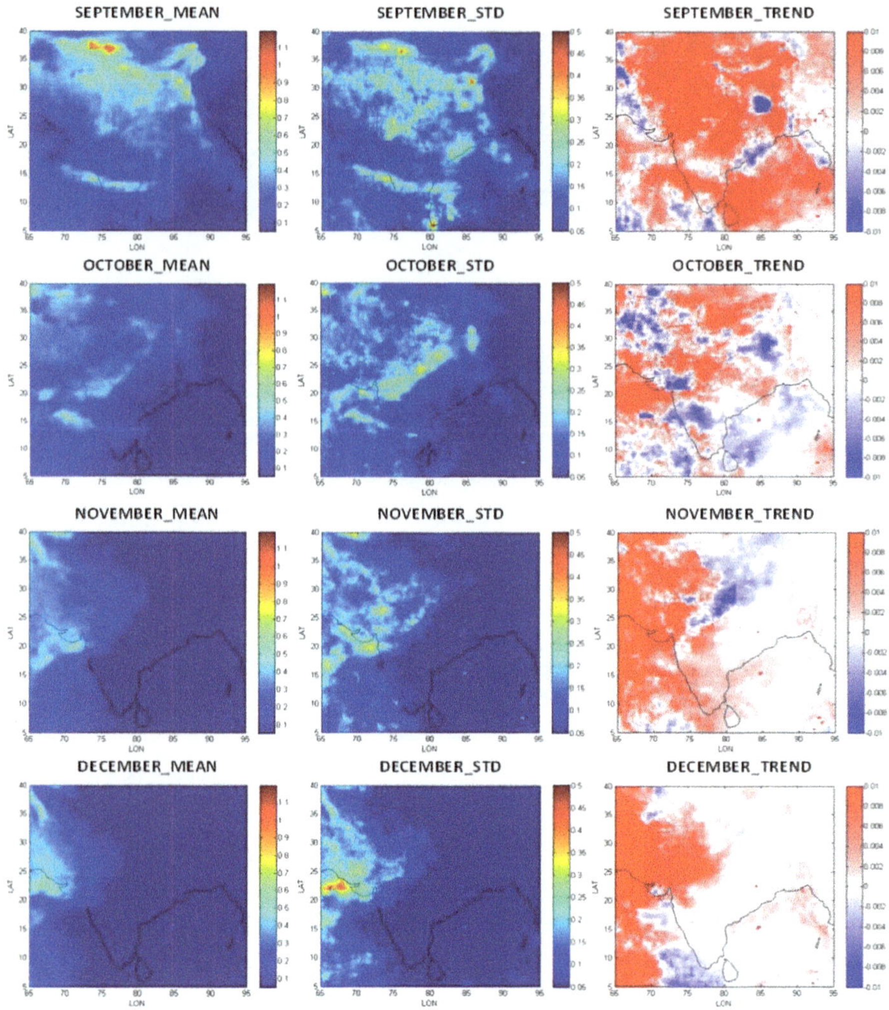

Fig. 11.6 Spatial variation of monthly (September–December) mean precipitation, associated standard deviation, and trend

precipitation received is more than 1.1 mm/h throughout the period over the Northern and some part of the Southern region but interestingly the negative trend is also observed where maximum precipitation was found and a positive trend was observed where less precipitation was observed.

Though the deviation of >0.5 mm/h from the mean precipitation has been observed over the places where maximum precipitation has received as this scenario is noticed in monsoon and post-monsoon season as compared to the pre-monsoon where no clear positive trend was found rather a negative trend has been observed over the whole Indian region. This deviation or reduction in monsoon strength might be due to the significant increase in atmospheric solar heating (Sharma et al. 2016). With the finding of this research, it is clear that the precipitation deteriorates with time from 1998 to 2014. Only in the pre-monsoon season, the trend is convincingly positive while all in the rest of the season the trend is dominated as negative. The months of July, August, and September are having a positive trend, and this trend has shifted toward the Western part of India. The study output is very fascinating and establishes an interesting relation between climatology and trend. The research output could be further analyzed for significance tests and precipitation forecast. Further looking at the trend analysis, the outcome could be used by the policymakers for the formulation of better strategies for natural disasters management and sustainable development as well.

Acknowledgements The authors are grateful to Dr. Amit Kumar Mishra, Assistant Professor, Jawaharlal Nehru University, New Delhi for providing the research space and peer guidance throughout the preparation of this manuscript.

References

Arjun KM (2013) Indian agriculture-status, importance and role in Indian economy. Int J Agric Food Sci Technol 4(4):343–346

Bagla P (2006) Controversial rivers project aims to turn India's fierce monsoon into a friend. Science 313:1036–1037

Bhatt BC, Nakamura K (2005) Characteristics of monsoon rainfall around the Himalayas revealed by TRMM precipitation radar. Mon Weather Rev 133(1):149–165

Blasco F, Bellan MF, Aizpuru M (1996) A vegetation map of tropical continental Asia at scale 1:5 million. J Veg Sci 7(5):623–634

Cao Y, Zhang W, Wang W (2018) Evaluation of TRMM 3B43 data over the Yangtze River Delta of China. Sci Rep 8(1):1–12

Changnon SA (1971) Note on hailstone size distributions. J Appl Meteorol (1962–1982) 10(1): 168–170

Endris HS, Hirons L, Segele ZT, Gudoshava M, Woolnough S, Artan GA (2021) Evaluation of the skill of monthly precipitation forecasts from global prediction systems over the greater horn of Africa. Weather Forecast 36(4):1275–1298

FAO (2022) Food and Agriculture Organization. FAO in India. https://www.fao.org/india/fao-in-india/india-at-a-glance/en/

Fleming K, Awange J, Kuhn M, Featherstone W (2011) Evaluating the TRMM 3B43 monthly precipitation product using gridded rain-gauge data over Australia. Australian Meteorol Oceanogr J 61:171–184

Guha S, Govil H, Gill N, Dey A (2021) A long-term seasonal analysis on the relationship between LST and NDBI using Landsat data. Quat Int 575:249–258

Guhathakurta P, Rajeevan M (2008) Trends in the rainfall pattern over India. Int J Climatol 28(11): 1453–1469

Hashemi H, Nordin M, Lakshmi V, Huffman GJ, Knight R (2017) Bias correction of long-term satellite monthly precipitation product (TRMM 3B43) over the conterminous United States. J Hydrometeorol 18(9):2491–2509

Heiblum RH, Koren I, Altaratz O (2012) New evidence of cloud invigoration from TRMM measurements of rain center of gravity. Geophys Res Lett 39(8):1158

Kalyan S, Sharma D, Sharma A (2021) Spatio-temporal variation in desert vulnerability using desertification index over the Banas River Basin in Rajasthan, India. Arabian J Geosci 14(1): 1–13
Kumar V, Jain SK, Singh Y (2010) Analysis of long-term rainfall trends in India. Hydrol Sci J J Sci Hydrol 55(4):484–496
Li X, Li Z, Lin Y (2020) Suitability of TRMM products with different temporal resolution (3-hourly, daily, and monthly) for rainfall erosivity estimation. Remote Sens 12(23):3924
Loucks DP, van Beek E (2017) Water resources planning and management: an overview. In: Water resource systems planning and management. Springer, Berlin, pp 1–49
Mitra AK, Bohra AK, Rajeevan MN, Krishnamurti TN (2009) Daily Indian precipitation analysis formed from a merge of rain-gauge data with the TRMM TMPA satellite-derived rainfall estimates. J Meteorol Soc Jpn Ser II 87:265–279
Mohapatra GN, Beham AR, Kavyashree C, Kavya S, Singh ML, Vaishnavi J (2018) Study of Indian summer monsoon rainfall trend during the period 1901–2013 through data mining. Int J Res Appl Sci Eng Technol 6(5):1701–1705
Ohba M (2021) Precipitation under climate change. In: Precipitation. Elsevier, Amsterdam, pp 21–51
Pai DS, Rajeevan M, Sreejith OP, Mukhopadhyay B, Satbha NS (2014) Development of a new high spatial resolution (0.25 × 0.25) long period (1901–2010) daily gridded rainfall data set over India and its comparison with existing data sets over the region. Mausam 65(1):1–18
Parthasarathy B, Kumar KR, Munot AA (1993) Homogeneous Indian monsoon rainfall: variability and prediction. Proc Indian Acad Sci Earth Planet Sci 102(1):121–155
Peel MC, Finlayson BL, McMahon TA (2007) Updated world map of the Köppen–Geiger climate classification. Hydrol Earth Syst Sci 11(5):1633–1644
Pradhan RK, Sharma D, Panda SK, Dubey SK, Sharma A (2019) Changes of precipitation regime and its indices over Rajasthan state of India: impact of climate change scenarios experiments. Clim Dyn 52(5):3405–3420
Rajeevan M, Bhate J, Jaswal AK (2008) Analysis of variability and trends of extreme rainfall events over India using 104 years of gridded daily rainfall data. Geophys Res Lett 35(18):143
Ramanathan V, Chung C, Kim D, Bettge T, Buja L, Kiehl JT, Wild M (2005) Atmospheric brown clouds: impacts on South Asian climate and hydrological cycle. Proc Natl Acad Sci 102(15): 5326–5333
Seneviratne S, Nicholls N, Easterling D, Goodess C, Kanae S, Kossin J, Zwiers FW (2012) Changes in climate extremes and their impacts on the natural physical environment. Cambridge University Press, Cambridge
Sharma S, Nunez L, Ramanathan V (2016) Atmospheric brown clouds. In: Oxford research encyclopedia of environmental science. Oxford Univeristy Press, Oxford
Sharma A, Sharma D, Panda SK, Dubey SK, Pradhan RK (2018) Investigation of temperature and its indices under climate change scenarios over different regions of Rajasthan state in India. Glob Planet Change 161:82–96
Tarek MH, Hassan A, Bhattacharjee J, Choudhury SH, Badruzzaman ABM (2017) Assessment of TRMM data for precipitation measurement in Bangladesh. Meteorol Appl 24(3):349–359
Wagholikar NK, Ray KS, Sen PN, Kumar PP (2014) Trends in seasonal temperatures over the Indian region. J Earth Syst Sci 123(4):673–687

Chapter 12
Timberline and Climate in the Indian Western Himalayan Region: Changes and Impact on Timberline Elevations

Priyanka Sah, Subrat Sharma, Avantika Latwal, and Rehana Shaik

12.1 Introduction

Climate, as a primary factor, controls geographic distribution of plants, influences ecology of vegetation, and changes are responsible for range shifts of plants (Forman 1964; Kumar 2012; Telwala et al. 2013). Climatic conditions in mountains are related to the elevation and locations, where altitude causes steep environmental gradients, particularly in the Himalaya (Singh et al. 2018). In mountains, heat and water conditions directly affect tree growth along the altitudinal gradients, and growth of trees has been strongly correlated with the environmental conditions (Nedlo et al. 2009; Liang et al. 2010). The position of the timberline or tree line depends on multi-factors, especially, the temperature is commonly held to be the main abiotic factor caused constrains the growth and regeneration of tree species (Körner 1998, 2003, 2012; Holtmeier and Broll 2005; Jump et al. 2007; Harsch et al. 2009).

High altitudes of Himalaya not only encase the highest timberline in Northern Hemisphere but also witnesses many processes of natural and anthropogenic events (Fig. 12.1) which makes Himalayan timberline sensitive to climatic and anthropogenic actions. In the recent decades, Himalaya have warmed up more rapidly than other areas of the globe (Field and Barros 2014). The warming effect is seen in the tree line ecotone where trees struggle to survive under cold stress in mountain summits (Butler et al. 2009). Worldwide, seasonal mean air temperature between 5 and 7 °C is the most consistent predictor of the altitudinal position of tree lines

P. Sah · S. Sharma (✉)
G.B. Pant National Institute of Himalayan Environment (NIHE), Ladakh Regional Centre, Leh, Ladakh (UT), India

A. Latwal · R. Shaik
International Institute of Information Technology, Hyderabad, Hyderabad, Telangana, India

B. Pathak, R. S. Dubey (eds.), *Climate Change and Urban Environment Sustainability*, Disaster Resilience and Green Growth,
https://doi.org/10.1007/978-981-19-7618-6_12

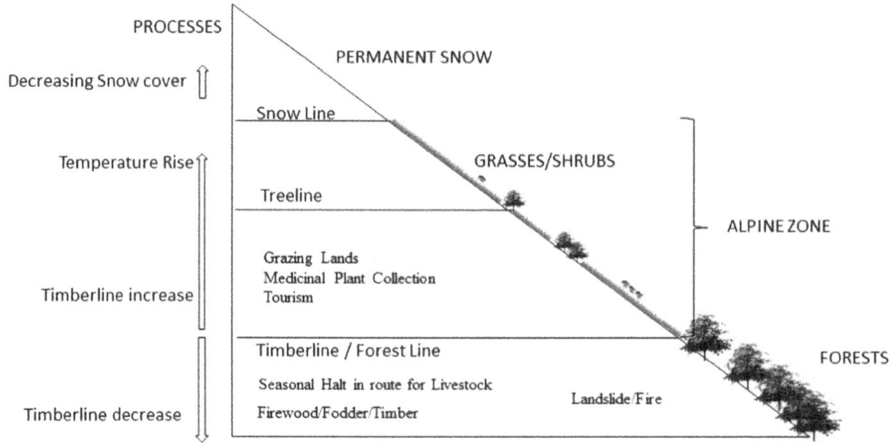

Fig. 12.1 A typical high-altitude profile of Himalaya showing various ecotones and processes

(Körner 1998, 2012). It has been established that tree growth is limited by the isotherm of 10 °C for the warmest months but sometimes other factors also limit their growth. For example, in southwestern Alaska, the isotherm stands ca. 400 km beyond the edge of the forest (Griggs 1937) and stunted and isolated forest stands can be formed at higher elevation in favorable microclimatic conditions.

Temperature reduction with increasing elevation is a primary driver of species biodiversity and the formation of tree lines (Mayor et al. 2017) in alpine ecosystems. The tree line is shifting upward to higher elevation primarily because of the warming (Grabherr et al. 1994; Walther et al. 2005; Schickhoff et al. 2015). Along the elevation gradients, the number of vascular plant species decreases as harsher cold environment prevails toward the nival zone (Mohapatra 2015). Temperature and precipitation are essential environmental parameters that govern the tree line dynamics. The tree line ecotones are of considerable ecological interest as they serve as indicators of response of plant species to environmental gradients, as loci for species diversity and as transition between two ecosystems reflected by spatial flux of species, matter, and energy (Malanson 1997). Thus, the alpine tree lines are ideally suited to study the impacts of climate change on vegetation and, conversely, recent changes in climate using climate-sensitive species and indicators. The impacts of climate change (particularly global warming) are evident through increase in air temperature and changing patterns of precipitation. Impacts are being felt on earth systems, viz. melting of the glaciers, etc.

In recent decades, the mountain ecosystem in Himalaya has warmed more rapidly than many other areas of the globe (IPCC 2014; Bhutiyani et al. 2010). Therefore, an increasing number of studies have focused on the Himalayan timberline due to its sensitive to climate change, and as field laboratory to impacts of global warming (Walck et al. 2011; Qi et al. 2015; Wielgolaski et al. 2017) where long-term studies are useful in assessing the responses of high-elevation ecosystems to climate change (Rogora et al. 2018). Along altitudinal and latitudinal gradients of vegetation distribution, global warming is supposed to affect the plant growth (Takahashi et al. 2011), and rise in temperature due to global warming is expected as causing factor for upward movement of tree line toward a higher altitude (MacDonald et al. 1998; Grace et al. 2002; Schickhoff et al. 2015).

In the worldwide, high-altitude climatic tree line is associated with a seasonal mean ground temperature of 6.7 °C ± 0.8 °C during the growing period (Körner and Paulsen 2004). Across most of the mountainous region of the world, air temperature is expected to rise significantly over coming decades (IPCC 2013) and also large warming trends in the Himalaya (up to 1.2 °C per decade at higher altitudes) have been observed in the past 30–40 years (Shrestha and Aryal 2011; Gerlitz et al. 2014; Hasson et al. 2016). The Himalaya during wintertime has already experienced a relatively high rate of warming with greater magnitude (Shrestha et al. 2012). In the high mountain range and cold deserts of the Trans-Himalayan region, warming episodes are even more intense, making it most vulnerable of all ecosystems (Sharma and Tsering 2009; Xu et al. 2009). Temperature and precipitation are the essential environmental parameters that govern tree line dynamics (Singh et al. 2018).

In the climate change scenarios, the rate of change in temperature with the rise in elevation (Temperature Lapse Rate, TLR) is the indication for elevation-dependent warming, which might have several implications on vegetation ecotone in Himalaya (Joshi et al. 2018). In addition, the response of tree line ecotone could vary significantly to climate change, which depends upon warmer temperatures of pre-monsoon months as well as increased precipitation of the intensify droughts, hence, climate warming may cause the upslope movement of tree line species, whereas grazing pressure may depress tree lines (Singh et al. 2018). High-elevation tree line is expected to be particularly sensitive to changes in climate as it receives low precipitation (Aryal et al. 2012), hence, it provides a unique opportunity to study the different climatic parameters (temperature, precipitation) response on vegetation in high-altitude environment. Sah and Sharma (2022) provided information on timberline in the Western Himalayan region by identifying new locations of timberline occurrence in the outer Himalayan ranges. This study explores (1) change detection in spatially distinct Western Himalayan timberline elevations in a time frame of nearly four decades (upward/downward movement or no change), (2) patterns of climatic parameters along altitudinal gradient of timberline elevations in inner and outer ranges of timberline occurrence, and (3) impacts of temperature and precipitation change on timberline elevations as indicator of climate change.

12.2 Study Area and Timberline

Western Himalayan state of Himachal Pradesh is spread in an area of 55,673 km^2 between 30°22′N–33°12′N latitude and 75°45′E–79°04′E longitude. This region has influence of both the Indian Summer Monsoon system in summer and mid-latitude westerlies in the winter (Pandey et al. 2017). The influence of Indian

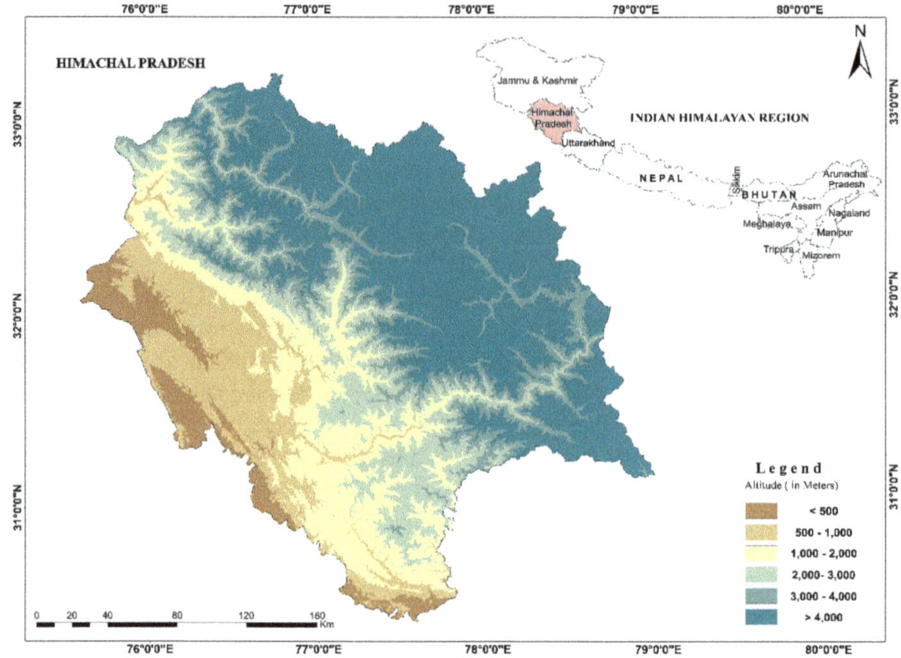

Fig. 12.2 Location map of Himachal Pradesh state and altitudinal gradient

Summer Monsoon can be seen predominantly over eastern and southern part of Himalaya, whereas mid-latitude westerlies strongly influence north-western Himalaya and Karakoram region (Bolch et al. 2012; Pandey et al. 2017). About one-third of the area in the state is permanently under snow, glaciers, and cold desert. The average annual rainfall is about 1800 mm. The temperature varies from sub-zero to 35 °C. Altitudinal zonation of the state is given in Fig. 12.2. Forest cover accounts for about 27% of the total geographical area of the state with a preponderance of moderate dense cover (7126 km^2) followed by open forest (5195 km^2, FSI 2019).

A timberline is the upper edge of a high-altitude forest with at least 30–40% crown density which is a parallel line to permanent snowline, referred to as continuous timberline (CTL). But some timberlines enclosed or encircled small islands of alpine meadows near the summits in other areas where summits are part of the alpine meadows, and this summit enclosing timberline is referred as ITL (Sah and Sharma 2018). Details of different spatial arrangements of two types of timberlines (CTL and ITL) have been described recently by Sah et al. (2021) for Himachal Pradesh. The total timberline length (sum of CTL and ITL) in Himachal Pradesh was ~3256 km, of which CTL accounted for about 91% (2972 km) and the remaining was ITL.

12.3 Methodology

12.3.1 Mapping of Timberline

Following Sah and Sharma (2018), Latwal et al. (2018), Sah et al. (2021) and Sah and Sharma (2022) timberlines were delineated for Himachal Pradesh state using Landsat data (Band 2, 3, 4, 5, and 7). The long-term availability of Landsat (Since 1972) makes it possible to realize spatio-temporal variability of timberline at larger scale. In order to map the longest spatio-temporal dynamics of timberline in Himachal Pradesh Landsat-8 and Landsat-2 were used. Oldest required suitable scene was available in 1976. We opted to use this image, which was resampled at 30 m to make it comparable with image of 2015. Different images were co-registered with latest image of 2015 to do change analysis of timberline. Spatial tools of ERDAS IMAGINE 2016 were used for these purposes (geometric correction, image enhancement, techniques, etc.). The satellite images were then subjected to knowledge-based interpretation technique and timberline was delineated by applying visual interpretation.

Realizing complex heterogeneous landscape features in high altitudes and limitations of auto-extraction methods in mountains, knowledge-based manual interpretation (visual) was preferred to delineate timberline. Working in Sikkim, Singh et al. (2018) realized that auto-extraction methods overlook/leave certain complex areas, which requires manual corrections. Visual interpretation appeared more appropriate in mountains (rugged terrain) where complex topography (steep, including shadow) challenges auto extraction of features. Change in timberline was recorded as a function of shift in altitude from the past (1976) to the current (2015) position. Sah et al. (2021) studied the timberline shift and Latwal et al. (2021)) (in press) studied variations in the air temperature and precipitation at timberline for Sikkim Himalaya.

The timberline data (vector) were used to create point data (30 m separation between two points in a line) for corresponding years (1976–2015) using Pixel to ASCII Converter feature of ERDAS IMAGINE 16. Thirty-meter spatial resolution of points (i.e., timberline elevation) were in tune to resolution of ASTER DEM (30 m) which was used to extract altitudinal information. Points at every 30 m were generated over the entire timberlines (past and present) to match the spatial attributes of DEM, and differences (elevation and distance) were recorded.

12.3.2 Climate Data at Timberline Elevations

Following the Latwal et al. (2021)) (in press), temporal change analysis (1976–2015) was done for these two types of timberlines and changes were marked as “shift” (upward or downward) and “no change” (stationary) in timberline position with respect to the base year of 1976. Himalaya is generally data deficient and it is truer for meteorological data, particularly in the higher altitudes. Gridded/modelled data sets were explored for the present study as many global and regional climatic datasets are available, such as APHRODITE (Asian Precipitation—Highly-Resolved Observational Data Integration Towards Evaluation). Daily gridded precipitation is the only long-term (1951 onward) continental-scale daily product that contains a dense network of daily rain-gauge data for Asia including the Himalayas, South and Southeast Asia and mountainous areas in the Middle East (Yatagai et al. 2012). Number of valid stations ranges between 5000 and 12,000, representing 2.3–4.5 times the data available through the Global Telecommunication System network. These were used for most of daily grid precipitation products. Climatologically daily mean precipitation and temperature data are available for Monsoon Asia region at 0.05° × 0.05° resolution (WWW1 2020). This product contributes to studies related to diagnosis of climate changes, evaluation of Asian water resources, statistical downscaling, forecast improvements, and verification of numerical model simulation and satellite precipitation estimates. APHRODITE precipitation data can be used as a benchmark for various estimations of gridded precipitation. Due to its long-term availability of daily products (1951 and 1961 for precipitation and temperature, respectively) in longer time frame, APHRODITE was preferred over high resolution (e.g., 1 × 1 km of WorldClim) and lower resolution data of IMD (gridded data of 1°).

Air temperature (daily mean) and daily precipitation (rainfall) at timberline altitude in Himachal Pradesh State were extracted from gridded data (resolution of 0.25°) of APHRODITE for two different years (1976 and 2015) using MATLAB R2019a. This dataset is available at http://www.chikyu.ac.jp/. Annual mean temperature and total annual rainfall were derived from the daily data set. To extract various points of annual mean temperature and annual precipitation at timberline altitude, bilinear interpolation technique was used to harmonize the different resolutions of two data sets, which uses four near neighbour grids and estimates the distance average with closer the grid being given higher weights. Bilinear interpolation algorithm is popular due to its computational efficiency and quality (Dilip et al. 2014). It is particularly useful for downscaling meteorological input data which are already gridded (Schulla and Jasper 2007). Temperature and precipitation (rainfall) were recorded for both the years (1976 and 2015).

To account for orographic effects, comprehensive interpolation methodology was used (WWW1 2020). It is believed to be one of the most realistic datasets for Asia because it uses the largest number of gauge observations (Ménégoz et al. 2013). For two spatially distinct timberlines, total data points (at 30 m interval) generated in different years were (1) ITL = 8613 and CTL = 92,113 for year 2015 and (2) ITL = 8576 and CTL = 92,113 for year 1976. For each data point latitude, longitude, altitude, daily air temperature, and precipitation were recorded. This data set was further categorized on spatial arrangements of timberline (ITL and CTL) and elevation class (100 m interval). All the values in an elevation class were averaged to represent that class. Comparisons were made for stationary and shifted timberline in each category. Statistical analysis (average, standard deviation) was performed using Microsoft Excel 2016.

12.4 Results

12.4.1 Changes in Timberline Elevation During Last Four Decades (1976–2015)

The mean elevation of the timberline position in Western Himalaya (i.e., state of Himachal Pradesh) in the year 2015 was 3442 m while almost four decades back (year 1976), this value was much below, i.e., 3248 m. An upward shift of 145 m was observed in the mean timberline elevation with a rate of 38 m upward shift per decade (or 3.8 m per year), however maximum occurrence elevation (4262 m) of timberline remained the same between this time frame. This indicates that entire timberline has not been moved upward. Change detection between the studied time frame (38 years) is presented in Fig. 12.3. The timberline position and their upward shift are variable throughout the mountains because of other factors (viz. topography, macro- and micro-climate, and other site conditions) responsible for the growth of tree species. Upward movement of *Pinus wallichiana*, a timberline tree species of Western Himalaya, through dendrochronological studies indicates different upward shift on different slopes between 1860 and 2000 [19 m decade^{-1} on the southern and 14 m decade^{-1} shift on the northern slopes; Dubey et al. (2003) and Singh et al. 2018)]. The changes at the timberline have occurred and it is significant in terms of

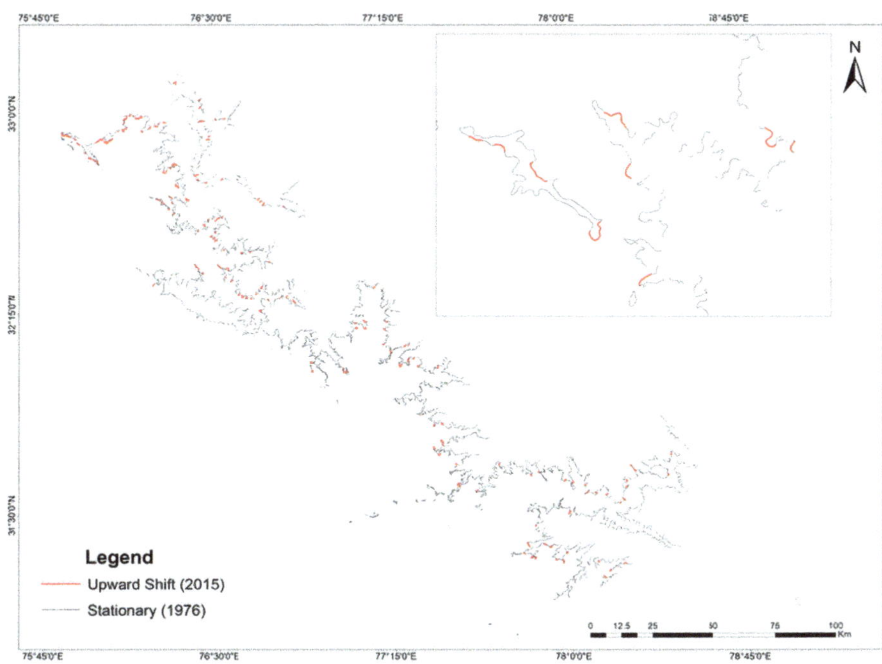

Fig. 12.3 Changes in timberline between 1976 and 2015 in Himachal Pradesh. In frame depicting details in changes in a part and fragmentary nature of timberline elevations

numbers. Upward shift observed through remote sensing-based timberline elevations was higher than the shift observed through field studies. More points of timberline species location throughout the high-elevation belt in Himalaya will lend more support to the remotely sensed derived timberline position.

Present study observed that there is a marginal increase in total timberline length (approx. 2 km, 3253 km in 1976) in four decades, and a large portion (94.6%) of this timberline length remained stationary, while only 5.4% (177.4 km in length) moved upwards. Total 65 island type timberline (ILT) locations were mapped in 1976, and most of them (60 ITL locations) remained stationary. Upward movement in 5 ILTs was observed, among them entire timberline was changed at two locations, while partial changes were observed in three ILT locations. In island type timberline locations, 11.5 km length of timberline (6.5% of total shifted timberline) was shifted.

12.4.2 Annual Mean Temperature and Total Rainfall at Timberline—Present (Year 2015)

For the year 2015, annual mean temperature and total annual rainfall in different altitudinal zones of two timberline types (ITL&CTL) are presented in Fig. 12.4a, b. ITL altitudes (away from the permanent snow) were warmer than the same timberlines altitudes close to permanent snowline (CTL) as reflected from average annual mean temperature (ITL = average 15.7 °C ± 2.9, range 7.8–20.7 °C among the different elevation zones), while this value was 10.6 °C ± 3.3 (average) at CTL (range −0.6 to 17.8 °C). Along ITL (outer ranges), the minimum temperature observed was 14.5 °C in the altitudinal band of 3500–3600 m and maximum temperature was 18.7 °C in the altitudinal band of 2800–2900 m. Along CTL (inner ranges), minimum temperature was 5.2 °C toward highest altitudinal band 4200–4300 m and maximum temperature was 16.4 °C in the lowest altitudinal band of 2400–2600 m (Fig. 12.4a).

Between elevational bands of ITL, total annual rainfall (range 966–2016 mm) was 1274 ± 219 mm, however, in CTL elevational bands, it was observed between 677 and 2130 mm with an average of 1286 ± 293 mm. Along ITL elevations, the minimum rainfall was observed in altitudinal band of 2900–3000 m (1190 mm) and maximum rainfall (1320 mm) was observed in the two altitudinal bands (3300–3400, 3500–3600). Along CTL elevations, minimum rainfall was 712 mm in highest altitudinal band (4200–4300 m) and maximum rainfall (2020 mm) was in lowest altitudinal band (2500–2600 m) (Fig. 12.4b). This observation supports the argument that inner Himalayan ranges are drier than the outer Himalayan ranges and rainfall decreases with increase in elevation.

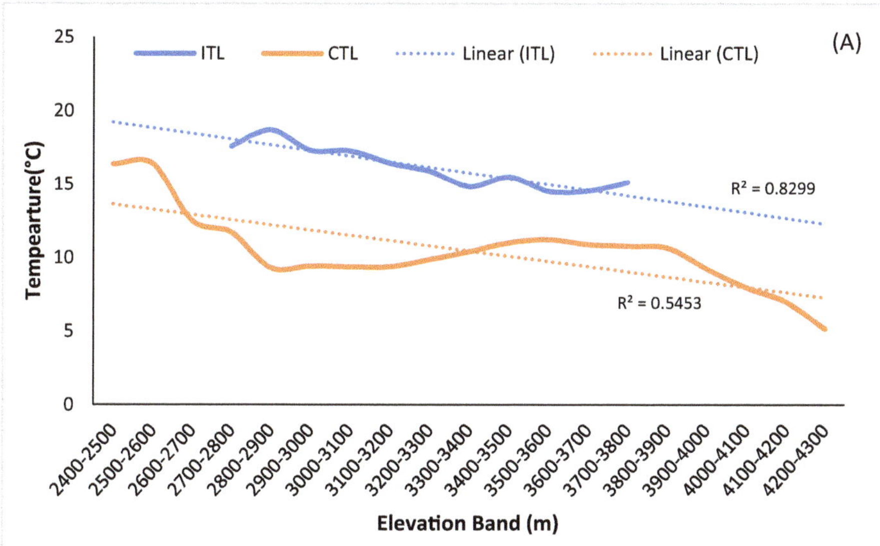

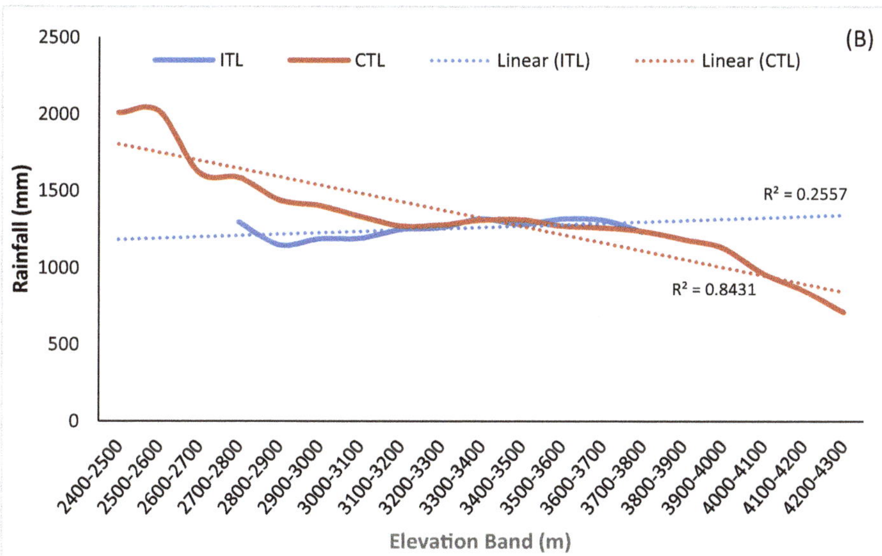

Fig. 12.4 Annual mean temperature (**a**) and annual precipitation (**b**) along elevational gradient of timberline altitudes (year 2015). Values averaged for each 100 m elevation band

12.4.3 Annual Mean Temperature and Total Rainfall at Timberline Altitudes—Past (Year 1976) and Rate of Changes

For the year 1976, annual mean temperature and total annual rainfall in different altitudinal zones in two spatially distinct timberlines (ITL/CTL) are presented in Fig. 12.5a, b. Similar to present time (year 2015), ITL altitudes (away from the permanent snow) were warmer (range of annual mean temperature from 6.83 to 19.92 °C; average 15.09 °C ± 3.13) than the CTL altitudes (close to permanent snow) where the annual mean temperature between different elevation bands of timberline elevation ranged between −1.7 and 17.0 °C (average 9.8 °C ± 3.4). Annual mean temperature at timberline elevations was decreasing with an increasing altitude (Fig. 12.5a) but was more prominent in ITL elevations ($r^2 = 0.8$) than the CTL elevations ($r^2 = 0.5$).

It was observed that annual mean temperature at timberline elevations is increasing at different rates in different spatial arrangements of timberline. For example, at ITL elevations the rate of increase was 0.15 °C decade^{-1} while at CTL altitudes the rate was slightly higher 0.22 °C decade^{-1}. This observation indicates that inner Himalayan ranges warming faster than the outer Himalayan ranges.

Total annual rainfall in different elevation bands of ITL varied from 782.78 to 2678.68 mm (average 1349.44 ± 318.76 mm) while in CTL elevations it ranged from 484 to 2988 mm (average 1236 ± 488 mm) (Fig. 12.5b). A slight increase in rainfall along the timberline elevations was observed in outer Himalayan ranges (ITL locations) while it was decreasing in inner Himalayan ranges (CTL locations, $r^2 = 0.77$). Similar to air temperature, it was observed that the rate of rainfall is increased from 1976 and was more in inner Himalayan timberline altitudes CTL, 13 mm decade^{-1} but was decreasing in outer Himalayan timberline elevations (ITL = 19.75 mm decade^{-1}).

12.4.4 Changes in Timberline Elevations Between 1976 and 2015 and Corresponding Changes in Climate at Different Timberline Locations

Shift in timberline has been attributed to the climate change/anthropogenic activities at high altitudes of the Himalaya. Timberline dynamics between studied years measured as changes through shift, i.e., upward/downward movement or no changes, i.e., stationary. Therefore, changes in climatic parameters in timberline segments of 1976 were analyzed for (1) changed (timberline segment moved upward) and (2) stationary (no changes in altitudes of timberline segments since 1976).

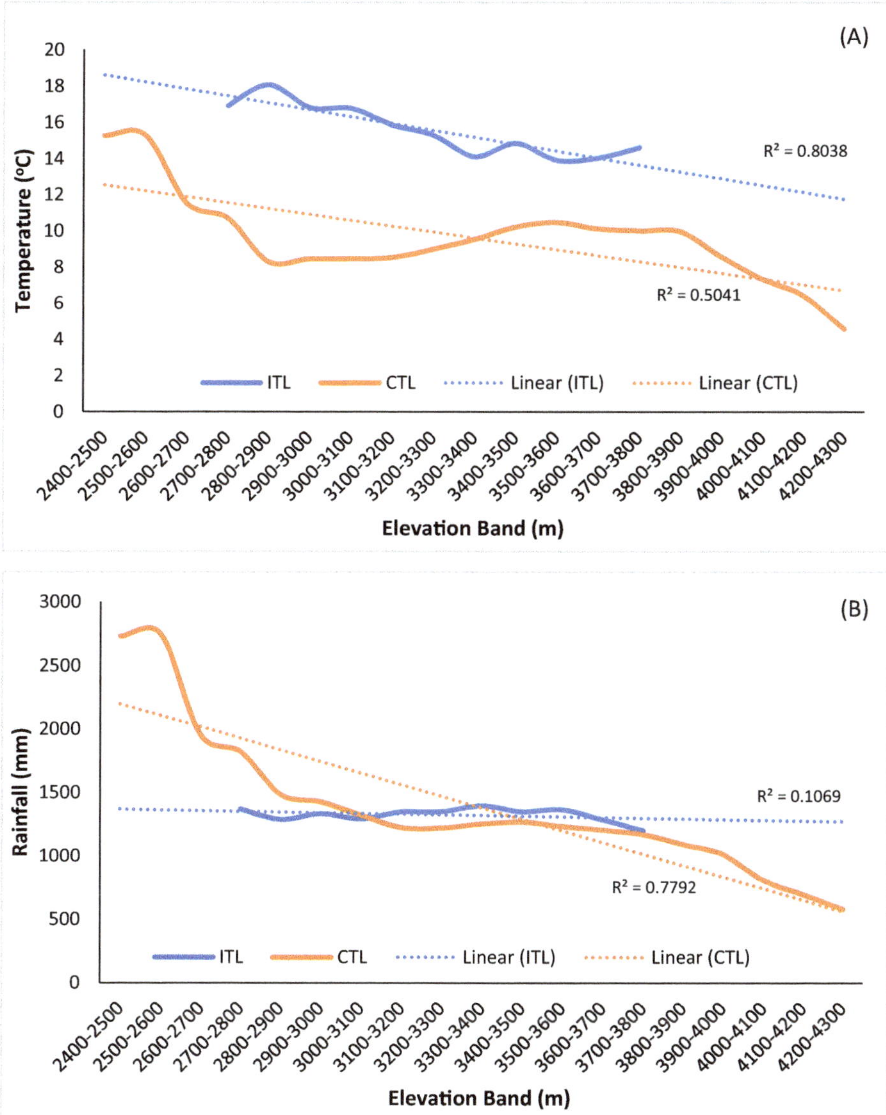

Fig. 12.5 Annual mean temperature (**a**) and annual precipitation (**b**) along elevational gradient of timberline altitudes (year 2015). Values averaged for each 100 m elevation band

12.4.4.1 No Change in Timberline Altitude Between the Years

At stationary timberline (no change) in different elevation bands of ITL, less than 1 °C increase (range 0.4–0.7 °C) in annual mean temperature (average 0.5 °C) was observed whereas in CTL, it was 1 °C (higher range 0.6–1.1 °C) in almost four decades. Difference in temperature between two timeframes shows a decreasing

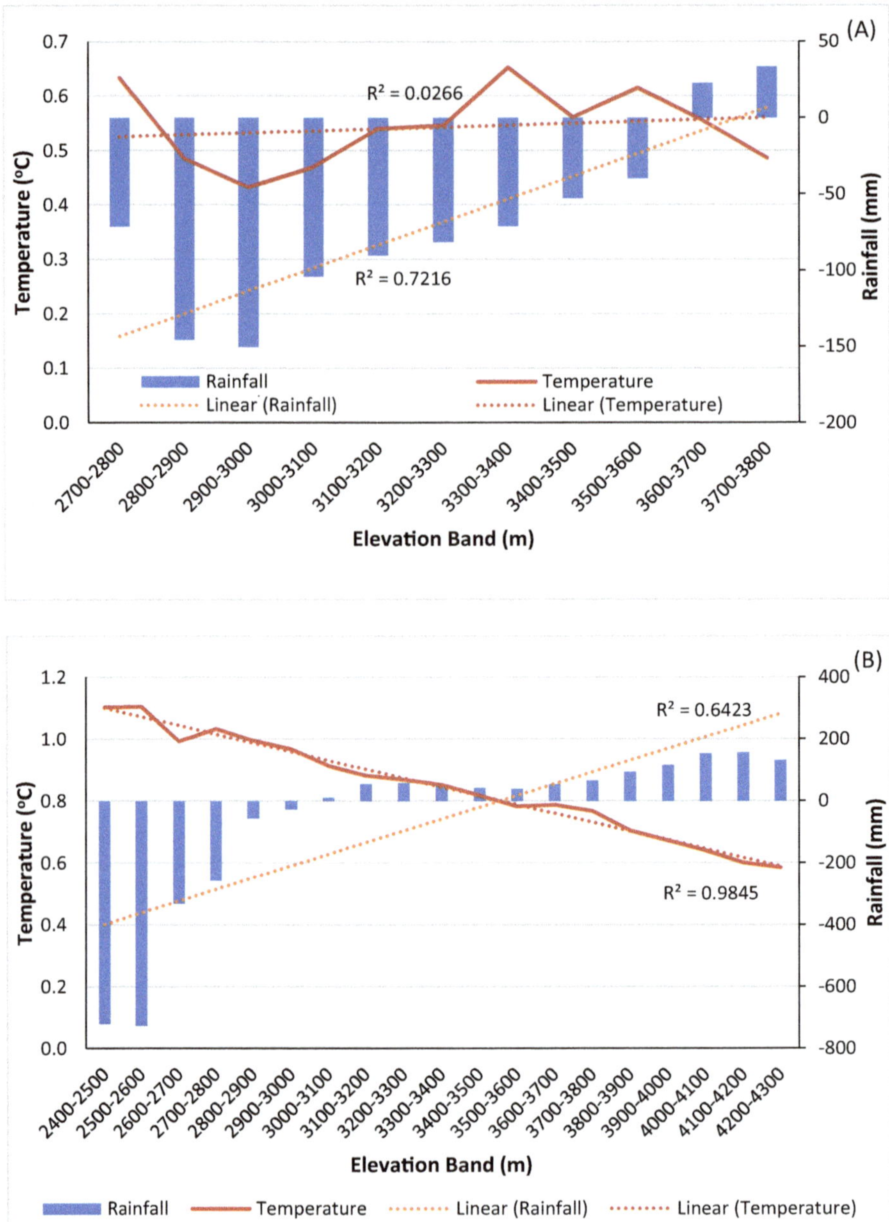

Fig. 12.6 Difference between 1976 and 2015 in annual mean temperature and total annual rainfall at altitude of stationary timberline of (**a**) ITL, (**b**) CTL

trend toward high altitudes of ITL ($r^2 = 0.9$; Fig. 12.6a); however, this relationship was weak in the case of CTL (Fig. 12.6b). This could be due to rugged topography in inner Himalayan ranges where timberline elevations are dispersed in valleys and ridges.

In this time period (38 years), decrease in rainfall was observed below 3000 m altitude of stationary timberline of ITL and below 3600 m altitude of stationary CTL (Fig. 12.6a, b) while above these mentioned altitudes, respectively, it increases in both types of timberlines. In stationary ITL, this difference ranged between −727 and 157 mm (average −68 mm across all elevation bands) and in stationary CTL, this range was from −126 to −7 mm (average −57 mm across all elevation bands). Overall, an increasing trend toward higher elevations of ITL ($r^2 = 0.6$) and CTL (weak relationship) was observed (Fig. 12.6a, b). Thus, it appears that both temperature and rainfall are important in making large part of timberline stationary.

12.4.4.2 Upward Shift in ITL Locations

The annual mean temperature of the upward moved ITL (present position in 2015) varied from 11.1 to 17 °C with an average of 14.6 °C ± 2.3, whereas in the 1976 (past position) this range varied from 9.9 to 16.1 °C with an average of 13.4 °C ± 2.4. It was observed that average temperature difference from 1976 to 2015 year was 1.2 °C ± 0.4. Same as the temperature, the annual precipitation (present position in 2015) ranged from 1268 to 2015.9 mm (average 1491.8 mm ± 222.3), whereas this range in the 1976 (past position) was between 1266.8 and 2678.7 mm (average 1636.2 mm ± 442.4). Average difference in annual rainfall between 1976 and 2015 was 132.8 mm ± 199.5 (Fig. 12.7).

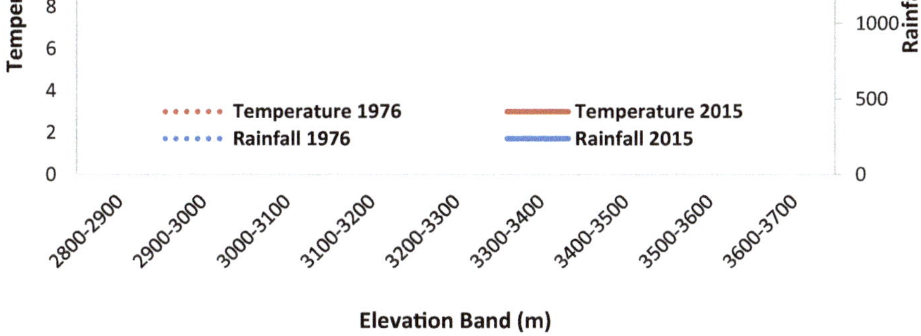

Fig. 12.7 Annual mean temperature and total annual rainfall between 1977 and 2015 at timberline locations shifted to upper altitudes ITL

The rate of changes in the last four decades (1976–2015) for climatic parameters at upward moved ITL locations were +0.32 °C decade^{-1} (annual mean temperature) and −38 mm decade^{-1} (annual precipitation).

12.4.4.3 Upward Shift in CTL Locations

The annual mean temperature at upward moved CTL locations (present position in 2015) varied from 1.5 to 17.3 °C (average 10.0 °C ± 3.2) whereas this range was from 0.4 to 16.6 °C (average of 9.1 °C ± 3.3) in the year 1976 (past position). It was observed that average temperature difference from 1976 to 2015 year was 0.8 °C ± 0.3 in CTL elevations (Fig. 12.8). Similar to temperature, annual precipitation at present elevations (2015) ranged from 740 to 1808 mm (average 1299 mm ± 218), whereas in the past (1976) this range was from 517 to 2412 mm (average 1219 mm ± 351). Since 1976, average difference in rainfall at present CTL locations (2015) was 83 mm ± 55.

The rate of changes in the last four decades (1976–2015) for climatic parameters at upward moved CTL locations were +0.23 °C decade^{-1} (annual mean temperature) and +221 mm decade^{-1} (annual precipitation).

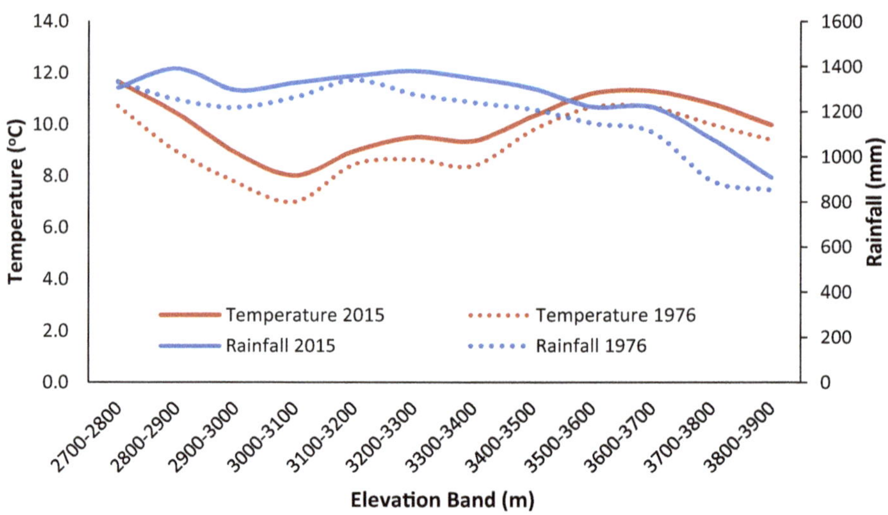

Fig. 12.8 Annual mean temperature and total annual rainfall between 1976 and 2015 at timberline locations shifted to upper altitudes CTL

12.5 Discussion

12.5.1 Comparison with Previous Studies

In the eastern Himalaya, Kumar (2012) concluded that 1 °C increase in temperature led to 250–300 m elevation shift of tree line species. Changes (shift) in Himalayan timberline/tree line have been reported by various workers (Singh et al. 2012, 2018, 2021; Mohapatra et al. 2019; Bharti et al. 2011; Sah et al. 2021). For entire Indian Himalayan region, overall vertical shift in the alpine tree line ecotone over four decades (1976–2014) has been reported (381 ± 73 m at a rate of c. 95 m decade^{-1}, Singh et al. 2021). For different geographical regions, this mean upward shift of tree line was as following—North-western Himalaya (Jammu & Kashmir UT) = 441 ± 71 (@ 116 m decade^{-1}), Western Himalaya (Himachal Pradesh State) = 301 ± 77 (@ 79 m decade^{-1}), Indian Central Himalaya (Uttarakhand state) = 388 + 60 m (@ 114 m decade^{-1}), Eastern Himalayan states of Sikkim = 301 ± 66 m (@ 81 m decade^{-1}), and Arunachal Pradesh = 452 + 74 m (@ 113 m decade^{-1}) in different time frames. Singh et al. (2021) also reported a fundamental niche shift of c. 109.9 m decade^{-1} for Sikkim tree line in response to climate change scenario (IPCC5 RCP8.5 for year 2061–2080). Sah et al. (2021) reported first time a downward shift in timberline of Sikkim Himalaya, the mean upward shift was 100 m ± 89 (@ 2.71 m year^{-1}) and mean downward shift was 56 m ± 54 (@ 1.52 m year^{-1}).

For a part of Uttarakhand and Himachal Pradesh states, Rai et al. (2013, 2019) concluded that 81.5 km^2 timberline ecotone area changes in the three decades (1980–2010). The magnitude of change was higher (44.1 km^2) than the preceding decades in the last 10 years. There was also an increase in sub-alpine forest and alpine scrub. From a dendrochronological study [Dubey et al. (2003), an upward shift of tree line (dominated by blue pine, *Pinus wallichiana*)] but with lesser magnitude was observed in Himachal Pradesh, on a long-time scale (1860–2000) for Himalayan pine with different rates of upward shift (19 m decade^{-1} and 14 m decade^{-1}). The present study also observes a close value (@ 3.8 m per year) than the other studies and highlights the importance of baseline data and remote sensing methods. A repeat photography study between 1923 and 2003 in Yunnan region of China shows an increasing trend of 8–15 m decade^{-1} (Baker and Moseley 2007).

Alpine tree line dynamics is an indicator of climate change mainly due to long-term changes in temperature regime (Zhang et al. 2009). Timberline/tree line is climate-sensitive transition zone, especially to the temperature, that are expected to respond to climate warming by advancing beyond their current position (Harsch et al. 2009). In the present study, the variation of climatic parameters (annual mean temperature and annual precipitation) was analyzed for spatially different timberline elevations (ITL and CTL) and their relationship with timberline elevations indicates that both the temperature and precipitation are important climatic parameters that govern timberline dynamics. The entire Hindu Kush Himalayan (HKH) region is experiencing (Sabin et al. 2020; 1901–2014) an increase in annual mean surface–air–temperature (rate ~0.1 °C per decade) and a faster rate after 1951 (~0.2 °C per decade). Rate of increase, per decade, in whole HKH region for mean temperature was 0.104 °C, for mean minimum temperature was 0.176 °C, and mean maximum temperature was 0.077 °C (Ren et al. 2017). In Nepal Himalaya, mean annual maximum temperature increase (1977–1994) in different ranges varies from >0.06 °C year^{-1} (northern and central/western mountain ranges of Middle Mountains) to >0.12 °C year^{-1} in some pockets, however rate of increase in annual regional mean for Himalaya is 0.09 °C year^{-1} ($p > 0.01$; Shrestha et al. 1999).

The high-altitude climatic tree lines are associated with a seasonal mean ground temperature; however, this is higher (7–8 °C) in the temperate and mediterranean zone tree lines, and is lower in subarctic Boreal zone (6–7 °C) and Equatorial tree lines (5–6 °C; Körner and Paulsen 2004). In the present study, the mean temperature at timberline elevation of western Himalayan region was 10.3 °C (inner Himalayan ranges) and 15.3 °C (outer Himalayan ranges).

In central and north-western Tibetan plateau, annual precipitation amounts to about 50–300 mm (Liao 1990; Wang et al. 2011; Zhang et al. 2002; Zheng and Li 1990); therefore, the highest tree line in the Northern Hemisphere occurs in the south-eastern region of the Tibetan plateau (Yao and Zhang 2015). In a mountain ecosystem, rainfall is also an important factor that governs the growth and phenology of tree line species. It is evident from tree line distribution in Tibetan plateau where at least 500 mm of annual precipitation is required for tree growth (Hou 1982a, b). Odland (2015) studied the timberline of *Betula* in central Scandinavian mountains, where latitude explained approximately 71% of the timberline variation but mountain height also contributed significantly with an explanatory rate of nearly 20%, which means if a mountain height is low, an increase in temperature may result in no or small timberline uplift. The mass elevation effect (mountain height) on both air temperatures and timberline elevation for the Tibetan Plateau also studied (Yao and Zhang, 2015) which showed that the mass elevation effect of the central high mountain areas pushed the 10 °C isotherm upward in the warmest month up to elevations of 4600–4700 m, resulting shift the tree line altitude to 500–1000 m higher than along the eastern edge where the mountains reached only 1000 m asl. Therefore, the mass elevation effect contributes to the highest tree line in the Northern Hemisphere, which is present on the south-eastern Tibetan Plateau.

Ren et al. (2017) reported that in HKH region, the precipitation shows a slight decreasing trend in 114 years but it increased significantly (@ 5.28% per decade) between 1961 and mid-1980s. Pre-monsoon warming accompanied by rain is likely to increase seedling recruitments and upslope rise of tree lines in the Himalaya (Singh et al. 2019). Kumar (2012) concluded that the rate was higher than the rate observed between 1957 and 2005 (49.6 mm decade^{-1} annual mean precipitation) at lower altitude of 1765 m. Singh et al. (2018) observed decreasing trend (206.5 mm decade^{-1}) in annual cumulative precipitation from the period of 1977–2013 using 0.25° gridded data of IMD. Thus, Himalayan studies require more robust long-term data to study the response and position of timberline to climate change. European Alps are also experiencing a temperature increase that is stronger than the global mean of about +0.7 °C for the last century (Hansen et al. 2006). The position of the tree line is influenced by complex interactions with past and current climates and various disturbances (Hofgaard 1997; Körner 1999; Oksanen et al. 1995). The rapid warming of 3–4 °C at the Late Glacial/Holocene transition (11,600 cal. BP) caused significant altitudinal displacements of alpine species that were additionally affected by the rapid upward movement of trees and shrubs (Berthel et al. 2012).

12.6 Conclusion

The study concludes that in the Western Himalaya majority of timberline (94.6%) remained stationery and rest (5.4%) shifted upwards in last 38 years (1976–2015). It observed that the mean upward shift in the timberline is 145 m approximately 38 m per decade (@ 3.8 m per year). As timberline altitude increases, precipitation and temperature both decrease. These findings suggested that the timberline ecotone dynamics could refer to as the sensitive area under climate change, and climatic parameters (temperature and precipitation) are main drivers of the natural shift in timberline. The present study also observed that at timberline elevations in outer Himalayan ranges (away from permanent snow) have higher temperature and precipitation than the timberline elevations of inner Himalayan ranges, hence outer Himalayan timberline is more susceptible to the impacts of climate change. Climatic conditions may erode timberline habitats at some locations bringing timberline elevation down. Besides being an effective indicator of climate warming, structural and functional changes in timberline have implications to decline in biodiversity, wildlife habitats, provisioning of ecosystem services, such as medicinal plants, grazing sites for migratory livestock, recreational use, etc.

Acknowledgements Authors are thankful to Prof. S. P. Singh for guidance to conduct this research. Director, G. B. Pant National Institute of Himalayan Environment (NIHE), Almora for providing necessary facilities. IIIT, Hyderabad for computational and climate data support. Financial grant for this study was supported by National Mission on Himalayan Studies, Ministry of Environment, Forest and Climate Change, Govt. of India.

References

Aryal A, Hipkins J, Ji W, Raubeinhimer D, Brunton D (2012) Distribution and diet of brown bear in Annapurna conservation area, Nepal. Ursus 23(2):231–236

Baker BB, Moseley RK (2007) Advancing treeline and retreating glaciers: implications for conservation in Yunnan, P.R. China. Arctic, Antarctic, and Alpine Res 39(2):200–209

Berthel N, Schwörer C, Tinner W (2012) Impact of Holocene climate changes on alpine and treeline vegetation at Sanetsch Pass, Bernese Alps, Switzerland. Rev Palaeobot Palynol 174:91–100

Bharti RR, Rai ID, Adhikari BS et al (2011) Timberline change detection using topographic map and satellite imagery: a critique. Trop Ecol 52:133–137

Bhutiyani MR, Kale VS, Pawar NJ (2010) Climate change and the precipitation variations in the northwestern Himalaya: 1866–2006. Int J Climatol 30(4):535–548

Bolch T, Kutusov S, Li X (2012) Updated GLIMS glacier database for the Tian Shan. National snow and ice data center, Boulder

Butler DR, Malanson GP, Walsh SJ, Fagre DB (2009) The changing alpine treeline: the example of glacier National park, MT, USA. In: Developments in earth surface processes, 12. Elsevier, United Kingdom (UK)

Dilip PA, Rameshbabu K, Ashok KP et al (2014) Bilinear interpolation image scaling processor for VLSI architecture. Int J Reconfigurable Embedded Syst 3(3)

Dubey B, Yadav RR, Singh J, Chaturvedi R (2003) Upward shift of Himalayan pine in western Himalaya. India. Curr Sci 85(8):1135–1136

Field CB, Barros VR (2014) Climate change 2014: impacts, adaptation, and vulnerability. Part A: Global and sectoral aspects. In: Intergovernmental panel on climate change (IPCC). Cambridge University Press, Cambridge

Forman RT (1964) Growth under controlled conditions to explain the hierarchical distributions of a moss, *Tetraphis pellucida*. Ecol Monogr 34:2–25

FSI (2019) Indian state of forest report. Forest survey of India

Gerlitz L, Conrad O, Thomas A, Böhner J (2014) Warming patterns over the Tibetan Plateau and adjacent lowlands derived from elevation-and bias-corrected ERA-Interim data. Clim Res 58(3): 235–246

Grabherr G, Gottfried M, Pauli H (1994) Climate effects on mountain plants. Nature 369:448

Grace J, Berninger F, Nagy L (2002) Impacts of climate change on the tree line. Ann Bot 90(4): 537–544

Griggs RF (1937) Timberlines as indicators of climatic trends. Science 85(2202):251–255

Hansen J, Sato M, Ruedy R, Lo K, Lea DW, Medina-Elizade M (2006) Global temperature change. Proc Natl Acad Sci 103(39):14288–14293

Harsch MA, Hulme PE, McGlone MS, Duncan RP (2009) Are treelines advancing? A global meta-analysis of treeline response to climate warming. Ecol Lett 12(10):1040–1049

Hasson S, Gerlitz L, Schickhoff U, Scholten T, Böhner J (2016) Recent climate change in high Asia. In: Singh RB, Schickhoff U, Mal S (eds) Climate change, glacier response, and vegetation dynamics in the Himalaya. Springer, Cham

Hofgaard A (1997) Inter-relationships between treeline position, species diversity, land use and climate change in the central Scandes Mountains of Norway. Glob Ecol Biogeogr Lett 6:419–429

Holtmeier FK, Broll G (2005) Sensitivity and response of northern hemisphere altitudinal and polar treelines to environmental change at landscape and local scales. Glob Ecol Biogeogr 14(5): 395–410

Hou XY (1982a) China vegetation geography and dominant plant composition. Science Press, Beijing

Hou XY (1982b) Vegetational geography and chemical components of dominant plants in China (in Chinese). Science Press, Beijing

IPCC (2013) Climate change 2013: the physical science basis. In: Stocker TF, Qin D, Plattner GK, Tignor M, Allen SK, Boschung J, Nauels A, Xia Y, Bex V, Midgley PM (eds) Contribution of working group I to the fifth assessment report of the intergovernmental panel on climate change. Cambridge University Press, Cambridge

IPCC (2014) Climate change 2013: impacts, adaptation, and vulnerability. Part A: global and sectoral aspects. In: Field CB, Barros VR, Dokken DJ, Mach KJ, Mastrandrea MD, Bilir TE, Chatterjee M, Ebi KL, Estrada YO, Genova RC, Girma B, Kissel ES, Levy AN, MacCracken S, Mastrandrea PR, White LL (eds) Contribution of working group II to the fifth assessment report of the intergovernmental panel on climate change. Cambridge University Press, Cambridge

Joshi R, Sambhav K, Singh SP (2018) Near-surface temperature lapse rate for treeline environment in western Himalaya and possible impacts on ecotone vegetation. Trop Ecol 59(2):197–209

Jump AS, Hunt JM, Peñuelas J (2007) Climate relationships of growth and establishment across the altitudinal range of *Fagus sylvatica* in the Montseny Mountains, Northeast Spain. Ecoscience 14(4):507–518

Körner C (1998) A re-assessment of high elevation treeline positions and their explanation. Oecologia 115:445–459
Körner C (1999) Alpine plant life: functional plant ecology of high mountain ecosystems. Springer, Berlin
Körner C (2003) Alpine plant life. Springer, Berlin
Körner C (2012) Alpine treelines: functional ecology of the global high elevation tree limits. Springer Science & Business Media, Berlin
Körner C, Paulsen J (2004) A world-wide study of high altitude treeline temperatures. J Biogeogr 31(5):713–732
Kumar P (2012) Assessment of impact of climate change on Rhododendrons in Sikkim Himalayas using Maxent modelling: limitations and challenges. Biodivers Conserv 21(5):1251–1266
Latwal A, Sah P, Sharma S (2018) A cartographic representation of a timberline, treeline and wood vegetation around a Central Himalayan summit using remote sensing method. Trop Ecol 59(2): 177–187
Latwal A, Sah P, Sharma S et al (2021) Ecology of treeline ecotone. In: Relationship between timberline elevation and climate in Sikkim Himalaya. Springer Nature, Berlin
Liao K (1990) The atlas of the Tibetan plateau. Science Press, Beijing
Liang E, Wang Y, Xu Y, Liu B, Shao X (2010) Growth variation in *Abies georgei* var. *smithii* along altitudinal gradients in the Sygera Mountains, southeastern Tibetan Plateau. Trees 24(2): 363–373
MacDonald GM, Szeicz JM, Claricoates J, Dale KA (1998) Response of the Central Canadian treeline to recent climatic changes. Ann Assoc Am Geogr 88(2):183–208
Malanson GP (1997) Effects of feedbacks and seed rain on ecotone patterns. Landsc Ecol 12:27–38
Mayor JR, Sanders NJ, Classen AT, Bardgett RD, Clément JC, Fajardo A, Lavorel S, Sundqvist MK, Bahn M, Chisholm C, Cieraad E (2017) Elevation alters ecosystem properties across temperate treelines globally. Nature 542(7639):91–95
Ménégoz M, Gallée H, Jacobi HW (2013) Precipitation and snow cover in the Himalaya: from reanalysis to regional climate simulations. Hydrol Earth Syst Sci 17:3921–3936. https://doi.org/10.5194/hess-17-3921-2013
Mohapatra J (2015) The changing face of the alpine ecosystem in the Himalaya. ENVIS Newsl Himal Ecol 12(2):9
Mohapatra J, Singh CP, Tripathi OP, Pandya HA (2019) Remote sensing of alpine treeline ecotone dynamics and phenology in Arunachal Pradesh Himalaya. Int J Remote Sens 40(20): 7986–8009. https://doi.org/10.1080/01431161.2019.1608383
Nedlo JE, Martin TA, Vose JM, Teskey RO (2009) Growing season temperatures limit growth of loblolly pine (*Pinus taeda* L.) seedlings across a wide geographic transect. Trees 23(4):751–759
Odland A (2015) Effect of latitude and mountain height on the timberline (Betula pubescens ssp. czerpanovii) elevation along the central Scandinavian mountain range. Fennia Int J Geogr 193 (2):260–270
Oksanen L, Moen J, Helle T (1995) Timberline patterns in northernmost Fennoscandia. Relative importance of climate and grazing. Acta Bot Fenn 153:93–105
Pandey P, Ali SN, Ramanathan AL, Venkataraman G (2017) Regional representation of glaciers in Chandra basin region, western Himalaya. India. Geosci Front 8(4):841–850
Qi Z, Liu H, Wu X, Hao Q (2015) Climate-driven speedup of alpine treeline forest growth in the Tianshan Mountains, Northwestern China. Glob Change Biol 21(2):816–826
Rai ID, Bharti RR, Adhikari BS, Rawat GS (2013) Structure and functioning of timberline vegetation in the western Himalaya: A. High-altitude rangelands and their interfaces in the Hindu Kush Himalayas, p 91
Rai ID, Singh G, Pandey G, Rawat GS (2019) Ecology of treeline vegetation in western himalaya: anthropogenic and climatic influences. In: Garkoti SC et al (eds) Tropical ecosystems: structure, functions and challenges in the face of global change. Springer Nature Singapore Pte Ltd, pp 173–212. https://doi.org/10.1007/978-981-13-8249-9_9

Ren Y-Y, Ren GY, Sun XB et al (2017) Observed changes in surface air temperature and precipitation in the Hindu Kush Himalayan region over the last 100-plus years. Adv Clim Chang Res 8(3):148–156
Rogora M, Frate L, Carranza ML, Freppaz M, Stanisci A, Bertani I, Bottarin R, Brambilla A, Canullo R, Carbognani M, Cerrato C (2018) Assessment of climate change effects on mountain ecosystems through a cross-site analysis in the Alps and Apennines. Sci Total Environ 624: 1429–1442
Sabin TP, Krishnan R, Vellore R et al (2020) Climate change over the Himalayas. In: Krishnan R, Sanjay J, Gnanaseelan C, Mujumdar M, Kulkarni A, Chakraborty S (eds) Assessment of climate change over the indian region. Springer, Singapore, pp 207–222. https://doi.org/10.1007/978-981-15-4327-2_11
Sah P, Sharma S (2018) Topographical characterisation of high altitude timberline in the Indian Central Himalayan Region. Trop Ecol 59(2):187–196
Sah P, Sharma S (2022) Geo-spatial attributes of Western Himalayan Timberline over Himachal Pradesh. In: Handbook of Himalayan ecosystems and sustainability. CRC Press, Boca Raton
Sah P, Latwal A, Sharma S (2021) Ecology of treeline ecotone. Challenges of timberline mapping in the Himalaya: a case study of Sikkim Himalaya. Springer Nature, Berlin
Schickhoff U, Bobrowski M, Böhner J, Bürzle B, Chaudhary RP, Gerlitz L, Heyken H, Lange J, Müller M, Scholten T, Schwab N (2015) Do Himalayan treelines respond to recent climate change? An evaluation of sensitivity indicators. Earth Syst Dyn 6:245–265
Schulla J, Jasper K (2007) Model description waSiM-ETH. Institute for atmospheric and climate science, Swiss federal institute of technology, Zürich
Sharma E, Tsering K (2009) Climate change in the Himalayas: the vulnerability of biodiversity. Sustain Mt Dev 55:10–12
Shrestha AB, Aryal R (2011) Climate change in Nepal and its impact on Himalayan glaciers. Reg Environ Change 11(1):65–77
Shrestha UB, Gautam S, Bawa KS (2012) Widespread climate change in the Himalayas and associated changes in local ecosystems. PLoS One 7(5):e36741
Shrestha AB, Wake CP, Mayewski PA et al (1999) Maximum temperature trends in the himalaya and its vicinity: an analysis based on temperature records from Nepal for the Period 1971–94. J Clim 12:2775–2787
Singh CP, Panigrahy S, Thapliyal A et al (2012) Monitoring the alpine treeline shift in parts of the Indian Himalayas using remote sensing. Curr Sci 102:559–562
Singh CP, Mohapatra J, Pandya HA, Gajmer B, Sharma N, Shrestha DG (2018) Evaluating changes in treeline position and land surface phenology in Sikkim Himalaya. Geocarto Int 35:453–469. https://doi.org/10.1080/10106049.2018.1524513
Singh SP, Sharma S, Dhyani PP (2019) Himalayan arc and treeline: distribution, climate change responses and ecosystem properties. Biodivers Conserv 28(8–9):1997–2016
Singh CP, Mohapatra J, Mathew JR et al (2021) Long-term observation and modeling on the distribution and patterns of alpine treeline ecotone in Indian Himalaya. J Geomat 15(1):68–84. ISSN: 0976-1330
Takahashi K, Kobori H, Seino T (2011) Effects of temperature and light conditions on growth of current-year seedlings of warm-temperate evergreen tree species and cool-temperate deciduous tree species. In: Casalegno S (ed) Global warming impacts: case studies on the economy, human health, and on urban and natural environments. InTech, Rijeka
Telwala Y, Brook BW, Manish K, Pandit MK (2013) Climate-induced elevational range shifts and increase in plant species richness in a Himalayan biodiversity epicentre. PLoS One 8(2):e57103
Walck JL, Hidayati SN, Dixon KW, Thompson KEN, Poschlod P (2011) Climate change and plant regeneration from seed. Glob Change Biol 17(6):2145–2161
Walther GR, Beißner S, Burga CA (2005) Trends in the upward shift of alpine plants. J Veg Sci 16: 541–548
Wang C, Zhou S, Tang X et al (2011) Temporal and spatial distribution of heavy precipitation over Tibetan plateau in recent 48 years. Sci GeogrSin 31(4):470–477

Wielgolaski FE, Hofgaard A, Holtmeier FK (2017) Sensitivity to environmental change of the treeline ecotone and its associated biodiversity in European mountains. Clim Res 73(1–2): 151–166

WWW1, Maeda M, Yasutomi N, Yatagai A (2020) National center for atmospheric research staff (2020) the climate data guide: aphrodite: asian precipitation—highly-resolved observational data integration towards evaluation of water resources. https://climatedataguide.ucar.edu/climate-data/aphrodite-asian-precipitation-highly-resolved-observational-data-integrationtowards

Xu J, Grumbine RE, Shrestha A, Eriksson M, Yang X, Wang Y, Wilkes A (2009) The melting Himalayas: cascading effects of climate change on water, biodiversity, and livelihoods. Conserv Biol 23(3):520–530

Yao Y, Zhang B (2015) The mass elevation effect of the Tibetan plateau and its implications for Alpine treelines. Int J Climatol 35(8):1833–1846

Yatagai A, Kamiguchi K, Arakawa O, Hamada A, Yasutomi N, Kitoh A (2012) APHRODITE: constructing a long-term daily gridded precipitation dataset for Asia based on a dense network of rain gauges. Bull Am Meteorol Soc 93(9):1401–1415

Zhang BP, Chen XD, Li BL et al (2002) Biodiversity and conservation in the Tibetan plateau. J Geogr Sci 12(2):135–143

Zhang Y, Xu M, Adams J, Wang X (2009) Can landsat imagery detect tree line dynamics? Int J Remote Sens 30(5):1327–1340

Zheng D, Li BY (1990) Evolution and differentiation of the natural environment of the Qinghai-Tibet plateau. Geogr Res 9(2):1–10

Chapter 13
Design of Humanitarian Logistics Network Using Imperialist Competitive Algorithm

Ashish Kumar Kaushal and Vandna Devi

13.1 Introduction

High population growth, global trend in urbanism (Suarez 2009), climate change (Kovács and Spens 2011), and global warming (IPCC 2001) have resulted in tremendous increase in the number, rate, and impact of natural disasters. Since the 1950s, natural disasters like earthquakes, floods, tsunamis have seen an exponential growth and have resulted in significant loss of human lives and capital (Guha-Sapir et al. 2012; Özdamar and Ertem 2015). According to Natural Disaster Database, 13,758 natural disasters have occurred between 1900 and 2015 resulting in loss of 32,583,037 lives and causing total economic loss 2,799,377,224 ('000$) (EM-DAT, 2015). The Haiti earthquake of 2010 demonstrated the potential severity of events following a natural disaster. It killed 222,570 people and affected a total of 3.9 million others. The disaster caused an estimated US$ 8.0 billion worth of damages and led to the collapse of around 70% of buildings and homes (Guha-Sapir et al. 2012). In India, 24 flash floods have occurred from 1990 to 2016 and have resulted in deaths of more than 7471 people, affected 23,446,633 people and have caused an economic loss of 454,200 ('000$) (EM-DAT, 2016). Estimation of the forecasts is that over the next 50 years, there will be a five times increment in the amount and intensity of natural and man-made disasters (Thomas and Kopczak 2005). These figures substantiate the importance of providing aid of the appropriate kind and

A. K. Kaushal (✉)
Jindal Global Business School, O P Jindal Global University, Sonipat, Haryana, India

Department of Management Studies, Indian Institute of Technology Delhi, New Delhi, India

V. Devi
Gandhinagar, India

Swami Vivekanand Government College, Ghumarwin, Himanchal Pradesh, India

B. Pathak, R. S. Dubey (eds.), *Climate Change and Urban Environment Sustainability*, Disaster Resilience and Green Growth,
https://doi.org/10.1007/978-981-19-7618-6_13

amount to those affected in the most efficient and effective way possible, to prevent loss and suffering (Christopher and Tatham 2011). Due to the high magnitude and economical impact of disasters, advances in the management of disaster operations are imperative and will contribute to an improvement in readiness, increase response speed, ease recovery, and provide institutional learning over time (Altay and Green 2006; Christopher and Tatham 2011; de la Torre et al. 2012; Thomas and Kopczak 2005).

Humanitarian logistics refers to the process of transferring and storing relief materials and information from the source to the destination in a systematic and efficient manner to meet the necessities of the affected people (Sheu 2007; Thomas and Mizusjima 2005). It involves three stages: (a) planning the route; (b) implementing the plan; and (c) reviewing the plan (Thomas and Mizusjima 2005). Humanitarian operations comprised of the duration of a disaster together with preparedness, response, and recovery phases (Altay and Green 2006). Thus, the capability to coordinate efficient and effective humanitarian operations is a vital component of a disaster relief process (Kovács and Spens 2011).

Modeling of commercial and humanitarian logistics is quite different due to different types of decision variables, constraints, inventories, and objective functions involved (Sheu 2007; Whybark 2007; Beamon and Balcik 2008; Apte 2010; Holguín-Veras et al. 2012). Objective functions for commercial logistics are often restricted to classical economic and/or logistics measures such as minimization of cost, time, and inventory levels maximization of profit, service level (Beamon and Balcik 2008; Holguín-Veras et al. 2012). Objective functions in humanitarian supply chains are associated with humanitarian principles (Beamon and Balcik 2008; Holguín-Veras et al. 2013) such as (1) *the human need principle*, in which the main concern is off rescuing human lives and minimizing sufferings of the mankind to the maximum amount possible and (2) *the impartiality principle*, which necessitates that the execution of any deed should be based only on necessity, with no biases between populations (Hilhorst 2005). The two principles (i.e., human need and impartiality) should be contemplated while designing humanitarian supply chains (de la Torre et al. 2012; Holguín-Veras et al. 2013).

The absence of sufficient preparedness measures in major flood prone areas has increased the possibility of destruction and losses caused by floods. Catastrophic effects of disasters are unavoidable but can be scaled down by proactive approaches and designing apt preparedness plans (Kunz et al. 2014; Salmerón and Apte 2010; Duran et al. 2011). As communities are not able to cope up with the aftermath of a disaster, response activities within the first 72 h play an imperative role in mitigating the effects of disaster (Salmerón and Apte 2010). Thus, humanitarian logistics demands a reconsideration of its planning process due to increasing intricacy, an increase in uncertain parameters, lack of coordination, and resources scarcity.

Uncertainty and unpredictability characterize the surroundings of disasters, and the logistical activities are to be performed in rapidly changing environments. Knowledge of timing and location of events is substantially difficult, if not impossible, to predict with any significant degree of certainty (Christopher and Tatham 2011). The same applies to the total magnitude of events immediately following a disaster. This translates into a number of different aspects: uncertainty regarding the nature of demand, the capacity of facilities to be used in the distribution process, the transportation capacity, and the amount of supply available for the decision maker, along with several other factors. Adding to this is the implicit urgency of need. Not only is the decision maker in charge of the distribution process required to make decisions based on limited and unreliable information, he must also make these at the earliest possible point in time following a disaster in order to prevent lives from being lost (Altay and Green 2006; de la Torre et al. 2012).

This research is driven by the intricate problem of designing a network for efficient relief delivery during flash floods in Uttarakhand, India. The state is located in an earthquake, flash floods prone region with 3 fault lines and many rivers passing through its populated districts (Geographical Survey of India (GSI) 1999). Uttarakhand was hit by flash floods on 16 June 2013. It resulted in deaths of about 8400 people and rendered 21,000 people homeless (EM-DAT, 2015). Rainfall and aftershocks were main factors complicating the rescue efforts, with potential secondary effects like additional landslides and further building collapses being major concerns. Impassable roads and damaged communications infrastructure posed substantial challenges to rescue efforts. Therefore, carrying out efficient relief activities under these restrictions is one of the major concern of local establishments, humanitarian organizations, and governments (State and Centre both). This comprised of designing a network for efficient delivery of relief items, which is addressed in this paper.

The aim of this paper is to present an optimization model for a location and relief distribution problem with stochastic elements that can be used for planning the distribution of aid following a catastrophic event. A mathematical model is proposed for capturing reliability of facilities, coverage radius of the distribution centers, uncertainty in demand, and the state of the infrastructure.

The remainder of the paper is organized as follows. Section 13.2 reviews the main papers available in the literature related to the topic considered, highlighting the main differences with the approach proposed in this paper. Section 13.3 gives a detailed description of the disaster response problem approached, whereas the mathematical programming model for this problem is presented in Sect. 13.4. A brief description of imperialist competitive algorithm is shown in Sect. 13.5. The case description and extensive computational study with application of the developed model to instances based on the Uttarakhand flash floods, 2013 are presented in Sect. 13.6, while concluding remarks are given in Sect. 13.7.

13.2 Literature Review

Decision-makers can be aided by optimization-based decision models for designing efficient humanitarian relief network (HRF) so as to dispense the best possible relief among the victims in shortest time. For example, de Treville et al. (2006) applied operations management logic and tools to save lives from tuberculosis under the

World Health Organization (WHO) mission to exterminate tuberculosis worldwide. Various optimization models have been proposed in the literature for designing HRFs (Caunhye et al. 2012; Hoyos et al. 2015). However, analytical approaches and mathematical models relating to logistical problems in HRFs with reliability constraints and coverage radius are notably very limited in the literature.

In this section, review of the previous studies which explore the location, routing, inventory, and transportation decisions while designing humanitarian network in the pre- and post-disaster phases (Table 13.1) is carried out. Caunhye et al. (2012) reviewed various optimization models developed for decision-making in emergency logistics. Chang et al. (2007) proposed a decision-making tool for flood relief logistics based on scenario planning approach. The objective of the model was to minimize the expected shipping distance and the sum of facility setup costs for local rescue bases under various scenarios. Balcik and Beamon (2008) presented a maximal-covering model to determine optimal locations and number of distribution centers (DC) along with the quantity of stocked relief items at each DC. The objective of the model was to maximize the total expected demand coverage by the opened up DCs in a scenario-based approach. Rawls and Turnquist (2010) in contrast to Balcik and Beamon (2008) proposed an integrated model for facility location and stocking level of relief items using stochastic mixed-integer programming formulation. They considered uncertainty related to demand and existence of pre-positioned stocks, and the transportation network situation under a scenario-based stochastic approach. The objective of the model was to minimize the expected costs of the whole scenarios and total cost comprised of various decisions related to location selection of warehouses, stocking of relief items, transportation, and penalty costs and holding costs for unused relief items. Rawls and Turnquist (2011) proposed an extension to their existing model, developed service quality constraints, and created a reliable set of scenarios. An extension to the Rawls and Turnquist (2010) model was proposed by Noyan (2012) which included a risk measure CVaR associated with the total cost of the objective function.

Rawls and Turnquist (2012) developed a dynamic allocation model to optimize the preparedness planning. The objective of the model was the satisfaction of the short-range demands (during the initial 72 h of relief) of relief items. Mete and Zabinsky (2010) developed a two-stage stochastic programming model for establishing warehouses and calculating inventory levels of medical items in a distribution relief network. The objective of the model was to minimize the total costs of opening facilities, unfulfilled demands, and transportation time. Duran et al. (2011) introduced a new practical objective function and deviated from the majority of the models by developing a time-minimizing model to enhance the operations of an organization in which cost data were unavailable. Tofighi et al. (2011) developed a two-stage fuzzy stochastic model while taking into account the integral ambiguity of demand and transportation time for prepositioning and the supply of assistance in a humanitarian supply chain.

Angelis et al. developed a weekly plan for world food program in Angola for emergency deliveries of food aid. The objective of the model was to maximize the total satisfied demand. Lodree Jr and Taskin (2009) proposed a stochastic inventory control problem for pre-disaster phase and was formulated as an optimal stopping

Table 13.1 The literature review

Studies	Disaster stages	No. of levels	L/A/Ro	Relief distribution	Total Cost	Total distance/time	Maximum distance/time	Unfulfilled Demand	Coverage Radius	Network reliability
Balcik and Beamon (2008)	P, R	1	L	Y				Y		
Mete and Zabinsky (2010)	P, R	1	L	Y	Y	Y		Y		
Rawls and Turnquist (2010)	P, R	1	L,Ro	Y	Y					
Rawls and Turnquist (2011, 2012)	P, R	1	L,A	Y	Y					
Salmerón and Apte (2010)	P, R	1	L.A	Y				Y		
Gormez et al. (2011)	P	2	L			Y	Y			
Duran et al. (2011)	P, R	1	L	Y		Y				
Döyen et al. (2012)	P, R	2	L,A	Y	Y					
Bozorgi-Amiri et al. (2013a)	P, R	1	L	Y	Y			Y		
Lu and Sheu (2013)	P, R	2	L, Ro		Y	Y				
Ahmadi et al. (2015)	P	2	L, Ro		Y	Y				
Bozorgi-Amiri and Khorsi (2015)	P	3	L, Ro			Y	Y	Y		
Zhang et al. (2015)	P, R	3	L, A	Y				Y		
Duhamel et al. (2016)	P	2	L, Ro		Y		Y			
Moreno et al. (2016)	R	3	Ro	Y	Y		Y			
Tofighi et al. (2016)	P,R	2	L, Ro	Y	Y	Y	Y	Y		
Caunhye et al. (2015)	P, R	2	L, A	Y	Y	Y	Y			
The proposed model	P, R	3	L, A	Y	Y	Y	Y	Y	Y	Y

P preparedness, *R* response, *L* location, *A* allocation, *Ro* routing, *Y* yes problem with Bayesian updates on the hurricane forecasts and solved by dynamic programming algorithm. Gatignon et al. (2010) demonstrated the development of optimal humanitarian supply chain via a study of the International Federation of the Red Cross (IFRC) decentralized supply chain, and performance assessment of the developed optimal plan was carried out through a case study of the IFRC's operations during the Yogyakarta earthquake in 2006.

Peeta et al. (2010) proposed a two-stage stochastic model for pre-disaster planning phase that attempted to reinforce a highway network. The objective of the model was to maximize the post-disaster connectivity and minimize transportation cost when network links were subjected to random failures. Qin et al. (2012) proposed an integrated single-period resource model for solving optimal order quantity of the recovery resources subjected to stock shortage. Analytical properties of the model were studied and a genetic algorithm-based simulation approach was proposed to solve the model. Tricoire et al. (2012) proposed a bi-objective covering tour model with stochastic demand and applied it to data set from Senegal to validate the model. The objective of the model was minimization of costs and expected uncovered demand.

Lu and Sheu (2013) proposed a robust vertex p-center model for locating urgent relief distribution centers where uncertain travel times were represented by fixed intervals or ranges. The objective of the model was to minimize worst-case deviation in travel time from the optimal solution. Simulated annealing based heuristics was developed to solve the model, and comparison with scenario-based two-stage stochastic model was carried out. Kumar and Havey (2013) proposed a decision support risk assessment and mitigation framework for disaster relief during Tsunami in Japan in March 2011. Different risks were recognized and evaluated during various phases in the relief supply chain by fault tree analysis and failure mode effects, and critical analysis method was used to evaluate the reliability of a relief supply chain system and its critical components. They concluded that the development of a robust communication plan and system will aid in enhanced coordination among all groups during all phases of a disaster and will help in providing a more efficient response. Davis et al. (2013) proposed a stochastic programming model for positioning and distribution of relief items between cooperative warehouse network. The model considered fairness in service, time constraints and subsequent traffic congestion from evacuation behavior.

Chakravarty (2014) developed a two-stage method where reaction time and relief quantities and pre-positioned inventory levels are determined for pre- and post-disaster stages. They concluded that response quantity and time must be adjusted to a social value according to disaster intensities, cost structures, and groups, and an increase in disaster intensity decreases the effectiveness of relief plan. Liberatore et al. (2014) proposed a model for recovery of damaged elements of the distribution network and applied the model called RecHADS to Haiti earthquake 2010. They also demonstrated analytically the significance of coordinating recovery and distribution operations optimization. Rath and Gutjahr (2014) proposed a model to support international aid organizations following the contingency of a natural disaster. Their model had three objectives (a) a medium-term economic; (b) a short-term economic, and (c) a humanitarian. Epsilon-constraint method was used to develop Pareto frontier and an exact solution method and a math-heuristic was proposed to solve the model, and was compared with the results of NSGA-II. Abounacer et al. (2014) proposed a multi-objective emergency location–transportation problem with capacity and multiple resource constraints. The three conflicting objectives were to minimize the total transportation time, the total uncovered demand, and the number of responders needed to open and operate the facilities. They presented an epsilon-constraint method that generated the exact Pareto front. Rennemo et al. (2014) proposed a three-stage mixed-integer stochastic programming model for disaster response planning, taking into consideration the opening of local distribution facilities, initial allocation of supplies, and last mile distribution of aid.

Kelle et al. (2014) proposed a decision criterion and applied it for supply of relief items in the Louisiana Gulf Coast and synchronized both the stages in a hierarchical humanitarian supply chain. Salman and Yucel proposed a systematic model to measure the spatial impact of the disaster on network links by random failures. The objective was to maximize the expected demand coverage within a specified distance over all possible network realizations. A tabu search heuristic was proposed and applied to Istanbul earthquake preparedness case. A comprehensive study for comparing solutions found in no link failure, independent link failure, and dependent link failure cases was carried out and it showed that integrating dependent link failures increased covered demand considerably.

Verma and Gaukler (2015) proposed a deterministic and stochastic location model, which took into consideration the effect of the disaster on the response facilities and the population centers in adjacent areas. A novel solution based on bender's decomposition was proposed and the locations determined by the stochastic model for the California earthquake considerably reduced the expected cost and were notably convincing for the budget-constrained situations. Rancourt et al. (2015) proposed an effective last mile food aid distribution model to determine a set of distribution centers, where the food was directly distributed to the beneficiaries, for the region of Gairsain in Kenya. The interest of all stakeholders associated with the response system: the World Food Programme, the Kenya Red Cross, and the beneficiaries were considered and GIS data of the road network was used to establish a set of potential distribution centers. Roh et al. (2015) developed a robust multi-criteria decision framework for prepositioning of relief warehouses from both the macro (country, region) and micro (proximate area) perspectives for Dubai. Moreno et al. (2016) developed two stochastic mixed-integer programming models for integrating and coordinating facility location, transportation, and fleet sizing decisions in a multi-period, multi-commodity, and multi-modal context under uncertainty and one model also considered the option of reusing vehicles. They also developed relax-and- fix and fix-and-optimize heuristics based on decompositions by time, scenario, and stage. Results indicated that the integration of decisions and reuse of vehicles reduces the total cost and increases the overall performance of the system.

Duque et al. (2016) developed an exact dynamic programming (DP) algorithm and an iterated greedy-randomized constructive method to solve the problem of scheduling and routing of the crew for the emergency repair of damaged rural road network after the occurrence of a disaster. Ransikarbum and Mason (2016) proposed multiple objectives in an integrated network optimization model for making strategic decisions in the supply distribution and network restoration phases of humanitarian logistics operations. The model provided an equity-based solution for constrained capacity, budget, and resource problems in post-disaster logistics management. Duhamel et al. (2016) proposed a mathematical model for solving a multi-period location–allocation problem in post-disaster operations, with logistics restrictions and solved by a decomposition approach. Computational experiments were performed on various scenarios generated from real data of Belo Horizonte city in Brazil. Roni et al. (2016) proposed a hybrid inventory management model considering both emergency and regular replenishments. A search-based heuristics was developed to find near optimal inventory management policies and results revealed that proposed hybrid inventory policy with split deliveries outperforms the other policies when holding and shortage costs are low.

Paul and MacDonald (2016) developed a stochastic optimization model to decide upon the location of relief material and volumes of medical items for enhanced disaster preparedness in the event of a hurricane. A mixed-integer programming model was developed and solved by evolutionary optimization heuristic. The model integrated various features to reflect hurricane conditions such as facility damage and casualty losses, based upon their severity levels and remaining survivability time as a function of time variant. Gutjahr and Dzubur (2016) proposed a bi-objective, bi-level optimization model for the location of relief distribution centers (DCs) in humanitarian logistics. Upper-level decision-maker does a selection of locations for capacitated DCs with the objective of minimizing total costs and total uncovered demand. Lower level beneficiaries select the DC based upon distance and quantity of supply to be expected. The model was solved by integrating the adaptive epsilon-constraint method, a branch-and-bound procedure, and the Frank–Wolfe procedure.

Ferrer et al. (2016) proposed a model to address the last mile distribution problem in disaster relief operations subjected to uncertain conditions and takes into consideration the cost and time of operation, the safety and reliability of the routes, and the fairness of aid distributed. Camacho-Vallejo et al. (2015) proposed a bi-level mathematical programming model to optimize decisions related to the distribution of international aid after a catastrophic disaster. As an international aid organization aims to minimize the transport cost, whereas the affected country seeks to distribute the relief items as swiftly and efficiently as possible. A case study of the earthquake in Chile in 2010 was used to validate the model and guidance for aid distribution during future disaster was provided. Bozorgi-Amiri and Khorsi (2015) developed a multi-objective dynamic stochastic programming model to integrate strategic, tactical, and operational decisions for a humanitarian relief logistics problem. Objectives of the model were minimizing the maximum quantity of deficiencies among the affected areas in all periods, the total travel time, and the sum of pre- and post-disaster costs and solved by ε constraint method for disaster planning of earthquake scenarios in the megacity of Tehran. Ahmadi et al. (2015) proposed a multi-depot location-routing model considering network failure, multiple uses of vehicles, and standard relief time. The model determined the locations of local depots and routing of vehicles for last mile distribution after an earthquake and was solved by a variable neighborhood search algorithm.

Garrido et al. (2016) developed an inventory optimization model level for emergency supplies so as to supply sufficient supplies to fulfill demands with a given possibility and availability of vehicles. The studied example showed large variances among the effects of logistics factors such as the number of periods, number of products, inventory size on the logistics cost and time. Kim et al. (2015) utilized graph theoretic approach for conceptualizing the supply network disruption and resilience by exploring the structural relationships among units in the network. The results showed that the network structure considerably governs the likelihood of disruption and resilience increases when the structural relationships in a network follow the *power-law*. Rath et al. (2015) proposed a two-stage bi-objective stochastic programming for disaster relief operations with monetary and humanitarian objectives. Various scenarios like fixed of free assignment of vehicles to depots, fixed running budget, homogenous and heterogeneous vehicle fleet were generated, and influence of the different alternatives on the solutions was analyzed. Hong et al. (2015) presented a risk-averse stochastic modeling approach for a pre-disaster relief network design problem under uncertain demand and transportation dimensions. The objective of the model was to determine the dimensions and sites of the response facilities and inventory levels of relief items at each facility while assuring a definite level of network reliability. Zhang et al. (2015) developed a scenario-based mixed-integer programming model for reliable capacitated location-routing problem in which depots were randomly disrupted. The objective of the model was to optimize outbound distribution routing, location of the depot and develop a backup plan and was solved by a metaheuristic algorithm based on simulated annealing, maximum-likelihood sampling method, and two-stage neighborhood search and route-reallocation improvement.

Zhang et al. (2016) developed a non-linear integer programming model for facility location considering the interruptions of facilities and the cost savings resulting from the effect of inventory risk-pooling and economies of scale. The objective was to minimize the expected total cost across all possible facility failure scenarios and was solved using two methods, an exact approach using special ordered sets of type two and a heuristic based on Lagrangian relaxation. He and Zhuang (2016) developed a two-stage dynamic programming model for optimal allocation of preparedness and relief expenditures. Sensitivity analysis was also carried out between pre-disaster preparedness and the total loss for all parameters. Orgut et al. (2016) proposed and analyzed mathematical models to enable a food bank's impartial and effective distribution of donated food among a population at risk for hunger. The model was applied to a real dataset from a North Carolina food bank and various studies, and sensitivity analysis was also carried out. Stauffer et al. (2016) modeled a global vehicle supply chain of an International Humanitarian Organization (IHO) with a dynamic hub location model across monthly periods. They found that there was dissimilarity between the demand of vehicles and of water or food. The results also demonstrated that having a lean hub design with a possibility of temporary hubs in disaster locations could bring down the supply chain cost over a long-time horizon.

A research framework was developed by Leiras et al. (2014) for carrying out literature in humanitarian logistics through qualitative and quantitative content analysis. Their main findings were (a) the prominent need for filling the existing gap between academic world and humanitarian administrators for growth of the practical research in the area and (b) the need for carrying out more research in the disaster recovery phase. Gutjahr and Nolz (2016) reviewed recent literature on the application of multi-criteria optimization to disaster management. They studied and analyzed various optimization criteria and multi-criteria decision-making approaches applied in this area. Anaya-Arenas et al. (2014) presented a systematic review of contributions on relief distribution networks in response to disasters and identified many future research perspectives in the field. Altay and Green (2006), Haddow et al. (2005), and Tomasini and van Wassenhove (2009) provide a basic and general introduction to the field of emergency management and humanitarian logistics. For full reviews on optimization models in emergency logistics, see Caunhye et al. (2012), Hoyos et al. (2015), Özdamar and Ertem (2015), Habib et al. (2016) and Snyder et al. (2016).

13.2.1 Gaps in Literature

Various researchers have addressed location, allocation, and routing problem (Table 13.1). They have considered various aspects of location, distribution, inventory, transportation time, etc. Yet network planning with reliability of facilities, coverage radius, and capacity restrictions has not been adequately addressed in these models. We have developed a mathematical model to address this issue and applied it to a specific case situation of flash floods in Uttarakhand, 2013, India.

13.2.2 Contributions to the Literature

In order to integrate the aforementioned gaps in the literature, a multi-objective mathematical model is developed. Tofighi et al. (2016) have conducted a survey of related research till mid-2012. We extend it further for newer addition to this list (Table 13.1). The developed model considers potential network damages and differentiates itself from other existing models as mentioned in Table 13.1.The main contributions of the paper lie mainly in the following two aspects:

1. It presents an integrated design of a humanitarian network for response planning, which contemplates a practical network configuration with reliability and capacity constraints of the facilities and coverage radius of the distribution centers. It also renders flexibility in humanitarian operations decision-making such as in location of facilities and distribution of relief items.

2. The extensive numerical analysis imparts an understanding of how the proposed model attributes (i.e., network structure, flexibilities in allocation and distribution) influence the response planning decisions and the resulting costs demand coverage and time.

13.3 Problem Description

This section describes a planning problem for a humanitarian supply chain in the event of a disaster. In this context, the disaster is presumed to have affected a wide area and as a result, affected victims are in the need of humanitarian aid. The affected victims must be reached as soon as possible by using the accessible infrastructure (some infrastructure might get damaged by the disaster). The location of relief distribution centers, allocation of relief items, distribution plan, types of vehicles to be used for shipment, and the severity of damage depend on the types of disaster. To streamline the demonstration, we examine specifically the humanitarian response in relation to flash floods although the developed model can be applied to various other disasters such as earthquakes, hurricanes, tsunamis, and volcanic eruptions, with some minor variations.

The fundamental job is to determine local distribution center(s) (LDCs) at which supplies should arrive and be governed for distribution to the victims [point of distribution(s) (PODs) and also the regional distribution center(s) (RDCs)] from which these supplies emanate.

The planning problem is focused on the generation of a distribution plan, from RDCs to PODs via operational LDCs, in order to fulfill the urgent needs of the affected population. The problem is intricate in nature due to the limited availability or unavailability of the information and uncertainty which demands relief planning activities of very high level of intricacy.

13.3.1 Contributors

There are several contributors across the different echelons in the humanitarian supply chain which help in carrying out relief activities. The Federal Emergency Management Agency (FEMA) (2008) and the International Federation of Red Cross and Red Crescent Societies (IFRC) (2012) provide explanations of the main players, out of which mainly three are relevant for the problem considered in this paper. They comprise the International central depots (ICDs) or the Regional distribution centers (RDCs), the Local distribution centers (LDCs), and the different Points of Distribution (PODs).

The ICDs/RDCs are the first points from where supply originates. They provide relief items, fleet services, and logistics support for the relief operations. The main objective of setting up an RDC is to be able to provide relief items instantaneously to the affected region (a relief to 5000 families within 48 h of a request, and to further 15,000 families within 2 weeks).

The LDCs receive the supplies from the RDCs and are of two types: permanent or temporary. The permanent LDCs are established under disaster preparedness plan that is before disaster strikes and serve as emergency storage points. When a disaster strikes, they are replenished by supplies from RDCs. They hold permanent inventory of relief items which helps them to supply initial relief items very swiftly after disaster strikes. The permanent inventory incurs an inventory holding and operating costs (Zhu et al. 2008). Inversely, the temporary LDCs are non-stationary and non-operational until disaster strikes. They are normally situated at airports, playgrounds, schools, or other sites suitable for carrying huge influx and efflux of relief items and employees, and can work as drop sites.

The PODs serve as a centralized location where the relief material is supplied from the LDCs, and the victims can collect the relief items from the locations FEMA (2008). The quantity and kinds of items to be sent to different PODs and operations and disbandment of PODs are controlled by the LDCs.

13.3.2 Course of Events

Shortly after a flash flood has occurred, aid agencies or the government will decide whether it is necessary to initiate an emergency response or not. If it is necessary, then the native representatives will start collecting information about the consequences of the disaster. For an efficient and effective emergency response, regular update of the information has to be carried out constantly. A team of experts is sent to the disaster affected site for a preliminary estimation of the disaster and the requirements of the victims. The estimation will help to build a request for specific items and amount needed to supply immediate relief to the affected population (International Federation of Red Cross and Red Crescent Societies (IFRC) 2008). The team of experts then advises the native representatives where to build PODs based upon the location of demand points. Real-time information about the location of the PODs and the condition of the infrastructure existing prior to the flash floods serve as the foundation for decision-making for opening and operating temporary LDCs.

As soon as an initial distribution plan is designed at an LDC, the vehicles are loaded and dispatched according to this plan. In some scenarios, the vehicles will not be able to cover the entire route due to infrastructure damage. In these cases, a different route will be developed based upon discussion with the native representative and updated information on the network. After some time of initial relief operation, the information related to the state of the infrastructure will be accessible via satellite pictures and actual understanding provided by drivers and local reports. As soon as the information is available, some vehicles may then reach their planned endpoints (PODs) whereas others may have to change their route because of difficulties in the network.

13.3.3 Sources of Uncertainty

There are numerous issues that affect the strategies of the native residents and confine the amount of alternatives that may be contemplated. First, demand is uncertain. The areas in the necessity of urgent relief may be located in isolated areas and perhaps the disaster location maybe in a disordered condition rendering a total overview of the affected region (Thomas and Kopczak 2005). Second, the usability of the network after the disaster is uncertain. Third, the capacity of available relief items is restricted. To the extent that this study deals with uncertainty, its emphasis lays mostly on these three aspects and the consequences that they involve.

Erratic demand patterns surge the intricacy of the distribution plan. Demand can vary unpredictably due to a number of causes. These reasons comprised of aftershock destructions, victims shifting between different areas in expectation of finding better relief, or unforeseen challenges such as an outburst of epidemics (de la Torre et al. 2012).

The preliminary distribution plan is subject to numerous changes because of the severe damage to the local distribution network by flash floods. Roads, bridges, and airports are often demolished. The ease of access of various victims may reduce depending upon their location as compared to the flash flood's originating point, and the quality of the network infrastructure joining the LDCs and the PODs. The attributes of an LDC like (location, coverage radius, and damage intensity) may, in addition, restrict it from covering various PODs in the infrastructure.

In this problem, we consider RDCs, relief warehouses (LDCs + TWH), PODs, and multiple types of relief products as shown in Fig. 13.1. Figure 13.1 illustrates the flow of material and information over time and when different actions are taken during emergency response. This figure and the details presented before regarding contributors, course of events, and sources of uncertainty are based on information obtained from media sources, websites, and experts or from the India Meteorological Department, Ministry of Earth Sciences, Government of India (IMD) (2015),

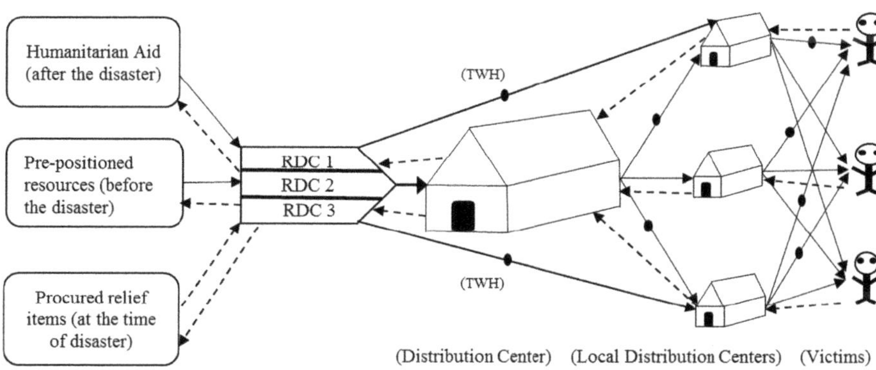

Fig. 13.1 Humanitarian logistics flow of materials. (• *TWH* temporary warehouses, → material flow, ---→ information flow)

National Center for Seismology (NCS) (2015), National Disaster Management Authority (NDMA) (2015), National Disaster Relief Force (NDRF) (2015), and International Red Cross Society (IRCS) (2015).

13.4 Mathematical Model

In this section, we formulate the reliable facility location and distribution (RFLD) problem. The objective of the developed model is to select, locate, and allocate a set of distribution centers and distribute relief items to victims so as to meet demands of the victims such that the total cost is minimized, maximum demand is covered, and total distribution time is minimized. We assume that each victim (POD) has an uncertain demand that follows a normal distribution. To make the problem more realistic, we have taken different capacity levels for each distribution center. During the response phase, some of the opened LDCs may be inoperative due to various aftershocks of the disaster-like landslides in the case of flash floods. Inoperative LDCs have zero capacity and as a result, network design should be carried out with respect to the active LDCs only. While designing a relief network, operative and inoperative LDCs are represented by a various set of scenarios to take account of all the events. The failure of an LDC is characterized by a group of scenarios, all linked with a possibility of realization $\left(ps = q^{n(D_r)}(1-q)^{n(D'_r)}\right)$. At the starting point, all LDCs are assumed to be operative ($D_1 = \emptyset$), and the total number of scenarios ($|R|$) is the complete permutation of operative and inoperative LDCs exclusive of a scenario in which all LDCs are inoperative ($|R| = 2^J - 1$). In reality, when one or more LDCs is/are inoperative, the demand of that region should be satisfied by other active LDCs but this is not completely achieved due to capacity restrictions of LDCs. With respect to this we formulate an objective function as minimization of shortage cost of unfulfilled demands (Eq. 13.2a). We extend upon the mathematical model of Jalali et al. and customize it to humanitarian settings and constraints. To customize a RFLD under the aforementioned assumptions used symbols are proposed as follows:

Notations:

I, J, K—Set of victims, LDC's and RDC's

i, j, k—Index set of victims, LDC's and RDC's

R—Set of scenarios

r—Index set of scenarios

P, H = Set of capacity levels for LDC's and RDC's

p, h = Index set of capacity levels for LDC's and RDC's

Q = Set of relief items

q = Index set of relief items ($q \in Q = \{1, 2, \ldots, q\}$)

CS = Set of crucial supplies ($\text{CS} = \{1, 2, \ldots, q^h\} \subset Q$)

D_r = Set of operative DC's during scenario r consisting of $n(D_r)$ elements

D'_r = Set of inoperative DC's during scenario r consisting of $n(D'_r)$ elements

a_i^r = Demand of victim i during scenario r

b_j^p = Nominal capacity of LDC's j at capacity level p

ω_{ij}^p = Coverage ratio of POD i by LDC j at capacity level p

$$L_{ij}^p = \begin{cases} 1, & \text{if } \omega_{ij}^p \neq 0 \\ 0, & \text{otherwise} \end{cases}$$

$$G_{r,j} = \begin{cases} 1, & \text{if } j \in D_r' \\ 0, & \text{otherwise} \end{cases}$$

$$b_J^{pr} = \begin{cases} 0, & \text{if } j \in D_r' \\ b_{j'}^p, & \text{if } j \in D_r \end{cases}$$

mh_q^r = Unit shortage cost of qth item under disaster scenario r

w_{qi}^r = Demand urgency for the qth item at the ith demand point under scenario $r \left(\sum_{i \in I} \sum_{q \in q} w_{qi}^s = 1 \right), \forall s \in S$

rc_k^h = Capacity of RDC's at capacity level h

FQ = Probability of failure of LDC's

RP_r = Probability of realization of scenario r

FL_j^p = Fixed cost of opening LDC j at capacity level p

TCL_{ji} = Carrying cost from jth LDC to ith victim

TCR_{kj} = Carrying cost from kth RDC to jth LDC

GR_k^h = Fixed cost of opening kth RDC at capacity level h

TT_{ij}^r = Total travelling time between jth LDC and ith demand point (POD) to consider various states of network infrastructure under disaster scenario r

TT_{jk}^r = Total travelling time between jth LDC and kth RDC to consider various states of network infrastructure under disaster scenario r

Decision Variables:

$$\mathrm{UQ}_j^p = \begin{cases} 1, & \text{if a DC with capacity level } p \text{ is opened in location } j \\ 0, & \text{otherwise} \end{cases}$$

$$\mathrm{VQ}_k^h = \begin{cases} 1, & \text{if a RDC with capacity level } h \text{ is opened in location } k \\ 0, & \text{otherwise} \end{cases}$$

XS_{ij}^r = Portion of ith satisfied victim (POD) demand by jth LDC during scenario r

YS_{jk}^{pr} = Portion of relief items sent to jth LDC with capacity level p by RDC k during scenario r

ZD_{iq}^r = Quantity of unfulfilled demand for the qth item at demand point i during scenario r

MT^r = Maximum transportation time under disaster scenario r

YD_{kjiq}^{pr} = Quantity of the qth item to be transported from RDC k to demand point i via LDC under disaster scenario r with capacity level p

XD_{ijq}^{r} = Quantity of the qth crucial relief item to be transported from LDC j to POD (demand point) i under disaster scenario r

Mathematical Model:

$$\mathrm{Min} f_1 = \sum\sum \mathrm{FL}_j^p \times \mathrm{UQ}_J^p + \sum\sum \mathrm{GR}_K^h \times \mathrm{VQ}_k^h + \mathrm{RP}_r\left(\sum_{i\in I}\sum_{j\in J}\mathrm{TCL}_{ji}\times a_i \times \mathrm{XS}_{ij}^r + \sum_{p\in P}\sum_{j\in J}\sum_{k\in K}\mathrm{TCR}_{kj}\times b_j^p \times \mathrm{YS}_{jk}^{pr}\right) \tag{13.1}$$

$$\mathrm{Max} f_2 = \frac{1}{\sum_{i\in I} a_i}\left[\sum_{r\in R}\sum_{i\in I}\sum_{j\in J}\mathrm{RP}_r \times a_i \times \mathrm{XS}_{ij}^r\right] \tag{13.2}$$

$$\mathrm{Min} f_2 = \left[\sum_{j\in J}\sum_{K\in RI} W_{qi}^p \, mh_q^r \mathrm{ZD}_{iq}^s\right] \tag{13.2a}$$

$$\mathrm{Min}\ f_3 = \mathrm{Min}\left(\sum_i\sum_j\sum_k\sum_{Q\in RI}\left(\mathrm{TT}_{ij}^r + \mathrm{TT}_{jk}^r\right)\mathrm{YD}_{kjiq}^{pr} + \sum\sum\sum \mathrm{TT}_{ij}^r \mathrm{XD}_{ijq}^r\right) \tag{13.3}$$

$$\mathrm{Min} f_4 = \mathrm{Min}\ (\mathrm{MT}^r) \tag{13.4}$$

Constraints:

$$\sum_{j\in J}\mathrm{XS}_{ij}^r = 1 \quad \forall i \in I, r \in R \tag{13.5}$$

$$\mathrm{XS}_{ij}^r \le \sum_{p\in P}\mathrm{UQ}_j^p G_{r,j}\ \omega_{ij}^p \forall i \in I, j \in J, r \in R \tag{13.6}$$

$$\sum_{p\in P}\mathrm{UQ}_j^p \le \sum_{i\in I}\sum_{p\in P} L_{ij}^p \ \forall\ j \in J \tag{13.7}$$

$$\sum_{i\in I} a_i^r \mathrm{XS}_{ij}^r \le \sum_{k\in K}\sum_{p\in P} b_j^p \mathrm{YS}_{jk}^{pr} \ \forall\ j \in J, r \in R \tag{13.8}$$

$$\sum_{k\in K}\mathrm{YS}_{jk}^{pr} \le \mathrm{UQ}_j^p, \forall j \in J, r \in R, p \in P \tag{13.9}$$

$$\sum_{j\in J}\sum_{p\in P} b_j^{pr}\mathrm{YS}_{jk}^{pr} \le \sum rc_k^h \mathrm{VQ}_k^h; \forall\ k \in K, r \in R \tag{13.10}$$

$$\sum_{p\in P}\mathrm{UQ}_j^p \le 1, \quad \forall j \in J \tag{13.11}$$

$$\sum_{h \in H} \mathrm{VQ}_k^h \leq 1, \quad \forall k \in K \tag{13.12}$$

$$\sum_{j \in J} \sum_{k \in K} \sum_{i \in I} \sum_{q \in Q} \mathrm{YD}_{kjiq}^{pr} + \sum \mathrm{XD}_{ijq}^r = a_i^r - \mathrm{ZD}_{iq}^r : \forall j \in J, Q \in \mathrm{CS} \tag{13.13}$$

$$\sum_{k \in K} \sum_{j \in J} \sum_{i \in I} \sum_{q \in Q} \mathrm{YD}_{kjiq}^{pr} = a_i^r - \mathrm{ZD}_{iq}^r : \forall j \in J, Q \neq \mathrm{CS} \tag{13.14}$$

$$\sum_{k \in K} \sum_{j \in J} \sum_{i \in I} \sum_{q \in Q} \mathrm{YD}_{kjiq}^{pr} \leq M \times D_r; \forall i \in I \tag{13.15}$$

$$\mathrm{MT}^r \geq \sum_{Q \in CR} \left(w_{qi}^r \sum_{i \in I} \mathrm{TT}_{ij}^r \mathrm{XS}_{ijq}^r \right); \forall j \in J \tag{13.16}$$

$$\mathrm{MT}^r \geq \sum_{Q \in RI} \left(w_{qi}^r \sum \sum \left(\mathrm{TT}_{jk}^r + \mathrm{TT}_{ij}^r \right) \mathrm{YD}_{kjiq}^{pr} \right); \forall j \in J \tag{13.17}$$

$$\mathrm{UQ}_j^p \in \{0, 1\} \forall j \in J, p \in P \tag{13.18}$$

$$\mathrm{VQ}_k^h \in \{0, 1\} \forall k \in K, h \in H \tag{13.19}$$

$$\left[\mathrm{XD}_{ijq}^r, \mathrm{YD}_{kjiq}^{pr}, \mathrm{ZD}_{iq}^r, \mathrm{MT}^r \geq 0, \forall i \in I, j \in J, k \in K, q \in Q. \right. \tag{13.20}$$

In RFLD model, the first objective function, Eq. (13.1), aims to minimize the fixed location costs and the carrying costs. The second objective function, Eq. (13.2), aims to maximize the demand coverage for all possible scenarios and in some cases it is also represented as minimization of the shortage cost of unfulfilled demands (Eq. 13.2a). The third objective function, Eq. (13.3), aims to minimize the total distribution time. The fourth objective function, Eq. (13.4), relates to the minimization of the maximum transportation time among each set of RDC/LDCs and POD.

According to constraint (13.5), allocation of extra demand to victims is prevented. Constraint (13.6) limits the fraction of allocation to victims based on opening, running, and covering the status of DCs. Constraint (13.7) inhibits from opening a DC which is not able to cover any customer. Constraint (13.8) refrain DCs from extra delivery, i.e., DCs cannot supply more than the received items. Constraint (13.9) restricts allocating a non-active LDC to an RDC. Constraint (13.10) restricts the transportation of items to RDCs in terms of capacities. Constraints (13.11) and (13.12) impede the consideration of multiple capacity levels for RDCs and LDCs. Unfulfilled demand for crucial and non-crucial relief items is calculated by constraints (13.13) and (13.14). Constraint (13.15) puts into effect that the transportation of relief items from RDCs to PODs must be carried through opened LDCs only. Constraints (13.16) and (13.17) calculate the maximum transportation time for the relief items at various PODs while taking demand priorities into consideration. Finally, constraints (13.18–13.20) represent the status of the decision of variables.

13.5 Proposed Algorithm for the Problem

Various algorithms and metaheuristics have been used in the literature to solve problems related to humanitarian logistics. Genetic-Algorithms (GA) are the most popular and widely used algorithm in the literature for solving production and operations management problems (Aytug et al. 2003) and various supply chain management problems and humanitarian relief planning (Mguis et al. 2014). Other widely used algorithms include: simulated annealing (SA) which has been used to determine the optimal allocation of emergency response services for managing disasters (Mitsakis et al. 2014); Tabu Search that has been used to investigate emergency facility location under random network damage, based on an earthquake in Istanbul (Salman and Yücel 2015); particle swarm optimization (PSO) which has been used by Uno et al. (2007) for solving a multi-objective programming model for optimal emergency facility location and Tian et al. (2011) for solving a multi-objective emergency transportation model; modified particle swarm optimization was applied by Bozorgi-Amiri et al. (2013b) for disaster relief logistics under uncertain environment; multi-objective PSO which has been used by Zheng et al. (2014a) for population classification in fire evacuation operations; ecogeography-based optimization algorithm (EBO) was developed by Zheng et al. (2014b) and applied to an emergency airlift problem in the 2013 Lushan Earthquake, China; multi-objective biogeography-based optimization (BBO) algorithm has been used by Zheng et al. (2014c) for an emergency engineering rescue scheduling problem; hybrid BBO which was used by Zheng et al. (2014a) for an emergency railway wagon scheduling problem. Ma et al. developed a genetic-SA algorithm for emergency service facility location problem. Zheng et al. (2016) presented a survey of the research advances in evolutionary algorithms (EAs) applied to disaster relief operations. They compared various state-of-art methods on a set of real-world emergency transportation problem cases, and some insights were derived from the experimental studies.

13.5.1 Imperialist Competitive Algorithm (ICA)

ICA is one of the latest approaches in the meta-heuristic field; it is a new socio-politically motivated global search strategy developed by Atashpaz-Gargari and Lucas (2007); Atashpaz-Gargari et al. (2008) for solving various optimization problems. Good performance has been achieved by the algorithm in both convergence rate and improved global optima attainment (Atashpaz-Gargari and Lucas 2007; Atashpaz-Gargari et al. 2008; Rajabioun et al. 2008). Hosseini and Al Khaled (2014) carried out an extensive review of emergence and implementations of ICA to various engineering areas such as design of structures, heat transfer, image processing and to numerous intricate discrete combinatorial optimization problems, such as Traveling Salesman Problem (TSP), Flow-shop Scheduling Problem (FSP), and Facility Layout Problem (FLP).

As compared to existing evolutionary algorithms, ICA also starts with a random initial population where each entity of the population stands for a country. Countries within the ICA are similar to chromosomes in the GA. In the first step, some of the top countries (less cost) are chosen as imperialist countries, and the leftovers are regarded as to be colonies of the imperialists. On the basis of imperialists' power, the colonies are then dispensed among the imperialists. After dispensing is complete, the colonies move in the direction of their pertinent imperialists inside the cultural space. A group of the imperialist country and several colonies forms an empire, which contends between empires and supersedes each other. The continued existence of an empire is dependent on its power to conquest colonies from other opponents and the power of superior empires multiplies, whereas the empires with a lesser amount of power crumble. After series of competitions between empires, the power of colonies comes close to that of the imperialist countries, specifying merging. In the end, when only one imperialist is leftover along with some colonies near to the imperialist country regarding the position, the competition ends. The pseudocode of the algorithm (Atashpaz-Gargari and Lucas 2007) is shown in Fig. 13.2.Various steps involved in the ICA are shown in Fig. 13.3 (compiled from Nazari-Shirkouhi et al. 2010). Flowchart of the ICA (Ghasemi et al. 2015) is presented in Fig. 13.4. The step-by-step illustration of ICA is described below.

13.5.1.1 Generation of Initial Empires (Fig. 13.3a)

The objective of optimization is to obtain an optimal solution in terms of the problem variables. An array of variable values is generated by the user which he wants to be optimized. While solving a N_{var} dimensional problem, a country is represented by an $1 \times N_{\text{var}}$ array as followed:

$$\text{Country} = [P_1, P_2, P_3 \ldots P_{N_{\text{var}}}], \tag{13.21}$$

where $P_i s$ are considered as the variables that should be optimized.

A mixture of a few socio-political factors such as religion, culture, language, and welfare are built-in in the prospective solutions of the problem, called country. Figure 13.5 depicts the understanding of country using some of the socio-political features (Javanmard et al. 2014). By calculating the cost function, f for variables $(p_1, p_2, p_3 \ldots p_{N_{\text{var}}})$, the cost of a country can be calculated by (Eq. 13.22)

$$\text{cost}\, t_i = f(\text{country}) = f(p_1, p_2, p_3 \ldots p_{N_{\text{var}}}). \tag{13.22}$$

In the initial stage of the algorithm, the preliminary population of size N_{country} is generated as shown in Fig. 13.3a. Empires are generated by selecting N_{imp} from the strongest population. The remaining N_{col} of the population will build the colonies

Imperialist competitive Algorithm

1. Initialize parameters;
2. Generate the population randomly;
3. Initialize the empire;
   For $i = 1$ to $(N_{pop} = N_{pop\ size})$
      Compute the evaluation cost c_i;
      Sort the computed cost c_i in descending order of entire population;
      Select N_{imp} (number if imperialist countries) out of N_{pop};
      Normalize the cost of each imperialist C_n;
      Compute the normalized power of each imperialist P_n;
      Assign N_{col} remained countries to the imperialists;
   End loop
4. Assimilation, Revolution, Imperialist Competition Process;
   For $j = 1$ to N_{imp}
      Move the colony toward the relevant imperialist (assimilation);
      Compute the cost of assimilated countries;
      Perform revolution on colony;
      If the cost of new colony is less than cost of imperialist
         Then exchange the position of colony and imperialist;
      Pick the weakest colony (colonies) from the weakest empire and assign it (them)
      to empire that has most likelihood to possess it;
   End loop
5. Elimination Process;
   If there is no imperialist with no colonies
      Then eliminate the imperialist;
6 Until stooping criterion is met

Figure 13.2 Pseudocode of the ICA

which are in command of one of the empires. Some of these colonies are given to each imperialist for distributing the early colonies between the imperialist according to their power. The normalized cost of an imperialist for proportionate division of colonies among the imperialists is explained as follows:

$$C_n = \max{}_i\{c_i\} - c_n. \tag{13.23}$$

In the equation stated above, cost of nth imperialist is c_n and normalized cost is C_n. As soon as the normalized costs of all imperialists are calculated, then the following equation is used to calculate the normalized power of an individual imperialist:

$$P_n = \left|\frac{C_n}{\sum_{i=1}^{N_{\text{imp}}} C_i}\right|. \tag{13.24}$$

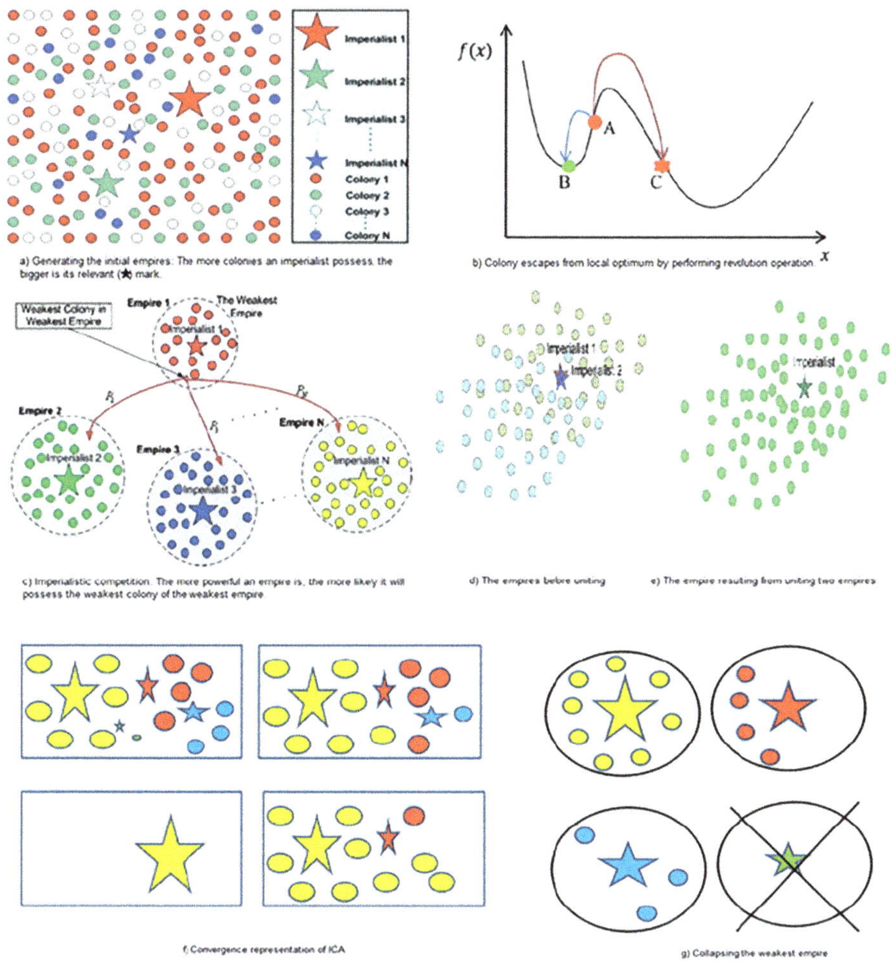

Fig. 13.3 Various steps involved in ICA (compiled from Nazari-Shirkouhi et al. 2010)

In the beginning, colonies are allocated among empires based on their power. Accordingly, in the beginning, number of colonies for nth empire will be:

$$N \times C_n = \text{round}\{P_n \times N_{\text{col}}\}, \tag{13.25}$$

where $N \times C_n$ factor signifies the number of colonies of the empire in the beginning and the entire number of current colonies countries in the preliminary countries groups are shown by N_{col}.

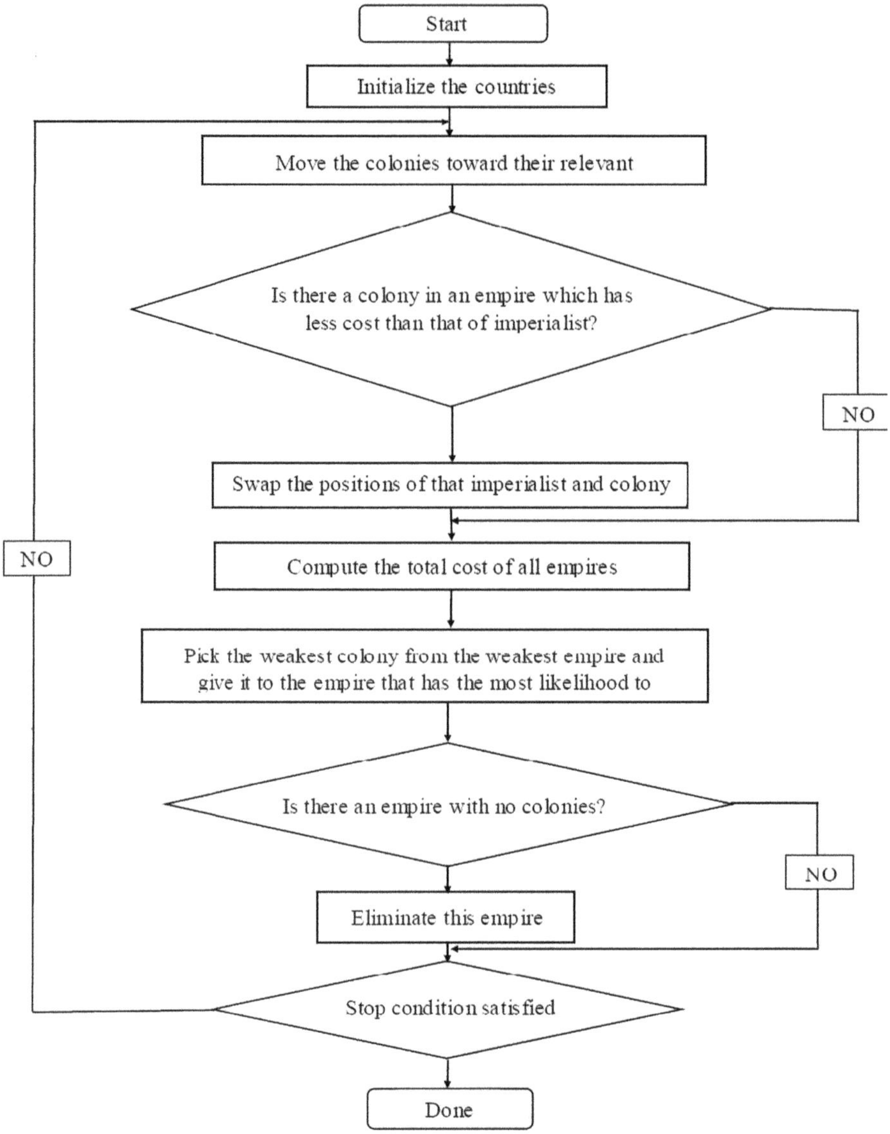

Fig. 13.4 Flowchart displaying the functioning of ICA algorithm

13.5.1.2 Assimilation: Movement of Colonies Toward the Imperialist

As described in the previous subsection, following assimilation policy, the imperialist states tried to captivate their colonies and make them a part of themselves through several socio-political axis, for example, culture and welfare (Seidgar et al. 2014). Considering a two-dimensional optimization problem, in Fig. 13.6, the

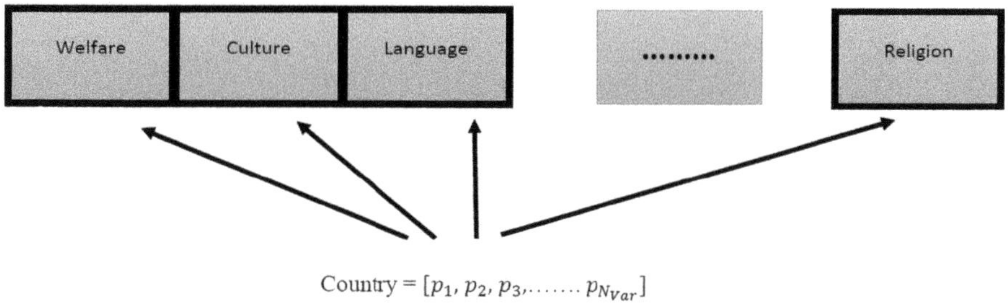

Fig. 13.5 Comprehension of a country by means of some of the socio-political features

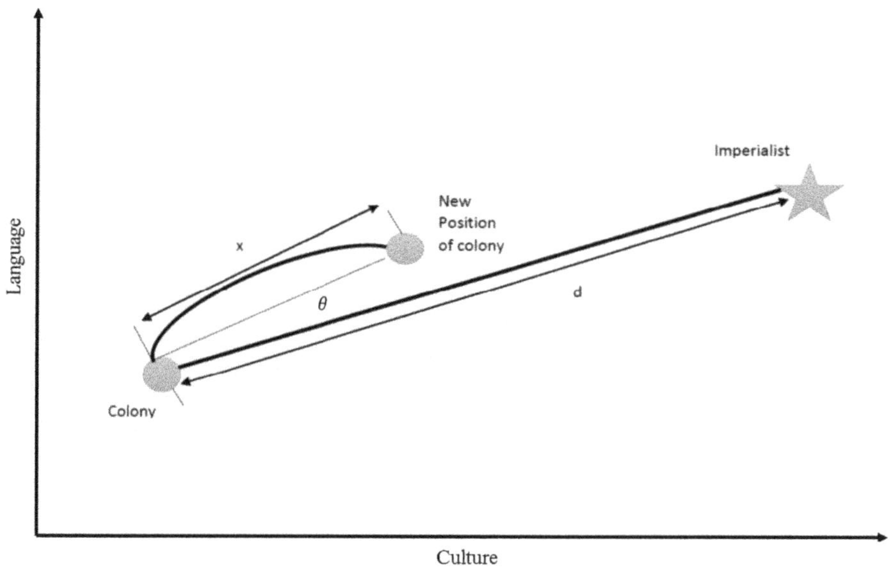

Fig. 13.6 Movement of colonies toward their relevant imperialist

colony is captivated by the imperialist in the culture and language axes. Then colony will get nearer to the imperialist in these axes. Continuance of this process will effect in all the colonies to be entirely assimilated into the imperialist. In Fig. 13.6, d and x variables represent the distance among colony countries and imperialist and random number with fixed distribution, correspondingly (Nazari-Shirkouhi et al. 2010).Therefore, x can be defined as follows:

$$x \sim U(0, \beta \times d), \tag{13.26}$$

where β is a number greater than 1. $\beta > 1$ induces the colonies to come nearer to the imperialist state from both sides. If no direct movement of the colonies in the direction of the imperialist is observed even after assimilating the colonies by the imperialist states, a random extent of deviation is added to the direction of movement. In Fig. 13.6, θ is a parameter with uniform distribution. Then:

$$\theta \sim U(-\gamma, +\gamma), \tag{13.27}$$

where γ is an optimal parameter that its increase in its value results in rise of examining area nearby imperialist and decrease results in colonies closing probably to the vector of joining colony to the imperialist. In maximum number of applications, the value of γ is random and about $\pi/4$ (Rad) and $\beta = 2$ effects in good merging of countries to the global minimum.

13.5.1.3 Revolution: A Sudden Change in Socio-Political Characteristics of a Country (Fig. 13.3b)

Revolution is an elemental modification in organizational structures or power that happen in a quite small amount of period. It causes a country to unexpectedly modify its socio-political features, viz. as an alternative of being assimilated by an imperialist, the colony randomly modifies its location in the socio-political axis. Figure 13.3b illustrates the revolution in the Culture-Language axis. The revolution intensifies exploration of the algorithm and helps in escaping the premature merging of the countries to local minimums. Percentage of colonies in each colony that will randomly alter their position is specified by the revolution rate. Exploitation power and convergence rate are reduced by a steep value of revolution (Nazari-Shirkouhi et al. 2010).

13.5.1.4 Movement of the Position of Imperialist and Colony

Substitution of the imperialist will take place if a colony achieves a superior point than an imperialist on the way to the imperialist country. This result in the continuation of the algorithm with an imperialist country in a new position and this time it is the new imperialist country which starts to apply integration policy for its colonies.

13.5.1.5 Total Power of an Empire

The total power of an empire is mainly determined by the power of the imperialist country. Though, the power of the colonies bears a very small effect on the total power of that empire. The sum cost (Eq. 13.28) of an empire is determined by taking into consideration both the above-stated factors (Afruzi et al. 2014)

$$T \times C_{\cdot n} = \text{Cost}(\text{Imperialist}_n) + \xi \text{mean}\{\text{Cost}(\text{colonies of empire}_n)\}, \tag{13.28}$$

where $T \times C_{\cdot n}$ is defined as the total cost of the nth empire and ξ is a positive number that has value between zero and one and near to zero. The value of 0.15 for ξ has shown good equitable results in most of the applications.

13.5.1.6 Imperialistic Competitions

All empires always tend to increase their control by dominating and taking possession of each other's colonies. This is the basis of imperialist competition in the algorithm. The imperialistic competition progressively results in a decrease of the power of weaker empires and an increase in the power of powerful ones (Fig. 13.3c). During the competition, every empire has a definite possibility of owning the stated colonies as compared to their total power. These powerful empires will try to control these weaker colonies. In the very first stage of the competition among the empires for owning these colonies, the feeblest empire is carefully chosen and then the ownership probability of every empire (P_p) is predicted as compared to the total power of the empire.For calculation of the total normalized cost of an empire we use (Eq. 13.29):

$$N \times T \times C_{\cdot n} = \max{}_i\{T \times C_{\cdot i}\} - T \times C_{\cdot n}, \tag{13.29}$$

where $T \times C_{\cdot n}$ is total cost of nth empire and $N \times T \times C_{\cdot n}$ is normalized cost of that nth empire. By having the normalized total cost, the ownership probability of each empire is defined by:

$$P_{p_n} = \left|\frac{N \times T \times C_{\cdot n}}{\sum_{i=1}^{N_{\text{imp}}} N \times T \times C_{\cdot i}}\right|. \tag{13.30}$$

In the following stage, the stated colonies will be separated inadvertently among the empires with a definite probability which is associated with the ownership probability of each empire. To carry out the distribution of the given colonies among the empires, vector P is formulated as stated below:

$$P = \left[P_{P_1}, P_{P_2}, \ldots, P_{P_{N_{\text{imp}}}}\right]. \tag{13.31}$$

After that, the vector R should be described with the similar size of vector P. The collections of this vector are inadvertent numbers with similar distribution as in [0, 1].

$$R = \left[r_1, r_2, \ldots, r_{N_{\text{imp}}}\right]. \tag{13.32}$$

Then, vector D is formulated by subtracting R from P.

$$D = R - P = \left[D_1, D_2, \ldots, D_{N_{\text{imp}}}\right]$$
$$= \left[P_{P_1} - r_1, P_{P_2} - r_2, \ldots, P_{P_{N_{\text{imp}}}} - r_{N_{\text{imp}}}\right]. \tag{13.33}$$

We give the stated colonies to the empires by having vector D so that associated index in vector D is greater than others. Progressively the outcome of the imperialist competition will result in an expansion of the power of great empires and a weakening of weaker ones. As a result, the feeble empires will slowly lose their power and get debilitated with the time (Fig. 13.3g). Carrying out these steps iteratively will induce the countries to congregate to the global minima or maxima of the chosen function as shown in Fig. 13.3f. Figure 13.7 shows the initial empires and empires at 50th iteration and final outcome.

We use multi-objective ICA (MICA) algorithm (Ghasemi et al. 2015) to solve the multi-objective RFLD problem. While applying this algorithm, an array of leading points is attained which are put in storage in a depository. Meanwhile, the magnitude of the depository should continue to be fixed, a fuzzy grouping technique is exploited for maintaining the fixed depository. Normalized value of membership function is calculated and successful answers are categorized based upon these values. After this, it splits the essential number of answers from them and hence by following this technique it maintains a fixed size of the depository.

13.6 The Case Description

The problem addressed in this research relates to a humanitarian relief organization carrying out relief activities during flash floods in Uttarakhand, India in 2013. Uttarakhand is popularly known as Devbhumi (Land of Gods) due to the presence of numerous Hindu pilgrimage sites. As a result, religious tourism forms a major portion of the tourism in the state. The services sector contributes around 53% to Gross State Domestic Product (GSDP) (*PHD Research Bureau, compiled from MOSPI*). The tourism business in Uttarakhand generated ₹ 23,000 crores during 2013–2014, however due to 2013 flash floods, it expected a 70% fall to ₹ 6900 crores during 2014–2015. The region was highly devastated due to heavy rainfall and flash floods, about 8400 people lost their lives, 21,000 people were rendered homeless, and an economic loss of 1,100,000 ('000 US$) was reported by EM-DAT, 2016.

The road network was highly damaged by the flash floods which in result affected the relief activities to a very large extent. Landslides made the situation worse by disconnecting the religious sites (Kedarnath, Badrinath, Gangotri, and Yamunotri) from rest of the state, as shown in Fig. 13.8. In Fig. 13.8, regions highlighted in red represent severely damaged areas which required air assistance for relief and blue shows sites from where relief activity was being carried out in the state.

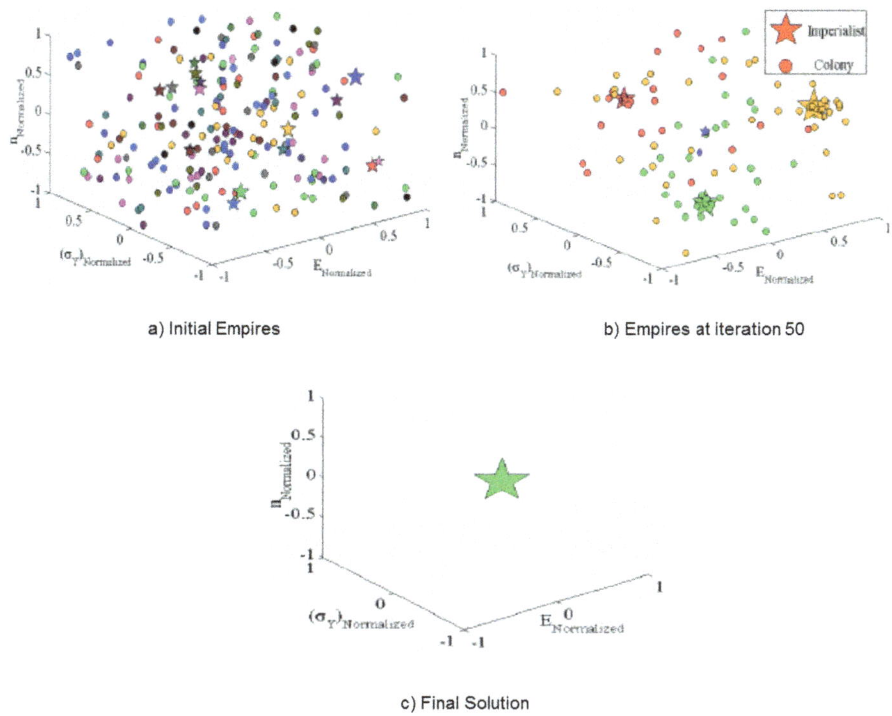

Fig. 13.7 Depicting the initial empire and final solution of ICA

In this section, we make available the particulars of the case study for the design of HRN in Uttarakhand, India. Uttarakhand consists of 13 districts, and 12 districts were affected by the flash floods (Indian Red Cross Society (IRCS) (2013)) and are taken as the demand points (PODs) for thorough development of the network. The priority for setting up PODs was calculated according to likely threats and damages (including average destruction, housing and land devastation ratio, and death ratio), social settings (population density), and existing network structure of the damaged location, which are derived from National Institute of Disaster Management (NIDM) (2013), India disaster report (2013). The relief network consists of 4 RDCs and 19 LDCs. Data for the case study has been taken or modified from the reports by India Meteorological Department, Ministry of Earth Sciences, Government of India, IMD (2015), National Center for Seismology (NCS) (2015), International Red Cross Society (IRCS) (2015), and Indian Red Cross Society (IRCS) (2013) and maps by Geographical Survey of India (1999). The data related to total demand in each district, damage of structures and road networks, and transportation times were calculated using information from these reports. The probability of devastation by a flash flood in each area is predicted for day and night individually since the destruction intensity of a flash flood is dependent on the time of occurrence.

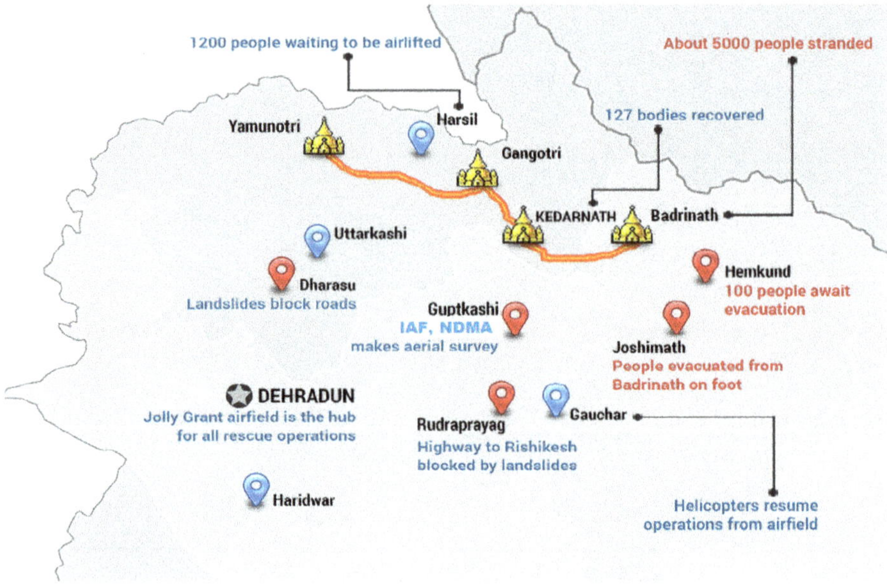

Fig. 13.8 Overview of disaster relief activities in Uttarakhand, India (2013). (http://uk.gov.in/home/index1)

RDCs of three types can be selected based upon their capacity levels, i.e., 50, 100, and 150 thousand families. We assume that an LDC has the potential of satisfying the necessities of 3000 families (of five as average family size). The location of LDCs was carefully chosen from the existing public facilities such as schools, playgrounds, stadiums, etc. throughout the state, which could be equipped to function like LDCs. Furthermore, in the present case, four candidate locations were selected for setting up required RDCs and 19 locations for setting up LDCs around Uttarakhand.

Various costs such as setup cost, operating cost, and transportation cost have been estimated by various experts from the relief organizations. The demand for various relief items has been estimated for each area based on the inhabitants and total affected people in each region. Relief items consist of essential requirements such as drinking water, food, medicines, first aid kits, clothes, and shelter for victims. Transportation times from RDCs to LDCs and LDCs to PODs have been estimated by considering the total distance as well as estimated destruction ratio and existing network structure. Percentages for the demolition of the pre-positioned relief items at each RDC and LDC have been taken into account by the subjective probability of damage of the warehouses during flash floods.

13.6.1 Computational Results for Large Instances

The algorithm described in Sect. 13.5 is coded in MATLAB R2013b and solved on Intel i3 processor, 3.10 GHz system with 32-bit Windows 7 operating system. Population size and the number of imperialists in the algorithm were taken as 180 and 16, respectively. A decent range can be $\beta = \beta_{\text{col}} = 2$ and $\gamma = \gamma_{\text{col}} = \frac{\pi}{4}$ (Rad). The developed model was first solved by MICA on a set of 5 small-sized test problems comprising of 2 RDCs, 8 LDCs, and 10 PODs with 1 relief item under 2 scenarios of the disaster as shown in Table 13.2.

To demonstrate the computational efficiency and performance of the MICA algorithm for the RFLD problem, we build five test cases. The total number of nodes comprising of RDCs, LDCs, and PODs are taken to be 40, 50, 60, 70, and 80, respectively. The supply amount and demand requirement are generated randomly from the interval $[200, 700]$ and $[300, 800]$, respectively. As shown in Table 13.3, the proposed MICA algorithm can solve our integrated model for large instances in a reasonable amount of time. For the case with a maximum number of nodes, the solution time is close to 8 min (last row in the table), which is a satisfactory response time during these types of situations. However, in real scenarios, the problem size and intricacy are dependent on the design of the time-space network, e.g., such as the number of nodes and arcs.

The first test considers varying the number of LDCs. Computational experiments were performed with two different data sets, first with 15 PODs and second with 40 PODs. LDCs are varied up to 90 and large number signifies more computational time and is reflected by the average values depicted in Fig. 13.9 for several number of LDCs. Only for one instance (with 90 LDCs), the algorithm was unable to find an optimal solution within 12 h. However, all remaining instances were solved far

Table 13.2 Computational results

Problem instance	Fixed costs ($)	Total transportation time (h)	Maximum transportation time (h)	Shortage cost ($)
1	2204	145	333	224
2	3614	178	378	386
3	4088	241	402	363
4	4604	292	460	442
5	5162	380	442	559

Table 13.3 Computational results for large instances

Nodes			MICA	
SN	TN	DN	Average CPU time (s)	Average error (e^{-4})
15	10	15	79.26	4.52
20	15	20	97.36	4.92
25	20	25	126.35	4.42
30	25	30	440.23	4.98

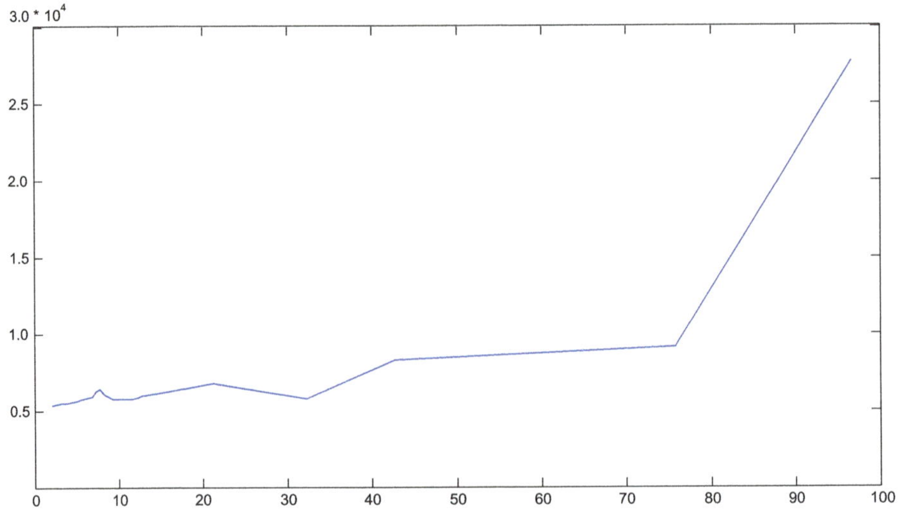

Fig. 13.9 Average solution time when varying the number of LDCs

within the acceptable time frame, indicating computational efficiency for varying levels of LDCs. By analysis of the trend line, we determined that the number of LDCs correlates with computational time.

The computational tests also reveal that the optimal objective function value, other parameters being held constant, does not increase substantially when the number of LDCs is increased beyond 25% of the number of PODs. This indicates that in practice, the number of LDCs should not be critical for the model.

The second test looks at the number of PODs. Computational experiments were performed by increasing the number of PODs and this resulted in an increase in computational time. Figure 13.10 illustrates the computational time required to solve problems when allowing up to 250 PODs. This shows that a decision-maker will still be able to obtain an optimal solution within an acceptable time frame of 45 min. However, only a small number of demand points for distribution are usually considered in the literature. Vitoriano et al. (2011) considered 9 demand points (PODs) while 6 and 30 PODs were considered by Barbarosoğlu and Arda (2004) and Rawls and Turnquist (2010), respectively.

The third test looks at the effect of increasing reliability factor on the cost as well as the demand coverage. Table 13.4 presents results for various problem settings related to the unreliability of location for (33 × 33) problem set. The plot of normalized values of the total cost at different reliability factors and demand coverage is shown in Fig. 13.11. As the coverage factor increases, the total cost also increases; and when the reliability is low, a high amount of cost is incurred maintaining the same demand coverage level.

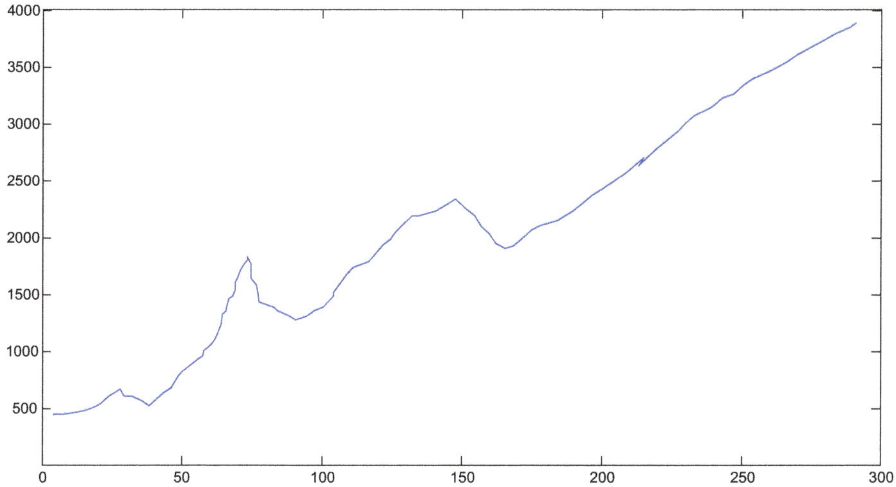

Fig. 13.10 Average solution time when varying the number of PODs

Table 13.4 Computational results for the unreliable location problem

Probability (P_j)	Fixed charge	Objective value
1.00	1000	14,832
0.90	1000	15,497
0.80	1000	16,163
0.70	1000	20,828
0.60	1000	24,494
0.50	1000	29,823
0.40	1000	32,627
0.30	1000	33,499
0.20	1000	(*)
0.10	1000	(*)

(*) No objective value was found

13.7 Conclusions and Future Research Directions

In this paper, we propose a multi-objective model for the design of humanitarian logistics network. In this model, we have integrated the reliability, capacity, and coverage radius constraints while designing the network. For demand satisfaction of various regions, we allow the supply of relief items from intermediate or temporary local distribution centers in contrast to most conventional network models which deal with the distribution of relief items from a single warehouse to various locations. The objective of the developed model is to find number and location of RDCs and LDCs that should be opened, the assignment of victims to PODs, allocation of victims to active routes, and the distribution plan so as to minimize total cost, minimize the shortage cost or maximize the demand coverage, minimize total

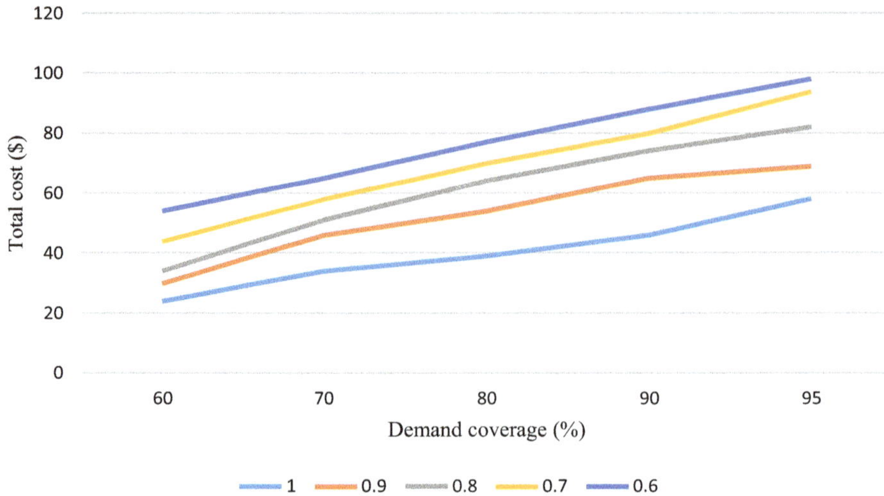

Fig. 13.11 Total cost for different reliability factor and demand coverage

distribution time, and minimization of maximum transportation time simultaneously. Since the proposed model is an NP-Hard problem, we have used a multi-objective imperialist competitive algorithm (MICA) to solve the model.

The main contributions of this paper are as follows. First, the formulation of an integrated mathematical model with consideration likes reliable locations, capacity levels, and coverage constraints for the design of a humanitarian network. Second, to the best of authors' knowledge, this is the first study that develops a multi-objective model for simultaneous optimization of total cost, distribution time, and demand coverage for humanitarian supply chain during flash floods.

Third, a sensitivity analysis has been carried out for a number of RDCs and total cost, the number of LDCs and total cost, and demand coverage and reliability. Fourth, employing multi-objective imperialist competitive algorithm (MICA) for RFLD problem is a contribution in the area of solution algorithms. The existing research can be extended in various dimensions. Various other demand patterns can be assumed and modeled such as fuzzy demand. Comparison of the proposed algorithm with other algorithms can also be a future research direction.

References

Abounacer R, Rekik M, Renaud J (2014) An exact solution approach for multi-objective location-transportation problem for disaster response. Comput Oper Res 41(1):83–93. https://doi.org/10.1016/j.cor.2013.08.001

Afruzi NE, Najafi AA, Roghanian E, Mazinani M (2014) A multi-objective imperialist competitive algorithm for solving discrete time, cost and quality trade-off problems with mode-identity and resource-constrained situations. Comput Oper Res 50:80–96. https://doi.org/10.1016/j.cor.2014.04.003

Ahmadi M, Seifi A, Tootooni B (2015) A humanitarian logistics model for disaster relief operation considering network failure and standard relief time: a case study on San Francisco district. Transport Res Part E Logist Transport Rev 75:145–163. https://doi.org/10.1016/j.tre.2015.01.008
Altay N, Green WG (2006) OR/MS research in disaster operations management. Eur J Oper Res 175(1):475–493. https://doi.org/10.1016/j.ejor.2005.05.016
Anaya-Arenas AM, Renaud J, Ruiz A (2014) Relief distribution networks: a systematic review. Ann Oper Res 223:53–79. https://doi.org/10.1007/s10479-014-1581-y
Apte A (2010) Humanitarian logistics: a new field of research and action. Found Trends (R) Technol Inf Oper Manag 3(1):1–100. https://doi.org/10.1561/0200000014
Atashpaz-Gargari E, Lucas C (2007) Imperialist competitive algorithm: an algorithm for optimization inspired by imperialistic competition. In: IEEE congress on evolutionary computation. CEC 2007. IEEE, New York, pp 4661–4667. https://doi.org/10.1109/CEC.2007.4425083
Atashpaz-Gargari E, Hashemzadeh F, Lucas C (2008) Designing MIMO PID controller using colonial competitive algorithm: applied to distillation column process. In: IEEE congress on evolutionary computation (CEC-2008). IEEE, New York, pp 1929–1934. https://doi.org/10.1109/CEC.2008.4631052
Aytug H, Knouja MJ, Vergara EF (2003) Use of genetic algorithms to solve production and operations management problems: a review. Int J Prod Res 41(17):3957–3979. https://doi.org/10.1080/00207540310001626319
Balcik B, Beamon BM (2008) Facility location in humanitarian relief. Int J Logist Res Appl 11(2): 101–112. https://doi.org/10.1080/13675560701561789
Barbarosoğlu G, Arda Y (2004) A two-stage stochastic programming framework for transportation planning in disaster response. J Oper Res Soc 55:1. https://doi.org/10.1057/palgrave.jors.2601652
Beamon BM, Balcik B (2008) Performance measurement in humanitarian relief chains. Int J Public Sector Manag 21(1):4–25. https://doi.org/10.1108/09513550810846087
Bozorgi-Amiri A, Khorsi M (2015) A dynamic multi-objective location-routing model for relief logistic planning under uncertainty on demand, travel time, and cost parameters. Int J Adv Manuf Technol 85:1633–1648. https://doi.org/10.1007/s00170-015-7923-3
Bozorgi-Amiri A, Jabalameli MS, Alinaghian M, Heydari M (2013a) A modified particle swarm optimization for disaster relief logistics under uncertain environment. Int J Adv Manuf Technol 60:357–371. https://doi.org/10.1007/s00170-011-3596-8
Bozorgi-Amiri A, Jabalameli MS, Mirzapour Al-e-Hashem SM (2013b) A multi-objective robust stochastic programming model for disaster relief logistics under uncertainty. OR Spectrum 35: 905–933. https://doi.org/10.1007/s00291-011-0268-x
Camacho-Vallejo JF, González-Rodríguez E, Almaguer FJ, González-Ramírez RG (2015) A bi-level optimization model for aid distribution after the occurrence of a disaster. J Clean Prod 105:134–145. https://doi.org/10.1016/j.jclepro.2014.09.069
Caunhye AM, Nie X, Pokharel S (2012) Optimization models in emergency logistics: a literature review. Socio Econ Plan Sci 46(1):4–13. https://doi.org/10.1016/j.seps.2011.04.004
Caunhye AM, Zhang Y, Li M, Nie X (2015) A location-routing model for prepositioning and distributing emergency supplies. Transport Res Part E Logist Transport Rev 90:161–176. https://doi.org/10.1016/j.tre.2015.10.011
Chakravarty KA (2014) Humanitarian relief chain: rapid response under uncertainty. Int J Prod Econ 151:146–157. https://doi.org/10.1016/j.ijpe.2013.10.007
Chang MS, Tseng YL, Chen JW (2007) A scenario planning approach for the flood emergency logistics preparation problem under uncertainty. Transport Res Part E Logist Transport Rev 43(6):737–754. https://doi.org/10.1016/j.tre.2006.10.013
Christopher M, Tatham P (2011) Humanitarian logistics: meeting the challenge of preparing for and responding to disasters. Kogan Page Series, Kogan
Davis BL, Samanlioglu F, Qu X, Root S (2013) Inventory planning and coordination in disaster relief efforts. Int J Prod Econ 141:561–573. https://doi.org/10.1016/j.ijpe.2010.01.003

de la Torre LE, Dolinskaya IS, Smilowitz KR (2012) Disaster relief routing: integrating research and practice. Socio Econ Plan Sci 46(2):88–97. https://doi.org/10.1016/j.seps.2011.06.001
de Treville S, Smith I, Rölli A, Arnold V (2006) Applying operations management logic and tools to save lives: a case study of the world health organization's global drug facility. J Oper Manag 24:397–406. https://doi.org/10.1016/j.jom.2005.03.004
Döyen A, Aras N, Barbarosoğlu G (2012) A two-echelon stochastic facility location model for humanitarian relief logistics. Optim Lett 6(6):1123–1145. https://doi.org/10.1007/s11590-011-0421-0
Duhamel C, Santos CA, Brasil D, Châtelet E, Birregah B (2016) Connecting a population dynamic model with a multi-period location-allocation problem for post-disaster relief operations. Ann Oper Res 247:693–713. https://doi.org/10.1007/s10479-015-2104-1
Duque MAP, Dolinskaya SI, Sörensen K (2016) Network repair crew scheduling and routing for emergency relief distribution problem. Eur J Oper Res 248:272–285. https://doi.org/10.1016/j.ejor.2015.06.026
Duran S, Gutierrez MA, Keskinocak P (2011) Pre-positioning of emergency items for care international. Interfaces 41(3):223–237
Federal Emergency Management Agency (FEMA) (2008) IS-26 Guide to points of distribution (PODs). Technical report. Emergency Management Institute, Emmitsburg
Ferrer MJ, Ortuño TM, Tirado G (2016) A GRASP metaheuristic for humanitarian aid distribution. J Heuristics 22:55–87. https://doi.org/10.1007/s10732-015-9302-5
Garrido AR, Lamas P, Pino JF (2016) A stochastic programming approach for floods emergency logistics. Transport Res Part E Logist Transport 75:18–31. https://doi.org/10.1016/j.tre.2014.12.002
Gatignon A, Van Wassenhove LN, Charles A (2010) The Yogyakarta earthquake: humanitarian relief through IFRC's decentralized supply chain. Int J Prod Econ 126:102–110. https://doi.org/10.1016/j.ijpe.2010.01.003
Ghasemi M, Ghavidel S, Ghanbarian MM, Massrur RH, Gharibzadeh M (2015) Application of imperialist competitive algorithm with its modified techniques for multi-objective optimal power flow problem: a comparative study. Inf Sci 281:225–247. https://doi.org/10.1016/j.ins.2014.05.040
Gormez N, Koksalan M, Salman SF (2011) Locating disaster response facilities in Istanbul. J Oper Res Soc 62:1239–1252. https://doi.org/10.1057/jors.2010.67
Guha-Sapir D, Vos F, Below R, Ponserre S (2012) Annual Disaster Statistical Review 2011: the numbers and trends. CRED, Brussels. http://www.cred.be/sites/default/files/ADSR_2011.pdf
Gutjahr JW, Dzubur N (2016) Bi-objective bilevel optimization of distribution center locations considering user equilibria. Transport Res Part E Logist Transport 85:1–22. https://doi.org/10.1016/j.tre.2015.11.001
Gutjahr JW, Nolz CP (2016) Multicriteria optimization in humanitarian aid. Eur J Oper Res 252: 351–366. https://doi.org/10.1016/j.ejor.2015.12.035
Habib SM, Lee HY, Memon SM (2016) Mathematical models in humanitarian supply chain management: a systematic literature review. Math Probl Eng 2016:3212095. https://doi.org/10.1155/2016/3212095
Haddow G, Bullock J, Coppola DP (2005) Introduction to emergency management. Butterworth-Heinemann, Oxford
He F, Zhuang J (2016) Balancing pre-disaster preparedness and post-disaster relief. Eur J Oper Res 252:246–256. https://doi.org/10.1016/j.ejor.2015.12.048
Hilhorst D (2005) Dead letter or living document? Ten years of the code of conduct for disaster relief. Disasters 29(4):351–369
Holguín-Veras J, Van Jaller M, Wassenhove LN, Pérezd N, Wachtendorf T (2012) On the unique features of post-disaster humanitarian logistics. J Oper Manag 30:494–506. https://doi.org/10.1016/j.jom.2012.08.003

Holguín-Veras J, Pérez N, Jaller M, Van Wassenhove LN, Aros-Vera F (2013) On the appropriate objective function for post-disaster humanitarian logistics models. J Oper Manag 31(5): 262–280. https://doi.org/10.1016/j.jom.2013.06.002

Hong X, Lejeune MA, Noyan N (2015) Stochastic network design for disaster preparedness. IIE Trans 47(4):329–357. https://doi.org/10.1080/0740817X.2014.919044

Hosseini S, Al Khaled A (2014) A survey on the Imperialist Competitive Algorithm metaheuristic: implementation in engineering domain and direction for future research. Appl Soft Comput 24: 1078–1094. https://doi.org/10.1016/j.asoc.2014.08.024

Hoyos MC, Morales RS, Akhavan-Tabatabaei R (2015) OR models with stochastic components in disaster operations management: a literature survey. Comput Ind Eng 82:183–197. https://doi.org/10.1016/j.cie.2014.11.025

India Online (2013) Uttarakhand tourism revenue loss pegged at Rs. 120 bn [online]. http://www.indiainfoline.com/article/news-top-story/uttarakhand-tourism-revenue-loss-pegged-at-rs-120bn-phdcci-113110810520_1.html

Indian Red Cross Society (IRCS) (2013). http://www.indianredcross.org/publications.htm. Accessed 23 Mar 16

International Federation of Red Cross and Red Crescent Societies (IFRC) (2008) Guidelines for assessment in emergencies. http://www.ifrc.org/Global/Publications/disasters/guidelines/guidelines-for-emergency-en.pdf. Accessed 23 Mar 2016

International Federation of Red Cross and Red Crescent Societies (IFRC) (2012) Federation procurement services. http://procurement.ifrc.org/. Accessed 23 Mar 2016

IPCC (2001) Climate Change 2001: the scientific basis. In: Houghton JT, Ding Y, Griggs DJ, Noguer M, van der Linden PJ, Dai X, Maskell K, Johnson CA (eds) Contribution of working group I to the third assessment report of the intergovernmental panel on climate change. Cambridge University Press, Cambridge, p 881

Javanmard S, Vahdani B, Tavakkoli-Moghaddam R (2014) Solving a multi-product distribution planning problem in cross docking networks: an imperialist competitive algorithm. Int J Adv Manuf Technol 70:1709–1720. https://doi.org/10.1007/s00170-013-5355-5

Kelle P, Schneider H, Yi H (2014) Decision alternatives between expected cost minimization and worst case scenario in emergency supply. Int J Prod Econ 157:250–260. https://doi.org/10.1016/j.ijpe.2014.06.009

Kim Y, Chen YS, Linderman K (2015) Supply network disruption and resilience: a network structural perspective. J Oper Manag 33:43–59. https://doi.org/10.1016/j.jom.2014.10.006

Kovács G, Spens MK (2011) Trends and developments in humanitarian logistics—a gap analysis. Int J Phys Distrib Logist Manag 41:32–45. https://doi.org/10.1108/09600031111101411

Kumar S, Havey T (2013) Before and after disaster strikes: a relief supply chain decision support framework. Int J Prod Econ 145:613–629. https://doi.org/10.1016/j.ijpe.2013.05.016

Kunz N, Reiner G, Gold S (2014) Investing in disaster management capabilities versus pre-positioning inventory: a new approach to disaster preparedness. Int J Prod Econ 157:261–272. https://doi.org/10.1016/j.ijpe.2013.11.002

Leiras A, De Brito I Jr, Peres EQ, Bertazzo TR, Yoshizaki HTY (2014) Literature review of humanitarian logistics research: trends and challenges. J Humanit Logist Supply Chain Manag 4(1):95–130. https://doi.org/10.1108/JHLSCM-04-2012-0008

Liberatore F, Ortuño MT, Tirado G, Vitoriano B, Scaparra MP (2014) A hierarchical compromise model for the joint optimization of recovery operations and distribution of emergency goods in humanitarian logistics. Comput Oper Res 42:3–13. https://doi.org/10.1016/j.cor.2012.03.019

Lodree EJ Jr, Taskin S (2009) Supply chain planning for hurricane response with wind speed information updates. Comput Oper Res 36:2–15. https://doi.org/10.1016/j.cor.2007.09.003

Lu CC, Sheu JB (2013) Robust vertex p-center model for locating urgent relief distribution centers. Comput Oper Res 40(8):2128–2137. https://doi.org/10.1016/j.cor.2013.02.019

Mete HO, Zabinsky ZB (2010) Stochastic optimization of medical supply location and distribution in disaster management. Int J Prod Econ 126(1):76–84. https://doi.org/10.1016/j.ijpe.2009.10.004

Mguis F, Zidi K, Ghedira K, Borne P (2014) Distributed and guided genetic algorithm for humanitarian relief planning in disaster case. In: 11th International symposium on distributed computing and artificial intelligence 2014, DCAI 2014. Springer, Salamanca, pp 149–156

Mitsakis E, Stamos I, Grau JMS, Aifadopoulou G (2014) Optimal allocation of emergency response services for managing disasters. Disaster Prev Manag 23(4):329–342. https://doi.org/10.1108/DPM-10-2013-0182

Moreno A, Alem D, Ferreira D (2016) Heuristic approaches for the multiperiod location-transportation problem with reuse of vehicles in emergency logistics. Comput Oper Res 69: 79–96. https://doi.org/10.1016/j.cor.2015.12.002

National Center for Seismology (NCS) (2015). https://www.moes.gov.in/programmes/national-centre-seismology

National Disaster Management Authority (NDMA) (2015) http://www.ndma.gov.in/en/

National Institute of Disaster Management (NIDM) (2013). www.nidm.gov.in/default.asp

Nazari-Shirkouhi S, Eivazy H, Ghodsi R, Rezaie K, Atashpaz-Gargari E (2010) Solving the integrated product mix-outsourcing problem using the imperialist competitive algorithm. Expert Syst Appl 37(12):7615–7626. https://doi.org/10.1016/j.eswa.2010.04.081

Noyan N (2012) Risk-averse two-stage stochastic programming with an application to disaster management. Comput Oper Res 39(3):541–559. https://doi.org/10.1016/j.cor.2011.03.017

Orgut SL, Ivy J, Uzsoy R, Wilson RJ (2016) Modeling for the equitable and effective distribution of donated food under capacity constraints. IIE Trans 48(3):252–266. https://doi.org/10.1080/0740817X.2015.1063792

Özdamar L, Ertem MA (2015) Models, solutions and enabling technologies in humanitarian logistics. Eur J Oper Res 244(1):55–65. https://doi.org/10.1016/j.ejor.2014.11.030

Paul AJ, MacDonald L (2016) Optimal location, capacity and timing of stockpiles for improved hurricane preparedness. Int J Prod Econ 174:11–28. https://doi.org/10.1016/j.ijpe.2016.01.006

Peeta S, Salman SF, Gunnec D, Viswanath K (2010) Pre-disaster investment decisions for strengthening a highway network. Comput Oper Res 37:1708–1719. https://doi.org/10.1016/j.cor.2009.12.006

Qin J, Xing Y, Wang S, Wang K, Chaudhry SS (2012) An inter-temporal resource emergency management model. Comput Oper Res 39:1909–1918. https://doi.org/10.1016/j.cor.2011.07.008

Rajabioun R, Hashemzadeh F, Atashpaz-Gargari E, Mesgari B, Rajaei Salmasi F (2008) Identification of a MIMO evaporator and its decentralized PID controller tuning using colonial competitive algorithm. IFAC World Congress, Yokohama, pp 11–12

Rancourt EM, Cordeau FJ, Laporte G, Watkins B (2015) Tactical network planning for food aid distribution in Kenya. Comput Oper Res 56:68–83. https://doi.org/10.1016/j.cor.2014.10.018

Ransikarbum K, Mason SJ (2016) Multiple-objective analysis of integrated relief supply and network restoration in humanitarian logistics operations. Int J Prod Econ 54(1):49–68. https://doi.org/10.1080/00207543.2014.977458

Rath S, Gutjahr WJ (2014) A math-heuristic for the warehouse location–routing problem in disaster relief. Comput Oper Res 42:25–39. https://doi.org/10.1016/j.cor.2011.07.016

Rath S, Gendreau M, Gutjahr WJ (2015) Bi-objective stochastic programming models for determining depot locations in disaster relief operations. Int Trans Oper Res 23:997–1023. https://doi.org/10.1111/itor.12163

Rawls CG, Turnquist MA (2010) Pre-positioning of emergency supplies for disaster response. Transport Res Part B Methodol 44(4):521–534. https://doi.org/10.1016/j.trb.2009.08.003

Rawls CG, Turnquist MA (2011) Pre-positioning planning for emergency response with service quality constraints. OR Spectrum 33(3):481–498. https://doi.org/10.1007/s00291-011-0248-1

Rawls CG, Turnquist MA (2012) Pre-positioning and dynamic delivery planning for short-term response following a natural disaster. Socio Econ Plan Sci 46(1):46–54. https://doi.org/10.1016/j.seps.2011.10.002

Rennemo SJ, Rø KF, Hvattum LM, Tirado G (2014) A three-stage stochastic facility routing model for disaster response planning. Transport Res Part E Logist Transport Rev 62:116–135. https://doi.org/10.1016/j.tre.2013.12.006

Roh S, Pettit S, Harris I, Beresford A (2015) The pre-positioning of warehouses at regional and local levels for a humanitarian relief organization. Int J Prod Econ 170:616–628. https://doi.org/10.1016/j.ijpe.2015.01.015

Roni SM, Eksioglu DS, Jin M, Mamun S (2016) A hybrid inventory policy with split delivery under regular and surge demand. Int J Prod Econ 172:126–136. https://doi.org/10.1016/j.ijpe.2015.11.015

Salman FS, Yücel E (2015) Emergency facility location under random network damage: Insights from the Istanbul case. Comput Oper Res 62:266–281. https://doi.org/10.1016/j.cor.2014.07.015

Salmerón J, Apte A (2010) Stochastic optimization for natural disaster asset pre-positioning. Prod Oper Manag 19:561–574. https://doi.org/10.1111/j.1937-5956.2009.01119.x

Seidgar H, Kiani M, Abedi M, Fazlollahtabar H (2014) An efficient imperialist competitive algorithm for scheduling in the two-stage assembly flow shop problem. Int J Prod Res 52(4): 1240–1256. https://doi.org/10.1080/00207543.2013.848490

Sheu JB (2007) Challenges of emergency logistics management. Transport Res Part E Logist Transport Rev 43(6):655–659. https://doi.org/10.1016/j.tre.2007.01.001

Snyder VL, Atan Z, Peng P, Rong Y, Schmitt JA, Sinsoysal B (2016) OR/MS models for supply chain disruptions: a review. IIE Trans 48(2):89–109. https://doi.org/10.1080/0740817X.2015.1067735

Stauffer J, Pedraza-Martinez JA, Van Wassenhove LN (2016) Temporary hubs for the global vehicle supply chain in humanitarian operations. Prod Oper Manag 25(2):192–209. https://doi.org/10.1111/poms.12427

Suarez P (2009) Linking climate knowledge and decisions: humanitarian challenges. The Pardee Papers. Boston University, Boston

Thomas AS, Kopczak LR (2005) From logistics to supply chain management: the path forward in the humanitarian sector. Fritz Institute, San Francisco, pp 1–15

Thomas A, Mizusjima M (2005) Logistics training: necessity or luxury? Forced Migr Rev 22:60–61

Tian J, Ma W-Z, Wang Y-L (2011) Emergency supplies distributing and vehicle routes programming based on particle swarm optimization. Syst Eng Theory Pract 31:898–906

Tofighi S, Torabi SA, Mansouri SA (2011) A fuzzy stochastic approach for pre-positioning and distribution of emergency supplies in disaster management. In: POMS 22nd Annual conference, Reno, Nevada, U.S.A. April 29 to May 2, 2011. www.pomsmeetings.org/ConfPapers/020/020-0550.pdf

Tofighi S, Torabi SA, Mansouri SA (2016) Humanitarian logistics network design under mixed uncertainty. Eur J Oper Res 250(1):239–250. https://doi.org/10.1016/j.ejor.2015.08.059

Tomasini RM, Van Wassenhove LN (2009 From preparedness to partnerships: case study research on humanitarian logistics. Int Trans Oper Res 16(5):549–559. https://doi.org/10.1111/j.1475-3995.2009.00697.x

Tricoire F, Graf A, Gutjahr WJ (2012) The bi-objective stochastic covering tour problem. Comput Oper Res 39(7):1582–1592. https://doi.org/10.1016/j.cor.2011.09.009

Uno T, Kato K, Katagiri H (2007) An application of interactive fuzzy satisficing approach with particle swarm optimization for multi-objective emergency facility location problem with a-distance. In: IEEE symposium on computational intelligence in multicriteria decision making. IEEE, New York, pp 368–373. https://doi.org/10.1109/MCDM.2007.369115

Verma A, Gaukler GM (2015) Pre-positioning disaster response facilities at safe locations: an evaluation of deterministic and stochastic modeling approaches. Comput Oper Res 62:197–209. https://doi.org/10.1016/j.cor.2014.10.006

Vitoriano B, Ortuno MT, Tirado G, Montero J (2011) A multi-criteria optimization model for humanitarian aid distribution. J Glob Optim 51(2):189–208. https://doi.org/10.1007/s10898-010-9603-z

Whybark DC (2007) Issues in managing disaster relief inventories. Int J Prod Econ 108:228–235. https://doi.org/10.1016/j.ijpe.2006.12.012

Zhang Y, Qi M, Lin W-H, Miao L (2015) A metaheuristic approach to the reliable location routing problem under disruptions. Transport Res Part E Logist Transport 83:90–110. https://doi.org/10.1016/j.tre.2015.09.001

Zhang Y, Synder VL, Qi M, Miao L (2016) A heterogeneous reliable location model with risk pooling under supply disruptions. Transport Res Part B Logist Transport 83:151–178. https://doi.org/10.1016/j.trb.2015.11.009

Zheng YJ, Ling FH, Xue YJ (2014a) Ecogeography based optimization: enhancing biogeography-based optimization with ecogeographic barriers and differentiations. Comput Oper Res 50:115–127. https://doi.org/10.1016/j.cor.2014.04.013

Zheng JY, Ling H, Xue J, Chen S (2014b) Population classification in fire evacuation: a multi-objective particle swarm optimization approach. IEEE Trans Evol Comput 18:70–81. https://doi.org/10.1109/TEVC.2013.2281396

Zheng JY, Ling FH, Shi HH, Chen SH, Chen YS (2014c) Emergency railway wagon scheduling by hybrid biogeography-based optimization. Comput Oper Res 43:1–8. https://doi.org/10.1016/j.cor.2013.09.002

Zheng YJ, Chen SY, Ling FH (2016) Evolutionary optimization for disaster relief operations: a survey. Appl Soft Comput 27:553–566. https://doi.org/10.1016/j.asoc.2014.09.041

Zhu J, Huang J, Liu D, Han J (2008) Resources allocation problem for local reserve depots in disaster management based on scenario analysis. In: The 7th International symposium on operations research and its applications. Springer, New York, pp 395–407. http://citeseerx.ist.psu.edu/viewdoc/download;jsessionid=668DF266CB99D4A9F45714A1645D9E03?doi=10.1.1.378.896&rep=rep1&type=pdf

Chapter 14
Impact of Climate Change Related Heat Stress on Thermophysiological Vulnerability Among Workers in Unorganized Sectors

P. Dutta and V. Chorsiya

14.1 Introduction

Climate change is leading to a rise in average temperatures and increased possibilities of severe heat waves. An increase in incidences of heat waves results in a deleterious effect on human health and well-being. Increased mortality due to increasing heat waves is predicted to be a major burden on health. Heat waves and resultant heat illness and heat stroke deaths are well-established facts (D'Ippoliti et al. 2010; Hess et al. 2014; Mazdiyasni et al. 2017). In Europe, during 2003 heat waves caused excess mortality estimates varying between 25,000 and 70,000 deaths (Maughan 2012; Lass et al. 2011). In 2010, the city of Ahmedabad, in the western India state of Gujarat, experienced a major heat wave. Locally, there was an excess of 1344 all-cause deaths in 1 month (May 2010) (Azhar et al. 2014). Following the 2010 heat wave, the Ahmedabad Municipal Corporation (AMC), partnered with national and international organizations, developed the Ahmedabad Heat Action Plan (HAP) to reduce the health impact of such extreme heat waves on vulnerable populations (Knowlton et al. 2014). Furthermore, NDMA supported the development of local heat action plans and issued its first issued guidelines in 2016 to provide a framework for implementation, coordination, and evaluation of extreme heatwave related activities (NDMA 2022). A 2018 study evaluated summer all-cause mortality before and after the HAP's implementation. The findings suggest

---

P. Dutta (✉)
Indian Institute of Public Health, Gandhinagar, India

V. Chorsiya
School of Physiotherapy, Delhi Pharmaceutical Sciences and Research University, New Delhi, India

B. Pathak, R. S. Dubey (eds.), *Climate Change and Urban Environment Sustainability*, Disaster Resilience and Green Growth,
https://doi.org/10.1007/978-981-19-7618-6_14

that the Ahmedabad HAP has helped avoid 1190 deaths annually since its 2013 launch (Hess et al. 2018). Still, these plans and guidelines lack their contributions toward the occupational fronts.

Moreover, human being are homoeothermic; that means their body temperature is regulated within narrow limits (Gagge and Gonzalez 2010; Kuht and Farmery 2014). The normal body "core" temperature at rest, measured for instance, in the rectum ranges between 36.5 and 37.5 °C and is maintained by a very efficient regulating system, even when the ambient temperature is high (Marui et al. 2017; Gagnon et al. 2010). The heat dissipating mechanisms involve peripheral vasodilatation, which increases heat convection and radiation, and water evaporation from the skin in the form of sweat. Inability to sustain heat results in a variety of discomforts, ill health, low productivity, and substantial increase in work accidents. The total heat load is the sum of heat generated in the body through routes (conduction, convection, and radiation) and the evaporative heat loss from the body to the environment. Through mechanisms for controlling heat balance by dissipation or production of body heat, as well as through protective measures like dressing and undressing, human can live and work in a variety of climatic environments. However, the optimum temperature at which the human body operates are only a very few degrees lower/higher from those at which thermal death occurs. Indeed, if the body core temperature drops to less than 34 °C, unconsciousness occurs, possibly followed by death; if it rises to 41 °C, unconsciousness also occurs and above 44 °C, denaturation of vital proteins begins (Huizenga et al. 2001; Arens and Zhang 2006).

Besides naturally occurring hot climates, occupational situations, such as open field tropical farming, glass and ceramic productions, molten metal operations, and many other hot industrial processes cause heat stress and strain. Occupational load represents over fourth/fifth of the effective heat load of the body.

On the occupational front, the concern areas are exposure of workers to heat stress, negative alteration in work performance, delivery, and productivity. In recent years, we are experiencing an increase in heat wave conditions in many regions of India, resulting in impact on human health and well-being. Temperature is an important factor associated with health risks to vulnerable people. The effect of heat stress becomes more life-threatening when occupational environments, especially in unorganized sectors expose the workers owing to improper infrastructure and poor thermal regulatory mechanism, for example, iron foundry and ceramic industry (indoor), construction sector (outdoor). This chapter will describe:

1. Thermophysiological vulnerability in unorganized sector workers based on thermophysiological variables.
2. Heat stress and health impacts in indoor and outdoor occupational groups.
3. Behavioral response of different occupational groups to perception of heat-related stress and strain.
4. Recommendations for human exposure to hot environments.

14.1.1 Thermophysiological Variables

1. The applicability of some of the frequently used thermal indices (Parsons 2019; Mutchler and Vecchio 1977) is discussed below:

   (a) Dry Bulb Temperature (DB): The directly measured dry bulb temperature of air (DB) is commonly used in estimating comfort zone for sedentary people in an indoor environment. The DB of 25 °C is considered comfortable for sedentary conditions. As the DB temperature and the metabolic heat production reach above the comfort zone, the level of heat stress and strain also increases when the use of DB alone is not justified.

   (b) Wet Bulb Temperature: This is considered as a useful measure for assessing heat stress and strain under high humid conditions, and when the radiant heat load and air velocity are minimal. The wet bulb temperatures have been used to predict risk of heat stroke in hot-humid conditions.

   (c) Effective Temperature (ET, CET, and ET*): The ET, widely used heat stress index, includes dry and wet bulb temperature and air velocity. The Corrected Effective Temperature (CET) is the modification of ET to replace ambient temperature by the black globe temperature to account for radiant heat. The ET/CET is based on subjective feeling of equivalent heat loads at 100% humidity and low air velocity, and an exposure of high ambient temperature and high air velocity, with low humidity. The new ET* uses 50% reference humidity in place of 100% humidity taken in ET and CET. This new ET* is useful for calculating ventilation of air-conditioning requirements in building enclosures, industries, etc.

   (d) Wet Bulb Globe Temperature (WBGT): The WBGT index was originally developed for environmental heat stress monitoring to control heat casualties at military training camps (Parsons 1993). The calculation for indoor use:

$$\mathrm{WBGT} = 0.7\,\mathrm{Tnwb} + 0.3\mathrm{Tg},$$

where nwb is the natural wet bulb temperature. For outdoor use:

$$\mathrm{WBGT} = 0.7\,\mathrm{Tnwb} + 0.2\mathrm{Tg} + 0.1\mathrm{Ta}.$$

The WBGT combines the effect of humidity and air movement (in natural wet bulb temperature), air temperature, and radiation as the factors in outdoor situations in the presence of sunshine. If there is a radiant heat load with no sunshine, the Tg reflects the effects of air temperature. The application of the WBGT index for determining training schedules for military recruits has resulted in a dramatic control or heat casualty incidents. Recommendation on hot environments is based on the WBGT index for wider application.

(e) Wet Globe Temperature (WGT): The WET (Botsball) thermometer is a portable environmental measuring device, consisting of a 3-inch copper globe covered by a black cloth. The globe and the cloth are kept 100% wet from a water reservoir. The temperature sensor is located at the center of the globe to read the temperature on a dial. Apparently the exchange of heat with the environment is similar in mechanism that a nude man with a total wet skin would exchange heat in an equivalent environment.

2. Physiological Indices

   Physiological indices provide an idea about the body's adjustment in response to changes in temperature like heart rate responses, blood pressure measurements, thermographic profile of the skin areas (T_{sk}), and deep body temperature (T_{cr}). The thermal profiling of the skin (T_{sk}) is conducted by ThermoCAM from different sites of the human body, i.e., head, hand, leg, and anterior and posterior trunk.

   (a) Weighted-Average Skin Temperature (WT_{sk}): The weighted average T_{sk} is an important thermal index based on the segmental surface area and sensitivity weighting of local skin areas. It can be determined by using an infrared thermographic camera (ThermoCAM, FLIR system, Sweden) that has automated calibration systems to adjust for ambient conditions, diurnal variations in temperature, and thermal drift. Varied region temperatures can be then digitized using the software ThermaCAM Reporter 2000 professional. The values then can be grouped as skin areas (T_{sk}) of head, trunk, back, upper arm, hand, thigh, and foot. Then the weighted average for skin temperature (T_{sk}) is calculated on the basis of the grouped areas in the skin area.
   (b) Skin Wettedness (%SWA): In addition to the sweat production required for temperature equilibrium (E_{req}) and the maximum amount of sweat that can be evaporated (E_{max}), as explained in HSI, the efficiency of sweat evaporation, with a greater body surface area will need to be wetted with sweat to maintain the required evaporative heat transfer. The ratio of wetted to non-wetted skin area is the new indicator of strain at high humidity and low air movement.
   (c) Predicted 4 h Sweat Rate (P4SR): This is a heat strain index based on both metabolic and environmental heat exchange. It has some limitations particularly for profuse sweating conditions when the predicted sweat rate is markedly underestimated. A P4SR of 4.5 L may likely cause serious heat disorders. A practical and safe upper limit of P4SR of 4 L has been adopted by Hall (1968) in studying outdoor manual workers in East Africa.

14.1.2 Physical Activity in Hot Environment

When a person is exposed to heat, the body temperature tends to increase depending upon the local climate, i.e., ambient temperature, humidity, and air movement. If the heat load is sufficiently large, the sweat glands are activated and the skin is cooled as the water in the sweat evaporates, depending upon the individual's capacity for sweating. Muscular activity increases the body temp as a linear function of workload. This is due to the exercise. This is due to the exercise-induced increase in body heat production, which may reach values 10–15 times the resting value, depending upon the intensity of activity. If the climatic conditions are unfavorable, e.g., air temperature above skin temperature (34–35 °C), lack of ventilation, and high humidity (close to saturation) heat dissipation from the body to the environment is seriously hampered and body temperature keeps increasing steadily. The thermoregulatory responses are fundamental to maintain a constant body temperature (ISO Standard 7243 1989). Since the climate of the tropics cannot be controlled, the thermoregulatory mechanism of a person is the only end where adjustment is possible through systematic acclimatization of persons in a standard condition and selective recruitment of persons depending on their capacity to sustain work in heat. The level of heat stress at which excessive heat strain results depends on the heat tolerance capabilities. Approximate repeated exposure to elevated heat stress causes a series of physiologic adaptations called acclimatization, whereby the body becomes more efficient in coping with the heat stress. Although people in the tropics are usually considered to be heat acclimatized, there is evidence of insufficient tolerance to fast climatic fluctuations. Because of the specificity of a person's heat acclimatization, predetermination of the exposure conditions as comfort zone, tolerable, just tolerable, and intolerable zones may be required to avoid heat casualty (World Health Organization 1990).

To cope with the high heat load, there is a considerable load on the cardiovascular system with increased demand for cardiac output and also accentuated by the opening up of superficial heat exchanging capillary vessels. Heat elimination mechanism by sweating leads to dehydration of the body, which may need to be replenished by water and salt supplement. At high humidity as well as a low air velocity condition, the sweating process often gets affected and exerts deleterious effects on man at work. In such situations, hidromeiosis (reduction of the sweat production and heat elimination) occurs. Therefore, the heat regulatory demands in turn influence and reduce the working capacity under hot conditions. Relatively low work capacity of the tropical population groups is to a large extent attributed to hot climate.

During prolonged exposure to a hot environment, a gradual decrease in the sweat may occur, owing to sweat gland fatigue. However, this condition is ordinarily quickly overcome, and after a week of heat exposure the sweat production increases up to twofold, while its salt content decreases, an acclimatization mechanism contributes to improved heat tolerance.

14.1.3 Acclimatization

Acclimatization is induced by repeated exposures to natural hot environments, while acclimation refers to experimentally induced physiological adaptation. With heat acclimatization, human beings demonstrate an increased ability to work in heat, reduce cardiovascular stress, and respond to acute heat stress with pronounced sudomotor response. Repeated exposure to heat results in improved work capacity. Increase in cardiac output after acclimation is attributable to an increased stroke volume. Acclimation reduced the threshold for vasodilatation. The splanchnic and renal blood flow, which are reduced during acute heat exposure, increases after acclimation, blood flow to exercising muscle also increases after acclimation. Even though these are hemoconcentration throughout heat exposure, the plasma volume in acclimatized state is equal to or in excess of the value of the initial heat exposure.

The principal physiological benefit appears to result from an increased sweating efficiency. This produces a drop in skin blood flow and reduces the need of skin blood flow for heat transfer. It is possible that heat acclimatization shifts the thermoneutral zone and sets a point in temperature regulation. It is observed that neither metabolic concentrations, nor the water and electrolyte regulating hormones (aldosterone and vasopressin) change with acclimation and therefore they cannot contribute to the entrance performance after acclimatization, nor to the delay in time to exhaustion. Acclimatization to heat is an unsurpassed example of physiological adaptation, it does not necessarily mean that individuals can work in widely different hot environments. This calls for attention that acclimatization may be specific to the level of heat exposure to which a person is exposed and may not respond well above the level of exposure. In spite of the fact that acclimatization is reasonably maintained for a few days of non-heat exposure, absence from work in the heat for a week or more results in a significant loss in adaptations. Normally the acclimatization to a hot environment job happens in around 2–3 days. For emergency placement of workers, sweat glands can be trained locally by repeated local exposures to heat, especially by daily hot water bath of a limb area (Hanna and Tait 2015; Sawka et al. 2011).

14.1.4 Human Tolerance to Work in Hot Environment

Management of human heat exposures is important to safeguard human health in industrial and community environments. Human strains increase as heat accumulates in the body. The upper limit of acceptable strains, often referred to as tolerance time, depends on the environment, characteristics of persons and one's thermal state of initiation of heat exposure. In an extremely hot environment, skin pain is usually the limiting factor. In a moderate hot environment, the skin temperature (T_{sk}) does not reach the painful level when heat tolerance is limited by the rise in body core temperature. In spite of the fact that the person in a moderately hot environment may not reach the critical level for skin pain and respiratory distress, they eventually become distressed and may show personality changes. These appear to be a close relation between subjective distress and total heat storage of the body. There is an obvious notion that it is always difficult to identify heat intolerant persons to avoid possible placement in high heat. In practice, the individuals who are least tolerant to heat are most vulnerable (Kjellstrom et al. 2009; Xiang et al. 2014).

14.1.5 Uses of Heat Stress and Strain Indices

An index for assessing heat stress and strain must satisfy the following criteria

1. Essential heat exchange factors (environmental, metabolic, clothing, physical work, etc.) must be considered.
2. The measuring instruments and techniques must derive data that reflect a person's heat exposure, without any interference with the person's activity.
3. Index must be applicable for setting exposure limits under a wide range of environmental and metabolic conditions.
4. Feasibility and accuracy of the index must be widely tested in actual exposure situations. Several schemes have been devised for assessing and/or predicting the heat stress and strain that a person might experience when working in hot environmental conditions.

To develop a better understanding about the heat load on workers in an outdoor and indoor environment, we present two case studies. They very well elucidate the interplay of the thermophysiological indices in order to provide the required thermoregulation to the worker's body.

14.2 Case Study 1: Heat Load on Workers in Open Cast Mines (Dutta et al. 2021)

For an outdoor occupation, an example of stone quarry (studied 934 men in the age range between 18 and 60 years) has been taken up, where it is difficult to control the environmental adversaries when compared to occupations like steel plant, steel plant, power plant, forge plant (Dutta et al. 2021). With the different seasons like summer, post-monsoon, and winter, the body's thermophysiology also behaves significantly differently. With this purpose, the environmental and thermophysiological indices were studied. The responses were the resultant of the combined effect of extreme temperature and work strenuousness that ranged from heavy to extremely heavy in the month of summer, post-monsoon, and heavy to moderately heavy in the month of winter.

The profiling of weighted T_{sk} indicates deviation from the thermoneutral reference, provoking a distinctive peripheral response for feedback and regulation in building up of body temperature. The association between WBGT outdoor and weighted T_{sk} is presented as Relative Risk (RR) with centering point 33 °C for the range from 20 to 43 °C of temperature. With increasing temperature, the curve increases gradually and then tend to plateau showing triggering of the thermoregulatory mechanism of the human body in all three defined seasons. It was evident that toward the body's thermoregulatory efforts, the weighted T_{sk} responses showed an increasing trend with the increasing WBGT. The temperature and humidity dominantly influenced the local T_{sk}.

As observed in the study, there was considerable difference in the tolerance time of stone quarry work in three different seasons, due to the differences in the environmental variables and workload. From the cross-sectional data on stone quarry workers, it was estimated that there was ~14% loss of tolerance time per degree increase of WBGT, from 33 to 35 °C WBGT. The loss of tolerance time might also indicate loss of productivity due to heat exposure, which Kjellstrom (2016) referred to as High Occupational Temperature Health and Productivity Suppression (Hothaps) effect, for loss of working ability or working capacity. The relative workload was higher during the winter season. It is likely that the workers might be adopting a self-adjustment strategy in the pace of work distributing the work and workload as per the varying environmental exposures. In repeated occupational exposures high heat load and strenuous physical activity, human's defense mechanism undergoes progressive changes for internal thermal stability (acclimatization), depending upon physiological adaptive capacity (Morioka et al. 2006). Data amply suggest that the workers during the summer months were at an unsafe zone of exposure and 14% of the workers were vulnerable to heat illnesses. Also, the workers lack awareness and measures to mitigate risks. In the month of May ~85% of the workers, the working heart rate was found to be more than 90 beats/min. The tolerance time level in summer months (71 ± 13 min at WBGT 35 ± 2.3 °C) was less than in other two seasons. The result of the cross-sectional survey study on open cast mine workers revealed that ~8% loss of tolerance time per degree increase of WBGT happens from 26 to 35 °C of WBGT (outdoor). The loss of tolerance time might also indicate loss of productivity due to heat exposure. Corresponding to observations of physiological and subjective responses to heat stress, the workers were vulnerable to heat illnesses. Around 30% of the workers reported diminished or no sweat along with reduction of urine output when exposed to summer conditions. During continuous and habitual exposure to hot climate during working along with strenuous physical activity, the human body acclimatizes to maintain thermal stability through progressive changes which is governed by physiological adaptation. The study concluded that the workers in the month of summer were found to be at an unsafe zone of exposure and ~17% of the workers were vulnerable to heat illnesses.

14.3 Case Study 2: Heat Load on Indoor Workers (Dutta and Chorsiya 2014)

To study the heat load vulnerability in indoor environment conditions, a cross-sectional survey study was done on a total of 1013 subjects including, iron foundry workers at Ahmedabad ($N = 587$) and ceramic workers at Morvi ($N = 426$). The workers surveyed during the working period between 11 a.m. to 4 p.m. in all seasons, i.e., summer (May and June), post-monsoon (September and October), and winter (December and January). The environmental measurements include

WBGT index measured by WBGT Monitor, Delta OHM (HD 32.1 Thermal Microclimate, Italy). The thermophysiological heat strain parameters include sweating response, heart rate and blood pressure measurements, weighted T_{sk} (WT_{sk}), and body's core temperature (T_{cr}).

The reactive heart rate in response to temperature was more in the summer months which then followed the winter months. The result indicated a varied range of seasonal variations, that is suggestive of workers' self-adjustment strategy with the pace of work. During summer, there might be a requirement for channeling more blood volume to the peripheral circulatory zones for thermoregulatory (cooling) purposes in contrast to winter which has a higher metabolic demand (Kristal-Boneh et al. 1995). In the study, the result showed a little increase in systolic blood pressure between different groups of workers in the winter months. The trend that the result showed is in accordance with the findings of Kristal-Boneh et al. (1995) which is suggestive that the average systolic BP during work was more during winter as compared to summer. Alpérovitch et al. (2009) demonstrated the association of blood pressure and ambient temperature and also suggested that the systolic blood pressure declines with increasing temperature, i.e., 8 mmHg decrease between the lowest (<7.9 °C) and the highest (>21.2 °C) temperature. The supporting mechanism probably could be that in the cooler environment, activation of the sympathetic nervous system and secretion of catecholamine might be increased that results in an increase in blood pressure through an increased heart rate and peripheral vascular resistance (Alpérovitch et al. 2009). In hot working environments, the primary importance is the thermoregulatory mechanism to dissipate excessive heat from the body via sweat production and its evaporation from the skin surface that results in loss of water and electrolyte (Galloway and Maughan 2000; Inbar et al. 2004). The result of the study revealed a statistically significant difference in iron foundry workers ($F = 31.2$, $p < 0.001$) and ceramic workers ($F = 392.3$, $p < 0.001$) in sweat loss between the three seasons studied. The workers in iron foundry and ceramic during summer were 12.6 ± 1.3 and 14.8 ± 2.0 g/min respectively (i.e., 3–3.7 L of P4SR). Consequently, performance may be impaired, and workers may pose a threat to health. The conditions of water requirements of informal workers are very much excruciating.

It was evident from the present study that toward the body's thermoregulatory efforts, the local T_{sk} responses showed an increasing trend with the increasing thermal stress. The indoor workers have no risk till 30 °C and their body's thermoregulatory mechanism starts to dissipate excess heat from the body. The association between WBGT outdoor and weighted T_{sk} is presented as Relative Risk (RR) with centering point 30 °C for the range from 20 to 43 °C of temperature. With increasing temperature, the curve increases gradually and then tends to plateau showing triggering of the thermoregulatory mechanism of the human body in all three defined seasons. It was evident that toward the body's thermoregulatory efforts, the weighted T_{sk} responses showed an increasing trend with the increasing WBGT.

The case studies elucidate the role of thermophysiological interplay toward peripheral thermoregulatory adjustability. The applicability of skin temperature (T_{sk}) is to assess vulnerability toward heat stress and to estimate human tolerance.

A heat gradient helps to transfer heat from the deep tissues (core) to the skin (shell), from where it dissipates to the ambient environment. Circulation of warm blood from central (core) body compartments to the skin results in an increase in skin temperature. Sometimes due to environmental conditions, it happens that sufficient heat is not dissipated to the environment from the skin that leads to an increase in core temperature of the body. As T_{cr} increases above 38 °C (100.4 °F), humans are prone to develop the risk of heat-related illness. The case studies mentioned above depict a difference of 3 °C in reaching to a centering point when comparing an outdoor occupation (33 °C) to an indoor occupation (30 °C). The difference is by the virtue of the fact that the environmental condition mainly, temperature and humidity dominantly influence the weighted T_{sk} in case of indoor occupation which hallmark the early initiation of thermoregulatory mechanism. The literature and the studies highlight the importance of thermophysiological vulnerability assessment to determine the magnitude of worker's thermal stress and physiological reaction to ensure optimal conditions for health in the hot and extreme environmental conditions.

14.4 Recommendations for Human Exposure to Hot Environment

- Engineering controls
  - Control the heat at source through the use of insulating and reflective barriers (insulate furnace walls).
  - Exhaust hot air and steam produced by specific operations.
  - Reduce the temperature and humidity through air cooling.
  - Provide air-conditioned rest areas/rooms or cool areas/rooms.
  - Increase air movement if temperature is less than 35 °C (fans).
  - Reduce physical demands of work task through mechanical assistance (hoists, lift-tables, etc.).
- Administrative controls
  - Organize training for employers, outdoor laborers, workers regarding health impacts of extreme heat and recommendations to protect themselves during high temperature.
  - Potentially overlay with irradiation map from IMD or heat island map to identify more high-risk outdoor workers.
  - Health and safety committees should assess the demands of all jobs and have monitoring and control strategies in place for hot days.
  - Change working hours of laborers. Encourage employers to shift outdoor workers schedules away from peak afternoon hours (1 p.m.–5 p.m.) during heat alert.
  - To schedule hot jobs to cooler times of the day.
  - Conduct publicity campaign during high-risk days to specific areas.

- To provide cool drinking water near workers and remind them to drink a cup every 20 min or so.
- Encourage individuals to discussion of early sign of heat exhaustion with their local doctor or Urban Health Centre.
- Workers should salt their food well, particularly, while they are acclimatizing to a hot job (workers with a low salt diet should discuss this with their doctor).
- Train workers to recognize the signs and symptoms of heat stress and start a "buddy system" since people are not likely to notice their own symptoms.
- Assign additional workers or slow down work pace.
- Make sure everyone is properly acclimatized.
- Pregnant workers and workers with a medical condition should discuss working in the heat with their doctor.
- Increase the frequency and length of rest breaks during heat wave period.

• Personal protective equipment

- Light clothing should be worn to allow free air movement and sweat evaporation.
- Outside, wear light-colored clothing.
- In a high radiant heat situation, reflective clothing may help.
- For very hot environments, air, water or ice-cooled insulated clothing should be considered.
- Vapor barrier clothing, such as acid suits, greatly increases the amount of heat stress on the body, and extra caution is necessary.

References

Alpérovitch A, Lacombe JM, Hanon O, Dartigues JF, Ritchie K, Ducimetière P, Tzourio C (2009) Relationship between blood pressure and outdoor temperature in a large sample of elderly individuals: the Three-City study. Arch Intern Med 169(1):75–80

Arens EA, Zhang H (2006) The skin's role in human thermoregulation and comfort. Woodhead Publishing, Cambridge

Azhar GS, Mavalankar D, Nori-Sarma A, Rajiva A, Dutta P, Jaiswal A, Sheffield P, Knowlton K, Hess JJ (2014) Heat-related mortality in India: excess all-cause mortality associated with the 2010 Ahmedabad heat wave. PLoS One 9(3):e91831

D'Ippoliti D, Michelozzi P, Marino C, de'Donato F, Menne B, Katsouyanni K, Kirchmayer U, Analitis A, Medina-Ramón M, Paldy A, Atkinson R (2010) The impact of heat waves on mortality in 9 European cities: results from the EuroHEAT project. Environ Health 9(1):1–9

Dutta P, Chorsiya V (2014) Thermophysiological vulnerability to heat stress among indoor workers. Public Health Rev Int J Public health 1(1):18–24

Dutta P, Chorsiya V, Nag PK (2021) Perceived thermal response of stone quarry workers in hot environment. Front Sustain Cities 3:640426

Gagge AP, Gonzalez RR (2010) Mechanisms of heat exchange: biophysics and physiology. In: Comprehensive physiology. Wiley, New York, pp 45–84

Gagnon D, Lemire BB, Jay O, Kenny GP (2010) Aural canal, esophageal, and rectal temperatures during exertional heat stress and the subsequent recovery period. J Athl Train 45(2):157–163

Galloway SD, Maughan RJ (2000) The effects of substrate and fluid provision on thermoregulatory and metabolic responses to prolonged exercise in a hot environment. J Sports Sci 18(5):339–351

Hall SA (1968) Heat stress amongst outdoor manual workers in East Africa. East Afr Med J 45(5): 351–357

Hanna EG, Tait PW (2015) Limitations to thermoregulation and acclimatization challenge human adaptation to global warming. Int J Environ Res Public Health 12(7):8034–8074

Hess JJ, Saha S, Luber G (2014) Summertime acute heat illness in US emergency departments from 2006 through 2010: analysis of a nationally representative sample. Environ Health Perspect 122(11):1209–1215

Hess JJ, Lm S, Knowlton K, Saha S, Dutta P, Ganguly P, Tiwari A, Jaiswal A, Sheffield P, Sarkar J, Bhan SC (2018) Building resilience to climate change: pilot evaluation of the impact of India's first heat action plan on all-cause mortality. J Environ Public Health 2018:7973519

Huizenga C, Hui Z, Arens E (2001) A model of human physiology and comfort for assessing complex thermal environments. Build Environ 36(6):691–699

Inbar O, Morris N, Epstein Y, Gass G (2004) Comparison of thermoregulatory responses to exercise in dry heat among prepubertal boys, young adults and older males. Exp Physiol 89(6):691–700

ISO Standard 7243 (1989) Hot environments—estimation of the heat stress on working man, based on the WBGT-Index (wet bulb globe temperature). International Standards Organization, Geneva

Kjellstrom T (2016) Impact of climate conditions on occupational health and related economic losses: a new feature of global and urban health in the context of climate change. Asia Pac J Public Health 28:28S–37S. https://doi.org/10.1177/1010539514568711

Kjellstrom T, Holmer I, Lemke B (2009) Workplace heat stress, health and productivity—an increasing challenge for low and middle-income countries during climate change. Glob Health Action 2(1):2047

Knowlton K, Kulkarni SP, Azhar GS, Mavalankar D, Jaiswal A, Connolly M, Nori-Sarma A, Rajiva A, Dutta P, Deol B, Sanchez L (2014) Development and implementation of South Asia's first heat-health action plan in Ahmedabad (Gujarat, India). Int J Environ Res Public Health 11(4):3473–3492

Kristal-Boneh E, Harari G, Green MS, Ribak J (1995) Seasonal changes in ambulatory blood pressure in employees under different indoor temperatures. Occup Environ Med 52(11): 715–721

Kuht J, Farmery AD (2014) Body temperature and its regulation. Anaesth Intensive Care Med 15(6):273–278

Lass W, Haas A, Hinkel J, Jaeger C (2011) Avoiding the avoidable: towards a European heat waves risk governance. Int J Disaster Risk Sci 2(1):1–14

Marui S, Misawa A, Tanaka Y, Nagashima K (2017) Assessment of axillary temperature for the evaluation of normal body temperature of healthy young adults at rest in a thermoneutral environment. J Physiol Anthropol 36(1):1–7

Maughan RJ (2012) Hydration, morbidity, and mortality in vulnerable populations. Nutr Rev 70 (suppl_2):S152–S155

Mazdiyasni O, AghaKouchak A, Davis SJ, Madadgar S, Mehran A, Ragno E, Sadegh M, Sengupta A, Ghosh S, Dhanya CT, Niknejad M (2017) Increasing probability of mortality during Indian heat waves. Sci Adv 3(6):e1700066

Morioka I, Miyai N, Kazuhisa M (2006) Hot environment and health problems of outdoor workers at a construction site. Ind Health 44:474–480. https://doi.org/10.2486/indhealth.44.474

Mutchler JE, Vecchio JL (1977) Empirical relationships among heat stress indices in 14 hot industries. Am Ind Hyg Assoc J 38(6):253–263

NDMA (2022) NDMA guidelines. https://ndma.gov.in/sites/default/files/PDF/Guidelines/heatwaveguidelines.pdf. Accessed 25 Jan 2022

Parsons KC (1993) Human thermal environments. Taylor & Francis, Abingdon, p 174

Parsons K (2019) Human heat stress. CRC Press
Sawka MN, Leon LR, Montain SJ, Sonna LA (2011) Integrated physiological mechanisms of exercise performance, adaptation, and maladaptation to heat stress. Compr Physiol 1(4): 1883–1928
World Health Organization (1990) Indoor environment: health aspects of air quality, thermal environment, light and noise. World Health Organization, Geneva
Xiang J, Bi P, Pisaniello D, Hansen A (2014) Health impacts of workplace heat exposure: an epidemiological review. Ind Health 52(2):91–101

Chapter 15
Seagrass of Southeast Asia: Challenges, Prospects, and Management Strategies

Akbar John, Bryan Nelson Raveen, Hassan I. Sheikh, Abu Hena Mustafa Kamal, Balaji Vedharajan, Ragavan Pandisamy, Kannan Sivakumar, P. Anantharaman, and Chandani Appadoo

15.1 Introduction

The Southeast Asian region, which has a population of 677 million people and accounts for more than 8.5% of the global population, is rich in biodiversity hotspots that are home to a vast diversity of endemic species and unique ecosystems in both terrestrial and marine habitats. The region covers 11 countries with various

A. John (✉)
Environmental Technology Division, School of Industrial Technology, Universiti Sains Malaysia, Penang, Malaysia

Renewable Biomass Transformation Cluster, School of Industrial Technology, Universiti Sains Malaysia, Penang, Malaysia

B. N. Raveen
Institute of Tropical Biodiversity and Sustainable Development, University Malaysia Terengganu, Kuala Nerus, Terengganu, Malaysia

H. I. Sheikh · A. H. M. Kamal
Faculty of Fisheries and Food Science, University Malaysia Terengganu, Kuala Nerus, Terengganu, Malaysia

B. Vedharajan
OMCAR Foundation, Velivayal, Tamil Nadu, India

R. Pandisamy
Ministry of Environment, Forest and Climate Change, New Delhi, India

K. Sivakumar · P. Anantharaman
Centre of Advanced Studies in Marine Biology (CASMB), Annamalai University, Parangipettai (Portnovo), India

C. Appadoo
Department of Biosciences and Ocean Studies, Faculty of Science, University of Mauritius, Réduit, Mauritius

B. Pathak, R. S. Dubey (eds.), *Climate Change and Urban Environment Sustainability*, Disaster Resilience and Green Growth,
https://doi.org/10.1007/978-981-19-7618-6_15

biogeographical niches and culturally distinct inhabitants that rely on the coastal region for their survival. Unlike the attention paid to the coral and mangrove ecosystems, the scientific community has paid less attention to the seagrass ecosystem. This is due to the large-scale ecosystem size of coral and mangrove habitats, as well as the economic and aesthetic values embedded in those ecosystems, as opposed to small-scale ecosystems like seagrass. Despite the fact that they only cover a small portion of the seabed, they have a more significant role in combating ocean acidification, global warming, and greenhouse gas footprints, which in return promote the recruitment of corals after bleaching episodes (Fortes et al. 2018).

Seagrass meadows are a prime example of a habitat that is largely understudied and poorly documented in the Southeast Asian region (Waycott et al. 2009), with only 62 ISI cited seagrass-related publications between the 1980s and 2010 (Ooi et al. 2011), the majority of which are on two specific sites in the Philippines' Northwest Luzon and Indonesia's South Sulawesi. Much of the information on the seagrasses of this region collected in the globally financed initiatives like the UNEP-GEF South China Sea Project (United Nations Environment Programme (UNEP) 2008) lies as grey literature. However, these studies focus only on specific locations, neglecting the larger Southeast Asian region, which is data deficient. Seagrass meadows are well-known for providing a variety of important ecosystem services, including habitat formation, nutrient cycling, carbon sequestration, and food provisioning (Cullen-Unsworth and Unsworth 2013), with many coastal populations in Southeast Asia directly reliant on these habitats for survival (William and Heck Jr. 2001; Masahiro et al. 2014). Seagrasses are also declining at a global rate of about 7% (Waycott et al. 2009). In Southeast Asia, the losses may be more severe due to increased coastal landform transformations and incomplete baseline information seagrasses. The socio-economic-cultural linkages between seagrass meadows and coastal inhabitants, which scholars now recognise, are maybe just as essential (Unsworth and Cullen 2010). Two previous reviews (Fortes 1995; Kirkman and Kirkman 2002) identified sediment circulation, eroded shores, and storms as reasons for seagrass loss in Southeast Asia, and these risks, as well as their accompanying issues, are still relevant today. Instead, this review paper will focus on the current level of knowledge about seagrass in Southeast Asia, with a particular emphasis on the breadth of seagrass within the varied biogeographic regions of the region's marine habitats. Based on the status review, we focus on the areal gaps in knowledge within Southeast Asia, followed by a broader thematic gap analysis. We also address the conservation and management problems posed by these gaps. We offer a comprehensive roadmap for seagrass conservation and study across Southeast Asia to improve the coastal environment's adaptive capability to natural and anthropogenic stressors and so as to help the people who rely on the coastal resources.

15.2 Seagrasses of Southeast Asia

Seagrass meadows are abundant in the tropical seascape, and their distribution relates closely to the habituation of corals and mangroves due to the availability of service agents. It is estimated that seagrass meadows contribute 15% of the ocean carbon storage worldwide and are identified to be the most vulnerable habitat due to exposure to various anthropogenic factors. Furthermore, many important herbivores on coral reefs use seagrass as an alternate grazing habitat; therefore, safeguarding fisheries in seagrass meadows contribute to coral reef fish assemblages' functional purpose. It means that supporting seagrass conservation activities does not compete but instead offers mutualistic support that works hand in hand with coral reef conservation efforts (Fortes et al. 2018).

Six seagrass bioregions span the globe's oceans, including tropical and temperate areas (Short et al. 2007). Southeast Asia is part of the Indo-West Pacific bioregion (Bioregion 5), which spans east Africa to the eastern Pacific Ocean and is known for being the world's largest and most biodiverse region (24 species). According to Fortes (1988)'s cluster study for the Indo-West Pacific seagrasses, Southeast Asia is potentially a separate biogeographic region within the Indo-West Pacific. The Philippines and Brunei Darussalam were also found to be slightly different from other locations due to high seagrass species counts in the former and low species numbers in the latter (Fortes 1988), implying that finer-scale zones within Southeast Asia may exist if distribution data is updated. In Malaysia, 16 species of seagrasses were recorded and distributed along 78 locations in a scattered fashion in coastal waters in peninsular and East Malaysia.

The Sunda Shelf and Western Coral Triangle have the highest species richness (15 species) at the province level. Individual ecoregions within these provinces have species richness ranging from 3 to 14 species (Fig. 15.1), but we omitted *Halophila gaudichaudii* and *Halophila tricostata* from our dataset due to locational uncertainty. The Cocos-Keeling/Christmas Island ecoregion (3 species) and the South China Sea Oceanic Islands ecoregion (3 species) had low species counts (4 species). Both ecoregions, which are made up of atolls with lagoon seagrass meadows, are relatively isolated. Low species richness in these lagoons is likely due to either a narrow dispersal pathway between meadows in Southeast Asia and these remote locations or a lack of acceptable biological drivers for the bulk of species. The number of seagrass species and the size of recognised meadows in Southeast Asian marine bioregions have been well recorded and discussed elsewhere (Fortes et al. 2018) and are shown in Fig. 15.1.

The Malacca Strait emerges as a distinct ecoregion in terms of species richness. It is one of the smaller ecoregions, although it is home to 14 seagrass species. Because it connects the Indian Ocean and the South China Sea, this small strait, spanning 926 km in length, is one of the busiest cargo channels in the world (Ibrahim and Nazery 2007). The placement of the strait in an area of overlap between Indian and the Pacific Ocean fauna was cited as a potential explanation for high species richness in a study of worldwide shore fish biodiversity (Carpenter and Springer 2005). This

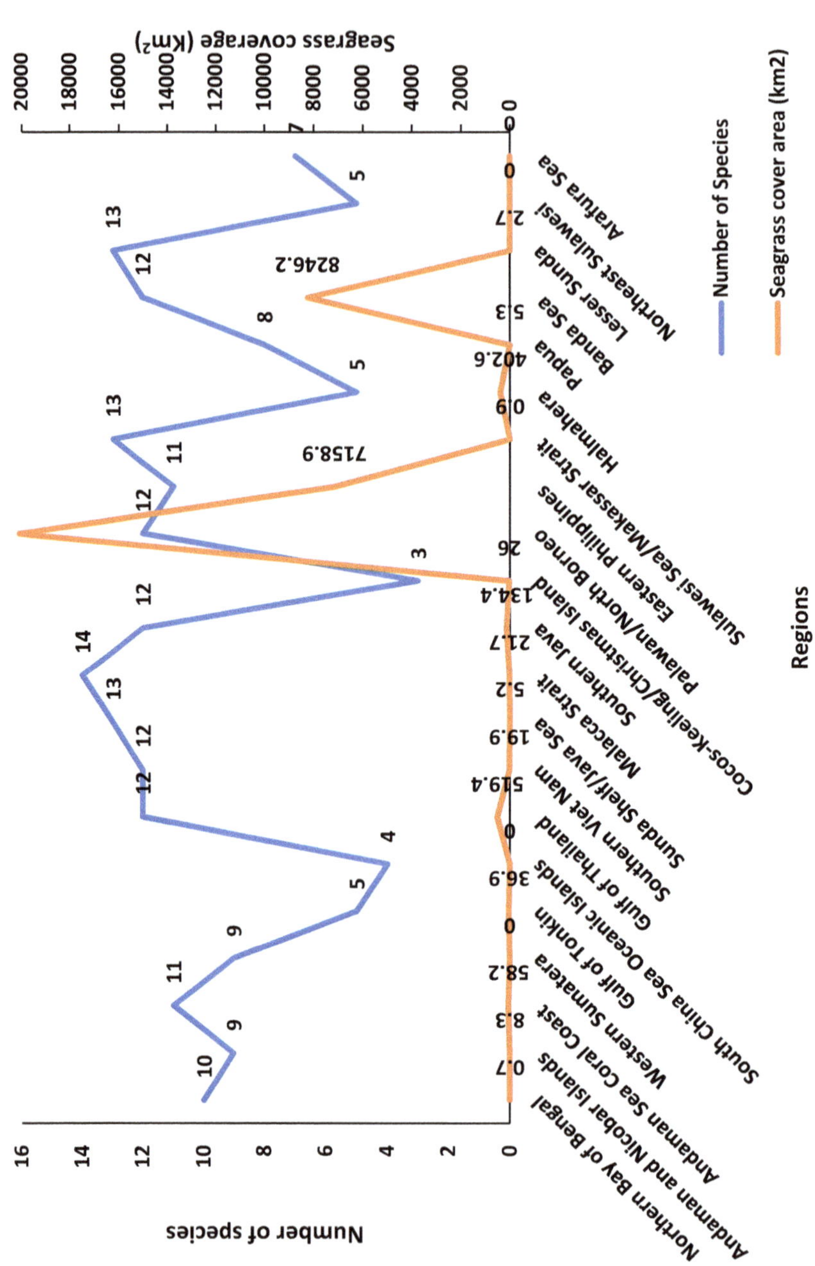

Fig. 15.1 Number of species and respective seagrass coverage area in (km^2) in representative regions are depicted. Value zero (0) in the secondary axis represents the data on seagrass coverage in those regions that are not available in the literature

ecoregion is known as the Indo-Pacific Barrier (Bowen et al. 2016), which is the equivalent of a marine Wallace line that separates populations of marine animals on both sides of the strait due to changes in sea level over glacial times. This knowledge helps guide the selection of sampling locations for seagrass phylogeographic studies that address questions about biodiversity distributions in Southeast Asia and test hypotheses about Southeast Asia as a centre of overlap, refuge, accumulation, or origin for seagrasses, such as those proposed by Mukai (1993) and Nguyen et al. (2013, 2014). However, we believe it is essential to highlight that development in the Malacca Strait due to shipping, port construction, and land reclamation is likely to influence how quickly seagrasses and other marine ecosystems change in the near future (Mokhzani 2004; Ibrahim and Nazery 2007).

Thalassia hemprichii is the most widespread species in Southeast Asia, with distribution records in all ecoregions, including the South China Sea Oceanic Islands. *Cymodocea serrulata* and *Cymodocea rotundata* are also frequent species, except for the South China Sea Oceanic Islands and the Cocos-Keeling/Christmas Islands, which exist in all ecoregions. *Halophila sulawesii* (Kuo 2007), with only one record on Samalona Island in the Spermonde archipelago (see also: Taxonomic Highlights, below), and *Zostera japonica* in the Gulf of Tonkin were two species that were unique to one ecoregion (Luong et al. 2012).

Seagrass meadows cover 36,762.6 km^2 in Southeast Asia, accounting for around 0.8% and 24% of the total area, respectively, compared to the entire landmass and coastline length. Palawan/North Borneo (20,115 km^2), the Eastern Philippines (7159 km^2), and the Banda Sea (8246.2 km^2) ecoregions, all of which are part of the Western Coral Triangle Province, have the most prominent areas. In terms of seagrass meadow estimates, not all ecoregions are equally well-represented. The Oceanic Islands of the South China Sea, Western Sumatera, Northeast Sulawesi, and the Arafura Sea ecoregions, for example, all have significant data gaps. These will be emphasised in greater detail in the section ‘Real knowledge and information shortages’.

15.3 Taxonomic Uncertainty

There are 21 species of seagrasses in Southeast Asia, but some are still considered taxonomically questionable. *Halophila* is the genus with the most significant number of unvalidated species. Although this genus has a lot of taxonomic diversity, its constituent species seem to have overlapped leaf morphologies, making confirmation difficult. The species’ plasticity in response to varied substratum, salinity, and light regimes has been studied (Japar Sidik et al. 2010). This is the main reason for taxonomic ambiguity based only on morphological features.

Halophila major has been investigated more thoroughly throughout Southeast Asia than other species in its genus, using a combination of morphological and molecular features. As a result, it has recently been recorded in Indonesia, Thailand (Tuntiprapas et al. 2015), Malaysia, Myanmar, Viet Nam, and the Philippines (Nguyen et al. 2013, 2014). Despite being recognised as two different species by

Kuo, *Halophila minor* and *Halophila ovata* were treated as synonyms for the seagrass flora of Singapore (Yaakub et al. 2013). Kuo (2000) offered *Halophila gaudichaudii* as a replacement for *H. ovata*, which was later deemed an illegitimate name (2006). As a result, all three confounding species, *H. minor, H. ovata, and H. gaudichaudii*, are found in this region's species records and are described in this review to maintain the transparency of these species listings until a taxonomic consensus is reached. However, *H. major* was found in records from Vietnam (Southern Viet Nam ecoregion) and Myanmar (Northern Bay of Bengal ecoregion); *H. gaudichaudii* was found in recordings from the Andaman and Nicobar Islands and the Philippines; and *H. ovata* was found in records from the Philippines (ecoregion undetermined). *Halophila tricostata* is native to Australia's east coast. Its presence in Southeast Asian species records suggests that it has the potential for long-distance migration from Australia's Great Barrier Reef to the Palawan/North Borneo ecoregion.

15.4 Fishes Are in Preference of Seagrass Ecosystem

Without a doubt, the seagrass ecosystem benefits the fishing industry by offering more suitable fishing grounds. Fishers will have to hunt elsewhere for resources when alternative dominating taxa replace hard corals (e.g., soft corals, corallimorphs, and sponges) and reef accretion declines. Seagrass ecosystems are an example of an alternative fisheries habitat that provides broad shallow-water fishing grounds that are often easily accessible (Murugan et al. 2014; Nordlund et al. 2018). When these seagrass systems are robust, they include an abundance of productive fish and invertebrate fauna (Al-Asif et al. 2020). Even with modest gear, it is often easy to catch fish and invertebrates in seagrass meadows. Seagrass meadows can host very productive fish assemblages of significant commercial and subsidence importance when healthy and effectively managed.

There is extensive evidence of high-intensity seagrass fishing efforts in Indonesia and the Philippines. Many of these sites have become severely deteriorated reefs. Fishermen target seagrass meadows in Indonesia where reef fisheries are fast declining (Unsworth et al. 2015). There is also evidence that fishers are rapidly working their way down the food chain, becoming more reliant on species that were originally thought to be highly unattractive, indicating unsustainable demand on the available seagrass resources (Abu Hena et al. 2015; Johan et al. 2020). In the Philippines, similar patterns have been seen. Seagrasses, as opposed to mangroves and coral reefs, have been proven to be the most visited fishing grounds in the Indo-Pacific region, delivering the most community benefits (Unsworth 2010). The growing reliance on seagrass meadows as a primary fisheries habitat has led to the widespread employment of ever more efficient and exploitative fishing tactics, such as static fish fences (Unsworth et al. 2015), which has resulted in increased food chain damage.

15.5 Feasibility in Seagrass Conservation

Tropical seagrass meadows receive very little conservation money, and seagrass research is minimal compared to other ecosystems like coral reefs and mangroves. Furthermore, the administration and management of seagrass ecosystems are almost non-existent in many regions of the planet.

Local and regional consequences such as deteriorated water quality, physical disturbance, and the collapse of food webs are putting a strain on seagrass meadows (Waycott et al. 2009; Unsworth et al. 2019). Seagrass decline has been rapid and widespread worldwide (Unsworth et al. 2019). The causes of seagrass loss, on the other hand, are mainly controllable, and dangers can be mitigated with targeted measures (Cullen-Unsworth and Unsworth 2016). Seagrass restoration is becoming more successful (Sudo et al. 2021), and a growing number of catchment management examples lead to long-term seagrass recovery. Although seagrass restoration is costly, and many major projects have failed in the past, there has been a significant shift in the tactics adopted in the last decade. As a result, there are now many successful projects carried out at a reasonable cost (Sudo et al. 2021). This includes recent multi-species tropical seagrass restoration, and in some conditions, restoration can now be done extremely cheaply by large-scale seed distribution (Sudo et al. 2021). Regardless, preserving extensive productive seagrass meadows rather than rebuilding or recreating them is the most efficient and viable conservation technique.

Seagrass protection must be addressed not only to boost seagrass viability, but also to prepare for fishermen's rising reliance on these areas. It is no longer adequate for marine protected area management plans to incorporate seagrass management as a 'check box exercise', if at all; rather, seagrass management must be intentionally included using best-practice science to increase fisheries productivity. The main challenge for coastal ecosystems in many parts of the world is watershed degradation (e.g., loss of river vegetation and deforestation). Therefore seagrass protection does not necessarily have to be the priority. Ridge-to-reef conservation programs can address more significant concerns like catchment degradation and poor water quality and serve as a platform for larger initiatives like seagrass protection.

15.6 Knowledge Gaps

Seagrass information in Southeast Asia is imbalanced spatially, with hotspots and cold spots of research activity. The degree of information varies within hotspots, with some reporting only species occurrence and others reporting both species presence and meadow size estimates. Because of the logistical challenges of mapping seagrass meadows, estimates of meadow size are rarely addressed. As a result of advancements in remote sensing technology and well-funded regional projects like the UNEP/GEF South China Sea Project (United Nations Environment Programme (UNEP) 2008), the Bay of Bengal Large Marine Ecosystem Project in which Myanmar represents from Southeast Asia (BOBLME 2015), and the JSPS-Asian CORE Project, areal estimates for seagrass meadows have only recently begun to emerge more rapidly. Meadow size data are essential for moving seagrass conservation and management forward in the region because they provide baselines for understanding ecosystem trajectories, either under natural conditions or in response to long-term environmental change, as in the case of Waycott et al. (2009)'s global analysis of seagrass trajectories.

To some extent, access to seagrass sites affects the distribution of data points in the region. Researchers cannot access remote locations because of distance, or the location itself lies within a conflict territory. The Spratly Islands are part of the South China Sea Oceanic Islands, which have been the subject of several overlapping maritime claims by the Philippines, Malaysia, Vietnam, Brunei, Taiwan, and the People's Republic of China for more than 50 years. These sites are natural laboratories for testing hypotheses concerning allopatric differentiation in seagrass and seagrass-associated species and elucidating seagrass dispersal routes in the region due to their physical isolation. In the case of coral reef fish larvae from the Spratly Islands, larval drift time and vector current charts show that the western Philippines, Taiwan, south-eastern China, Brunei, and Malaysia are natural sink habitats (McManus 2017). Long-distance dispersal has also been demonstrated in seagrasses using seeds, fruits, viviparous seedlings, and vegetative fragments, all of which can travel hundreds of kilometres (Kendrick et al. 2012). However, because of the quantity of recent island-building in this ecoregion as a result of aggressive territorial claims, urgent work is required if seagrass scientists in the region want to fill in this data gap (Southerland 2016). Sand dredging and land reclamation are frequently used in the construction of islands on shallow coral reefs and lagoons. Two-thirds of currently occupied atolls in the Spradley's have been demonstrated to have proportionally smaller reef extent than uninhabited atolls, showing that island-building has a negative impact on reef systems (Asner et al. 2017). Seagrasses, like these reefs, maybe similarly harmed before being documented and examined for research.

Despite the ongoing increase in research, it is clear that much work remains to be done in order to close the gaps in knowledge and information. There is almost minimal social science study on the social, cultural, and economic elements of human–seagrass interactions, for example, and there is emerging research that indicates critical linkages between coastal communities and seagrass beds (Cullen-Unsworth et al. 2014). Furthermore, while the importance of seagrass fisheries activity has been recognised internationally (Nordlund et al. 2017), there is a data gap in the Southeast Asian region regarding seagrass fishing resources and the flow-on benefits from ecological services it provides (Unsworth 2010).

15.7 Future Challenges

A number of challenges were extensively discussed to achieve better management of seagrass ecosystem. They are

1. Lack of trained researchers
2. Limited scope of work
3. Knowledge gaps
4. Misguided management efforts
5. Lack of implementation and enforcement of environmental laws
6. Socio-economic and cultural disconnect

The first three issues revolve around the absence of knowledge, competence, and information as fundamental issues. Collaborations and partnerships with research partners from outside the region have helped the region develop expertise in seagrass research and information on tropical seagrass environments. The region's actual capacity building and training of researchers appeared to be concentrated in a few institutions, with little to no information sharing outside of those institutions. Seagrass research conducted by researchers outside of centres of knowledge tends to be extremely descriptive, which is likely due to a lack of training, experience, and resources. This lack of information and training, particularly in basic procedures, exacerbates the absence of basic understanding about seagrass meadows in Southeast Asia, resulting in a scarcity of basic data like as spatial extent, species composition, and cover. To overcome these obstacles, a systematic and multi-pronged strategy is required, and it must originate from scholars stationed in Southeast Asia. Forming a (even if informal) network of scientists operating in the region fosters collaborative and supportive research and knowledge sharing. Workshops for young researchers should be promoted, as should inter-institutional collaboration and exchanges. Knowledge sharing can also help researchers avoid duplication of effort and resource and equipment sharing, and institutional diversity can help researchers compete for funding.

The following three difficulties concern policymaking and natural resource management. While management efforts have been launched at a number of locations around the region, they have generally focused on remedial or curative measures rather than addressing the underlying issues. Many of the causes of seagrass decrease in Southeast Asia have been extensively documented, and many of them may be traced back to anthropogenic pressures such as coastal development. To prevent continued degradation of seagrass meadows throughout the region, these issues must be addressed. Although seagrass relocation and restoration are periodically attempted, they are frequently not the most cost-effective alternative because success rates are low despite a substantial investment of time and money (Van Katwijk et al. 2015). Ineffective solutions are often exacerbated by a lack of effective links between science, government, and the commercial sector, resulting in poor management and conservation measures. There is a significant divergence between seagrass science, policy, and practice at the moment.

One of the most difficult aspects of enforcing environmental rules is dealing with opposition from local maritime communities (Bennett and Dearden 2014). The last major obstacle to seagrass protection is the social, economic, and cultural divide. TeamSeaGrass is a member of the Seagrass-Watch Network, and the information gathered by volunteers has been used in scientific papers (Yaakub et al. 2014; McKenzie et al. 2016), as well as shared with management from Singapore's National Parks Board (NParks).

15.8 Conclusion

Humanity cannot afford to let another marine ecosystem's integrity be jeopardised due to poor stewardship of our planet's resources. It is crucial to maintain essential ecosystem services. The moment has come for large conservation donors, government regulators, and conservation stakeholders to refocus their efforts on where ecosystem services will be needed now and in the future. Indeed, there is mounting evidence that global fisheries management regimes must adjust to changing climate conditions and adopt adaptive policy aims. It's also crucial to think about which conservation projects could benefit multiple aspects of the seascape, even if they aren't the most effective approach for one habitat in particular. Reducing land-based pollution, for example, will benefit seagrass meadows and coral reefs, as well as other nearby habitats. As a result, it's critical to analyse and argue for the seascape's overall benefits. Depending on the in-water circumstances, enhancing water quality may be the most significant activity for seagrass, and it would also benefit other systems. Importantly, we must prioritise ecosystem-related measures that promote ecosystem services. Some 'bright spots' in coral reef protection show the possibility of some coral reef survival. However, for our tropical seas to continue to feed fisheries and people, we must focus on maintaining ecosystems and species that provide the most vital ecosystem services while also being able to withstand future climate change. Seagrass meadows are one of these ecosystems, and preserving them is critical for the lives and food security of hundreds of millions of people. Seagrass ecosystems are in desperate need of protection, and now is the high time to act.

Acknowledgements Authors (PA & KS) would like to thank the Authorities of Annamalai University and Director and Dean, Faculty of Marine Sciences for offering facilities and support. They also thank Prof. L. Kannan, former Director of CAS in Marine Biology, Annamalai University for offering valuable comments.

References

Abu Hena MK, Zakaria MH, Japar Sidik B (2015) Seagrass ecosystems at Johor Straits, Malaysia: climate change impacts and fishermen outlook. In: Proceedings of climate change impacts on seagrass and salt marsh ecosystems in South and Southeast Asian Coasts, APN report, p 82

Al-Asif A, Hamli H, Abu Hena MK, Idris MH, Gerusu GJ, Ismail J, Karim NU (2020) Benthic macrofaunal assemblage in seagrass-mangrove complex and adjacent ecosystems of Punang-Sari Estuary, Lawas, Sarawak, Malaysia. Biodiversitas 21:4606–4615

Asner GP, Martin RE, Mascaro J, Williams R, Boyd D, D. (2017) Coral reef atoll assessment in the South China Sea using Planet Dove satellites. Remote Sens Ecol Conserv 3:57–65

Bennett NJ, Dearden P (2014) Why local people do not support conservation: community perceptions of marine protected area livelihood impacts, governance and management in Thailand. Mar Policy 44:107–116

BOBLME (2015) The Bay of Bengal: an overview of a large marine ecosystem, vol 36. BOBLME, Phuket

Bowen BW, Gaither MR, DiBattista JD, Iacchei M, Andrews KR, Grant WS, Toonen RJ, Briggs JC (2016) Comparative phylogeography of the ocean planet. Proc Natl Acad Sci U S A 113:7962–7969

Carpenter KE, Springer VG (2005) The center of the center of marine shore fish biodiversity: the Philippine Islands. Environ Biol Fish 72:467–480. https://doi.org/10.1007/s10641-004-3154-4

Cullen-Unsworth LC, Unsworth RKF (2016) Strategies to enhance the resilience of the world's seagrass meadows. J Appl Ecol 53:967–972

Cullen-Unsworth LC, Unsworth R (2013) Seagrass meadows, ecosystem services, and sustainability. Environ Sci Policy Sustain Dev 55:14–28

Cullen-Unsworth LC, Nordlund LM, Paddock J, Baker S, McKenzie LJ, Unsworth RK (2014) Seagrass meadows globally as a coupled social–ecological system: implications for human wellbeing. Mar Pollut Bull 83:387–397

Fortes MD (1988) Indo West Pacific affinities of Philippine seagrasses. Bot Mar 31:237–242

Fortes MD (1995) Seagrasses of East Asia: environmental and management perspectives. RCU/EAS Technical Report Series No. 6. United Nations Environment Programme, Bangkok, p 75

Fortes MD, Ooi JLS, Tan YM, Anchana P, Japar Sidik B, Yaakub SM (2018) Seagrass in Southeast Asia: a review of status and knowledge gaps, and a road map for conservation. Bot Mar 61(3):269–288

Ibrahim HM, Nazery K (2007) Growing shipping traffic in the strait of Malacca: some reflections on the environmental impact. In: Global maritime and intermodal logistics conference. Springer, Singapore

Japar Sidik B, Muta Harah Z, Fakhrulddin ML, Khairul Anwar MS, Arshad A (2010) Growth performance of Malaysian spoongrass, *Halophila ovalis* (R. Br.) Hooker f. under different substrate, salinity and light regime. Coast Mar Sci 34:103–107

Johan I, Abu Hena MK, Idris MH, Amin SMN, Denil NA, Kumar U, Karim NU (2020) Species composition and diversity of fishes from the seagrass habitat of Lawas, Sarawak, Malaysia. J Environ Biol 41:1382–1389

Kendrick GA, Waycott M, Carruthers T, Cambridge M, Hovey R, Krauss S, Lavery P, Les D, Lowe R, Mascaró i Vidal O, Ooi JLS, Orth RJ, Rivers D, Ruiz-Montoya L, Sinclair EA, Statton J, Van Dijk K, Verduin J (2012) The central role of dispersal in the maintenance and persistence of seagrass populations. Bioscience 62:56–65

Kirkman H, Kirkman JA (2002) The management of seagrasses in Southeast Asia. Bull Mar Sci 71: 1379–1390

Kuo J (2000) Taxonomic notes on *Halophila minor* and *Halophila ovata*. Biol Mar Medit 7:79–82

Kuo J (2007) New monoecious seagrass of *Halophila sulawesii* (Hydrocharitaceae) from Indonesia. Aquat Bot 87:171–175

Luong CV, Thao NV, Komatsu T, Ve ND, Tien DD (2012) Status and threats on seagrass beds using GIS in Vietnam. Proc SPIE Int Soc Opt Eng 8525:214–226

Masahiro N, Lee KS, Huang X, Almonte T, Japar Sidik B, Kiswara W, Ambo-Rappe R, Yaakub SM, Prabhakaran MP, Abu Hena MK, Hori M, Zhang P, Prathep A, Fortes MD (2014) Regional comparison of the ecosystem services from seagrass beds in Asia. In: Nakano S et al (eds)

Integrative observations and assessments. Ecological research monographs. Springer, Tokyo, pp 367–390
McKenzie LJ, Yaakub SM, Tan R, Seymour J, Yoshida R (2016) Seagrass habitats of Singapore: environmental drivers and key processes. Raff Bull Zool 34:60–77
McManus JW (2017) Offshore coral reef damage, overfishing, and paths to peace in the South China Sea. Int J Mar Coast Law 32:199–237
Mokhzani Z (2004) Strategic value of the Malacca Strait. In: Maritime Institute of Malaysia (MIMA). Kuala Lumpur
Mukai H (1993) Biogeography of the tropical seagrasses in the western Pacific. Tropical seagrass ecosystems; structure and dynamics in the Indo-West Pacific. Aust J Mar Freshw Res 44:1–17
Murugan S, Khan A, Lyla PS, Manokaran S, Raja S, John A (2014) Spatial and temporal variability in fish diversity of Vellar Estuary (South East Coast of India). Annu Res Rev Biol 4(13):2147–2162
Nguyen VX, Holzmeyer L, Papenbrock J (2013) New record of the seagrass species *Halophila major* (Zoll.) Miquel in Vietnam: evidence from leaf morphology and ITS analysis. Bot Mar 56: 313–321
Nguyen VX, Detcharoen M, Tuntiprapas P, Soe-Htun U, Japar Sidik B, Muta Harah Z, Prathep A, Papenbrock J (2014) Genetic species identification and population structure of Halophila (Hydrocharitaceae) from the Western Pacific to the Eastern Indian Ocean. BMC Evol Biol 14:92
Nordlund LM, Unsworth RK, Gullström M, Cullen-Unsworth LC (2017) Global significance of seagrass fishery activity. Fish Fish 19:1–14
Nordlund LM, Unsworth RKF, Gullström M, Cullen-Unsworth LC (2018) Global significance of seagrass fishery activity. Fish Fish 19:399–412
Ooi JLS, Kendrick GA, Van Viel KP, Affendi YA (2011) Knowledge gaps in tropical Southeast Asian seagrass systems. Estuar Coast Shelf Sci 92:118–131
Short F, Carruthers T, Dennison W, Waycott M (2007) Global seagrass distribution and diversity: a bioregional model. J Exp Mar Biol Ecol 350:3–20
Southerland M (2016) China's island building in the South China Sea. U.S.-China Economic and Security Review Commission, Washington, DC
Sudo K, Quiros TEAL, Prathep A, Luong CV, Lin H-J, Bujang JS, Ooi JLS, Fortes MD, Zakaria MH, Yaakub SM, Tan YM, Huang X, Nakaoka M (2021) Distribution, temporal change, and conservation status of tropical seagrass beds in Southeast Asia: 2000–2020. Front Mar Sci 8:637722. https://doi.org/10.3389/fmars.2021.637722
Tuntiprapas P, Shimada S, Pongparadon S, Prathep A (2015) Is Halophila major (Zoll.) Miquel a big H. ovalis (R. Brown) JD Hooker? An evaluation based on age, morphology, and ITS sequence. Sci Asia 41:79–86
United Nations Environment Programme (UNEP) (2008) National reports on seagrass in the South China Sea. UNEP/GEF/SCS Technical Publication No. 12. UNEP, Bangkok, p 207
Unsworth RKF (2010) Seagrass meadows of the Wakitobi National Park. In: Clifton J, Unsworth R, Smith D (eds) Marine research and conservation in the coral triangle. Nova Science Publishers Inc., New York, pp 101–126
Unsworth RKF, Cullen LC (2010) Recognising the necessity for Indo-Pacific seagrass conservation. Conserv Lett 3:63–73
Unsworth RKF, Collier CJ, Waycott M, McKenzie LJ, Cullen-Unsworth LC (2015) A framework for the resilience of seagrass ecosystems. Mar Pollut Bull 100:34–46
Unsworth RKF, Nordlund LM, Cullen-Unsworth LC (2019) Seagrass meadows support global fisheries production. Conserv Lett 12:e12566
van Katwijk MM, Thorhaug A, Marbà N, Orth RJ, Duarte CM, Kendrick GA, Althuizen IHJ, Balestri E, Bernard G, Cambridge ML, Cunha A, Durance C, Giesen W, Han Q, Hosokawa S, Kiswara W, Komatsu T, Lardicci C, Lee K, Meinesz A, Nakaoka M, O'Brien KR, Paling E, Pickerell C, Ransijn AMA, Verduin JJ (2015) Global analysis of seagrass restoration: the importance of large-scale planting. J Appl Ecol 53:567–578

Waycott M, Duarte CM, Carruthers TJB, Orth RJ, Dennison WC, Olyarnik S, Calladine A, Fourqurean JW, Heck KL Jr, Randall Hughes A, Kendrick GA, Judson Kenworthy W, Short FT, Williams SL (2009) Accelerating loss of seagrasses cross the globe threatens coastal ecosystems. Proc Natl Acad Sci U S A 106:12377–12381
William SL, Heck KL Jr (2001) Seagrass community ecology. In: Bertness MD, Gaines SD, Hay ME (eds) Marine community ecology. Sinauer Associates, Sunderland
Yaakub SM, Lim RLF, Lim WL, Todd PA (2013) The diversity and distribution of seagrass in Singapore. Nat Singap 6:105–110
Yaakub SM, McKenzie LJ, Erftemeijer PLA, Bouma T, Todd PA (2014) Courage under fire: seagrass persistence adjacent to a highly urbanised city–state. Mar Pollut Bull 83:417–424

Chapter 16
Effect of Climate Change and Urbanization on Mangrove Ecosystem

Priyanka Kumari and Bhawana Pathak

16.1 Introduction

Mangroves are one of the most vulnerable and fragile ecosystems and have socio-economic and ecological services. They provide goods and services to human societies and marine ecosystem. They have unique properties like aerial breathing roots and extensive supportive root systems (Kathiresan and Bingham 2001; Saddhe et al. 2016). These protective forests play a crucial role to provide physical protection by acting as natural ecological barrier against cyclones, storm, and high oceanic currents. One of the unique properties of their roots is that they act as sponges so that they absorb flood water and hence it helps in flood attenuation (Duke 2011). By performing these special characteristics, they save the most fragile coastal communities from devastating impact of climate change.

Mangrove forests play a significant role in mitigating climate change as they act as the sink of carbon, store them in large quantities in biomass and sediments for a long time. Because of excessive exploitation in coastal areas and increasing impact of climate change which includes sea level rise, alteration in weather patterns, mutations in dangerous microorganisms and species, acidification of the oceans, the mangrove cover and its health around the globe is degrading at a very alarming rate (3) (Mangroves for Coastal Defence Guidelines for Coastal Managers and Policy Makers Suggested Citation 2014). Most of the mangrove ecosystems have been exploited during the past few decades. Expansion of shrimp farming, urbanization, tourism, and industries are some of the main causes for mass clearing of

P. Kumari (✉) · B. Pathak
School of Environment and Sustainable Development, Central University of Gujarat, Gandhinagar, Gujarat, India
e-mail: priyanka15@cusb.ac.in

B. Pathak, R. S. Dubey (eds.), *Climate Change and Urban Environment Sustainability*, Disaster Resilience and Green Growth,
https://doi.org/10.1007/978-981-19-7618-6_16

mangrove patches (IUCN 2012). Mangrove forests are one of the world's threatened major tropical environments (Valiela et al. 2001).

Mangroves are ecologically important due to their unique features, such as high productivity, abundant detritus, and high levels of organic carbon, which may make them a preferential site for the uptake and preservation of PAHs from anthropogenic inputs (Bernard et al. 1996).

16.2 Causes of Mangrove Degradation

16.2.1 Climate Change

Climate is the average pattern of weather, if it is left to itself for centuries it remains uniform. Burning of fossil fuels, such as coal, oil, and natural gas has large impact on the climate, and these fossil fuels liberate carbon dioxide and other dangerous gases in the atmosphere. In the early 1800s, the percentage of carbon dioxide in the earth's atmosphere has increased by around 30%, as well as the average of global temperature also seems to have risen by 2 °F. This percentage increase of temperature is clearly linked to the basic property of the gas. As glass traps solar heat in a greenhouse or a sunroom partly in the same way carbon dioxide gas traps the heat from sun in the atmosphere. That's the reason sometimes the carbon dioxide is called a "greenhouse gases." As more carbon dioxide is added to the atmosphere, solar heat faces more trouble in getting out. Solar heat faces more trouble in getting out from earth's atmosphere as more carbon dioxide is added to the atmosphere.

The present trends of global warming, climate change related factors are the major cause of affecting the survival of mangrove ecosystems (Salik and Sehrish 2013; Di Nitto et al. 2014; Alongi 2015). During the past decades' global average air temperature near Earth's surface rose 0.74 ± 0.18 °C. Climate change factors like increasing temperature and atmospheric CO_2, change in precipitation patterns, sea level rise, and frequency and intensity of tropical storms are the serious concerns that are affecting mangrove ecosystems and livelihood across the mangrove ecosystem (McLeod and Salm 2006). Small warming effect from pre-industrial times to 1950 may also happen by natural phenomena such as solar variation combined with volcanoes. Effects of rise in temperature are very uncertain. The Earth's temperature has risen by 0.5 °C over the past century and on record the recent years have been among the hottest. It is expected by the year 2030 that the sea level would rise by 20 cm (Alongi 2015).

Mangrove decline is directly influenced by carbon levels in the atmosphere as these are the prominent source for carbon sequestration, absorbing carbon dioxide which helps in mitigation of climate change impact. Critical consequences occur when such type of ecosystems are damaged/decline. There are fewer sources for carbon sequestration of atmospheric carbon dioxide. Declining of such resources has long-term impact of climate change.

16.2.2 Effect of Rise in Temperature

Since in late nineteenth century, the earth has warmed by 0.6–0.8 °C and by the end of twenty-first century, earth is projected to warm by 2–6 °C due to very high rate of anthropogenic activities (Houghton et al. 2001). Maximum shoot diversity occurs at when mean air temperature increases up to 25 °C and mangroves stop producing leaves when the mean air temperature drops below 15 °C (Hutchings and Saenger 1987). Some species of mangroves show deformation in leaf pattern at temperature above 25 °C (Saenger 1985). Root structures and establishment of mangrove seedlings are affected by thermal stress temperature above 35 °C (UNESCO 1992). At temperature of 38–40 °C, almost no photosynthesis occurs (Clough and Attiwill 1982).

Photosynthesis in mangroves are affected by the oscillation of temperature. At higher leaf temperature, rate of photosynthesis is reduced (Andrews and Muller 1985). Photosynthesis in mangroves is similar to other C_3 plants and unlike that of C_4 plants. Reproduction patterns and the length of time between flowering are impacted by the change in temperature (UNEP 1994; Ellison 2000). Regeneration and nutrient recycling of bacterial growth and reproduction are increased by the enhancement in sediment temperature.

Mangrove plant parts show special or unique physiological and morphological adaptive features to overcome from several environmental stresses that are associated with their intertidal habitat (Halpern et al. 2007). The recently studied data reports estimate 0.61–1.1 m rise in the global mean sea level by the year 2100 which is expected to put many coastal cities in India at a high vulnerability rate of whole submergence or at least frequent inundation during conditions of high tide (Ramos 2017). It was reported by Gilman et al. (2008) that loss of mangrove area in Sundarbans is due to sea level rise. In the year 1996 and 1997, the Sundarbans mangrove forests were announced as "reserve forest" and "UNESCO world heritage site," respectively (Payo et al. 2016). According to forest survey report, the sea level rise and erosion in Bay of Bengal resulted in the loss of 2 Sq. km in mangrove cover between year 2017 and 2019 (DasGupta and Shaw 2013). Up to a saturated threshold, only the mangroves are well adopted to saline conditions both morphologically and physiologically. In recent decades, extreme variation in water and soil salinity has been recorded in Indian coastal states.

Lead researcher of Pichavaram Dr. V. Selvam and Muthupet in Tamil Nadu noted that in the year 2019 there was major increase in salinity level of soil and water which resulted in loss of five species of mangrove vegetation over just 70 years (Sengupta 2019). Increased salinity has adverse impacts on the species diversity of mangroves as well as even upon marine animal species. High salinity led to 90% reduction in freshwater flow (Murthy et al. 2015). In the Pichavaram mangroves, there is serious decline in the population of gastropods like snails, slugs, etc. and also in the population of several migratory birds.

16.2.3 Acidification

Carbon dioxide produced by living organisms is soaked up by the world's oceans, in the form of skeletons of small aquatic creatures that fall to become limestone or in the form of dissolved gas. As per earlier research, oceans absorb one ton of CO_2 per person per year. The CO_2 generated by anthropogenic activities since 1800. Carbon dioxide lowers the pH value of the ocean as in the aquatic phase, CO_2 converts into weak carbonic acid, which results into the lowering of pH level of ocean and increases the acidic value. Already in many studies, it has been stated that the increase in greenhouse gas during industrial revolution has degraded the average pH of seawater by 0.1 units. Estimated emissions could decrease it further about 0.5 units by 2100. Hence, decreasing of aquatic pH could have detrimental impact on coral reefs and whole aquatic system. Increased acidity may also impact the economical profit taken by the fishery and local community as there is increase in acidity then it directly affects the growth and reproduction of fish and planktons.

16.2.4 Effects on Coral Reefs

Coral reefs act as natural protective barrier for coastal regions. These are colorful underwater taxonomically diverse units. Coral reefs are affected by global warming in several ways. Coral bleaching has been rising in recent years. It occurs due to the rising temperature of seawaters. It occurs when the coral is stressed and expels the zooxanthellae. Without the zooxanthellae, the coral loses its color and becomes white. The optimum temperature for growth of coral reefs ranging between 66 and 86 °F. In several studies, it was noted that frequency is increased during El Nino events due to coral reef bleaching. Rise in sea level caused by melting of sea ice and thermal expansion of the oceans could also promote difficulties for survival of coral reefs. Almost a third of atmospheric CO_2 are absorbed by oceans. However, the acidity of the ocean increased due to the mixing of high level of CO_2. The point source of the carbon dioxide emissions are cars and factories. The skeletons of corals are made up of $CaCO_3$. Hence, it reacts with oceanic CO_2 and forms carbonic acid. This carbonic acid kind of substances slowed the growth of corals and the shells of oceanic organisms become thin and finally dissolved. 16% of the world's coral reefs were diminished just in 1 year as several studies reported. According to IPCC reports, there is a positive correlation between coral bleaching and global warming.

16.2.5 Urbanization

In current time, urbanization is one of the major social transformations, driving by multiple social, economic, and environmental processes. Urbanization impacts on the environment are profound and multifaceted at the local, regional, and global scale. Urban development induced climate change is main cause of the loss and subsequent fragmentation of mangrove forests. Urban development is considered as a driver for the lost and fragmentation of mangrove forests (Alongi 2002; Branoff 2017). Empirical knowledge linkages between urbanization and environment could understand by focusing on core aspects (air pollution, ecosystems functionality, land use change, biogeochemical cycles and water pollution, solid waste management, and biodiversity). Several emerging trends in urban environmental sustainability research like urbanization pattern and environmental impacts; changing distribution patterns; geographical and socio-economic gradients; urbanization impact on processes and mechanisms are going on for understanding of interactions and interlinkages between environmental, social, economic, and cultural aspect.

Empirical studies focused on the impacts of urbanization onto species richness show that this can both increase and decrease based on the intensity of urbanization (McKinney 2008).

For the livelihood, India's coastal communities are dependent on agriculture. Murthy et al. (2015) reported that around 1,50,000 hectares of mangrove area destruction over past 100 years in India and Bangladesh in order to make agricultural land. India coastal cities are facing concomitant development and rapid urbanization, land reclamation and development projects are the major reasons of mangrove loss. Mumbai is one of the example as it was built on cluster of seven island, surrounded by dense mangroves. According to recent news, for the construction of Navi Mumbai international airport and metro project, hundreds of acres of mangroves have been deforested. A very disappointing announcement was stated by the Maharashtra's transport minister in the year 2019 that around 32,044 mangroves would be cleared for Mumbai–Ahmedabad bullet train project (Chatterjee 2019). According to Mumbai-based non-profit organization "Conservation Action Trust" no land is available for the replantation of mangroves in Maharashtra. In megacities, such as Kolkata and Mumbai, pollution is another big challenge for the mangrove ecosystem. Mountains of solid waste and effluents from various point sources are dumped directly into mangrove ecosystems. Oil spills also badly impact on these fragile ecosystems.

16.2.6 Industrialization

Industrialization is one of the important sectors of development. The discharge of industrial effluents pollutes mangrove ecosystems. Most of the pollutants are adsorbed onto particulate material by mangroves (Baudo et al. 1990). Polycyclic aromatic hydrocarbons (PAHs) via industrial discharge, petroleum spills, the combustion of fossil fuels, automobile exhaust, and non-point sources: urban runoff and atmospheric deposition are also responsible for mangrove degradation (Macdonald et al. 2005). A number of PAHs are carcinogens and persist in the environment for long term will affect mangrove ecosystem.

16.2.7 Anthropogenic Activities

Mangrove ecosystem depletion is also due to natural disaster and anthropogenic activities (Fusi et al. 2016). Wood harvesting (Feka and Manzano 2008), aquaculture (Guimarães et al. 2010), urban infrastructures, sand extraction, petroleum exploitation, coastal erosion also contributing to be the important factors of mangrove degeneration. Other factors like collection of non-timber forest products, digging, landfill, dyke construction also contributed a lot in mangrove degradation. Oil spill also pose a major threat for mangrove survival (Duke 2016). Tourism sector also impacts of climate change and ecosystem degradation (WTO, UNEP 2008).

16.2.8 Mangrove Resistance to Climate Change

Mangrove ecosystems are one of the toughest ecosystems on the Earth. Mangrove plants have unique features to cope with different type if stresses. This fragile ecosystem acts as "bioshields." A 2004 Asian Tsunami was exacerbated clearing of mangroves. Unfortunately, no proper policy was designed to increase the resistance and resilience of mangroves. The elevation of mangroves sediment surface is influenced by several interconnected surfaces and subsurface processes. According to several studies, subsurface controls on mangrove sediment elevation can catalyze the sedimentation rates (Cahoon and Hensel 2006). The halophytes were capable to survive through quaternary despite substantial damage from major sea level fluctuations, which shows that mangroves are highly resilient to change (Woodroffe 1987, 1992; Salik et al. 2014).

16.3 Conclusion

Urbanization of the adjacent towns, climate change, and the intense anthropogenic activities (excessive harvest of mangrove trees for firewood, charcoal, clearing of mangrove areas for agricultural purposes, pollution) across coastal zone including mangrove ecosystem have led to a gradual degeneration of these ecosystems. Coastal plantation is important for climate change mitigation and adaptation to cope up with the mangrove deterioration (Duarte et al. 2013). For developing conservation strategies, people perception will contribute with major impact (Din et al. 2016; Nfotabong-Atheull et al. 2011). Awareness about the impacts is rising and it's changing how people approach infrastructure and engineering development. Initially, measuring and mitigating biodiversity net loss were the primary concern for sustainable development.

References

Alongi DM (2002) Present state and future of the world's mangrove forests. Environ Conserv 29(3): 331–349. https://doi.org/10.1017/S0376892902000231

Alongi DM (2015) The impact of climate change on mangrove forest. Curr Clim Change Rep 1:30–39

Andrews TJ, Muller GJ (1985) Photosynthetic gas exchange of the mangrove, *Rhizophora stylosa* Griff., in its natural environment. Oecologia 65(3):449–455. https://doi.org/10.1007/BF00378922

Baudo R, Giesy J, Muntau H (1990) Sediments: chemistry and toxicity of in-place pollutants. CRC Press, Boca Raton, pp 12–20

Bernard D, Pascaline H, Jeremie JJ (1996) Distribution and origin of hydrocarbons in sediments from lagoons with fringing mangrove communities. Mar Pollut Bull 32:734–773

Branoff BL (2017) Quantifying the influence of urban land use on mangrove biology and ecology: a meta-analysis. Glob Ecol Biogeogr 26(11):1339–1356. https://doi.org/10.1111/geb.12638

Cahoon DR, Hensel PF (2006) High-resolution global assessment of mangrove responses to sea-level rise: a review. In: Proceedings of the symposium on mangrove responses to relative sea-level rise and other climate change effects, 13 July 2006, pp 9–18

Chatterjee B (2019) 32K mangroves to be razed for bullet train, NHSRCL confirms—Hindustan times. https://www.hindustantimes.com/cities/32k-mangroves-to-be-razed-for-bullet-train-nhsrcl-confirms/story-FLUWtnYITKrcghylzIXyuL.html

Clough BF, Attiwill PM (1982) Primary productivity of mangroves. In: Mangrove ecosystems in Australia: structure, function and management. AIMS, Townsville, pp 213–222. https://ci.nii.ac.jp/naid/10008751540/

DasGupta R, Shaw R (2013) Cumulative impacts of human interventions and climate change on mangrove ecosystems of south and Southeast Asia: an overview. J Ecosyst 2013:1–15. https://doi.org/10.1155/2013/379429

Di Nitto D, Neukermans G, Koedam N, Defever H, Pattyn F, Kairo JG, Dahdouh-Guebas F (2014) Mangroves facing climate change: landward migration potential in response to projected scenarios of sea level rise. Biogeosciences 11:857–871

Din N, Ngo-Massou VM, Essomè-Koum GL, Kottè-Mapoko E, Emanè JM, Akusa DA, Tchoffo R (2016) Local perception of climate change and adaptation in mangrove areas of the Cameroon coast. J Water Resour Prot 8:608–618. https://doi.org/10.4236/jwarp.2016.85050

Duarte CM, Losada IJ, Hendriks IE, Mazarrasa I, Marbà N (2013) The role of coastal plant communities for climate change mitigation and adaptation. Nat Clim Change 3:961

Duke NC (2011) Mangrove floristics and biogeography. Wiley, New York, pp 63–100. https://doi.org/10.1029/CE041P0063

Duke NC (2016) Oil spill impacts on mangroves: a global review with recommendations. Mar Pollut Bull 109(2):700–715. https://doi.org/10.1016/j.marpolbul.2016.06.082

Ellison JC (2000) How south pacific mangroves may respond to predicted climate change and sea-level rise. Springer, Dordrecht, pp 289–300. https://doi.org/10.1007/0-306-47981-8_16

Feka NZ, Manzano MG (2008) The implications of wood exploitation for fish smoking on mangrove ecosystem conservation in the South West Province, Cameroon. Trop Conserv Sci 1:222–235

Fusi M, Beone GM, Suciu NA, Sacchi A, Trevisan M, Capri E, Daffonchio D, Din N, Dahdouh-Guebas F, Cannicci S (2016) Ecological status and sources of anthropogenic contaminants in mangroves of the Wouri River Estuary (Cameroon). Mar Pollut Bull 109(2):723–733. https://doi.org/10.1016/j.marpolbul.2016.06.104

Gilman EL, Ellison J, Duke NC, Field C (2008) Threats to mangroves from climate change and adaptation options: a review. Aquat Bot 89(2):237–250. https://doi.org/10.1016/J.AQUABOT.2007.12.009

Guimarães AS, Travassos P, Filho PWMES, Gonçalves FD, Costa F (2010) Impact of aquaculture on mangrove areas in the northern Pernambuco Coast (Brazil) using remote sensing and geographic information system. Aquacult Res 41:828–838

Halpern BS, Selkoe KA, Micheli F, Kappel CV (2007) Evaluating and ranking the vulnerability of global marine ecosystems to anthropogenic threats. Conserv Biol 21(5):1301–1315. https://doi.org/10.1111/J.1523-1739.2007.00752.X

Houghton JT et al (eds) (2001) Climate change 2001: the scientific basis: contribution of Working Group I to the third assessment report of the Intergovernmental Panel on Climate Change. Cambridge University Press

Hutchings P, Saenger P (1987) Ecology of mangroves. University of Queensland Press, St. Lucia

IUCN (2012) Annual report Nature+Towards Nature-based Solutions

Kathiresan K, Bingham BL (2001) Biology of mangroves and mangrove ecosystems. Adv Mar Biol 40:81–251

Macdonald RW, Harner T, Fyfe J (2005) Recent climate change in the Arctic and its impact on contaminant pathways and interpretation of temporal trend data. Sci Total Environ 342:5–86

Mangroves for Coastal Defence Guidelines for Coastal Managers and Policy Makers Suggested Citation (2014). www.nature.org

McKinney ML (2008) Effects of urbanization on species richness: a review of plants and animals. Urban Ecosyst 11:161–176. https://doi.org/10.1007/s11252-007-0045-4

McLeod E, Salm RV (2006) Managing mangroves for resilience to climate change. IUCN, Gland, p 64

Murthy IK, Prasad RC, Sahu SC, Suresh HS, Murthy IK, Ravindranath N (2015) Mangrove area assessment in India: implications of loss of mangroves. J Earth Sci Clim Change 6:280. https://doi.org/10.4172/2157-7617.1000280

Nfotabong-Atheull A, Din N, Essome-Koum GL, Satyanarayan B, Koedam N, Dahdouh-Guebas F (2011) Assessing forest products usage and local residents' perception of environmental changes in peri-urban and rural mangrove of Cameroon, Central Africa. J Ethnobiol Ethnomed 7:41. https://doi.org/10.1186/1746-4269-7-41

Payo A, Mukhopadhyay A, Hazra S, Ghosh T, Ghosh S, Brown S, Nicholls RJ et al (2016) Projected changes in area of the Sundarban mangrove forest in Bangladesh due to SLR by 2100. Clim Change 139(2):279–291. https://doi.org/10.1007/S10584-016-1769-Z/FIGURES/4

Ramos J (2017) On world mangrove day, a new strategy to protect the world's most important ecosystem. https://www.conservation.org/blog/on-world-mangrove-day-a-new-strategy-to-protect-the-worlds-most-important-ecosystem

Saddhe AA, Jamdade RA, Kumar K (2016) Assessment of mangroves from Goa, west coast India using DNA barcode. Springerplus 5(1):1–10. https://doi.org/10.1186/S40064-016-3191-4/TABLES/5

Saenger P (1985) Vegetative phenology of mangroves along the Queensland Coastline. https://agris.fao.org/agris-search/search.do?recordID=AU2019D12402

Salik KM, Sehrish J (2013) Coastal population more at risk; socio-economic vulnerability assessment of Indus delta under climate change. Sustainable Development Policy Institute (SDPI), Islamabad, Pakistan, Research & News Bulletin, vol. 20 no. 2

Salik KM, Mohammad MH, Sadia I (2014) Socio-economic vulnerability of mangrove ecosystems to climate change in South Asia: a case study of the Indus and Ganges deltas. APN Science Bulletin Issue http://www.apn-gcr.org/resources/items/show/1932. Accessed 4 Mar 2014

Sengupta A (2019) TN loses 5 mangrove species as soil salinity increases in coastal areas—DTNext. https://www.dtnext.in/News/TopNews/2019/06/06060527/1141087/TN-loses-5-mangrove-species-as-soil-salinity-increases-.vpf

UNEP (1994) Assessment and monitoring of climatic change impacts on mangrove ecosystems. Regional Seas Reports and Studies No. 154. https://wedocs.unep.org/xmlui/handle/20.500.11822/25973

UNESCO (1992) Coastal systems studies and sustainable development. UNESCO, Paris. https://digitallibrary.un.org/record/133454

Valiela I, Bowen JL, York JK (2001) Mangrove forests: one of the world's threatened major tropical environments. Bioscience 5:807–815

Woodroffe CD (1987) Pacific Island mangroves: distribution and environmental settings. http://scholarspace.manoa.hawaii.edu/handle/10125/1031

Woodroffe C (1992) Mangrove sediments and geomorphology, vol 7–41. Wiley, New York. https://doi.org/10.1029/ce041p0007

WTO, UNEP (2008) Climate change and tourism-responding to global challenges. WTO and UNEP, Nairobi

Chapter 17
Impacts of Urbanization and Climate Change on Habitat Destruction and Emergence of Zoonotic Species

Rama Shanker Dubey, Sandeep Kalyan, and Bhawana Pathak

17.1 Introduction

Animal migration is a natural phenomenon in the entire ecosystem, and it helps in the regulation of the ecosystem structure and services. The behaviours such as migration due to food scarcity, avoiding predation during reproduction and seeking physiologically optimal climate (Altizer et al. 2011) are commonly associated with zoonotic pathogens which sometimes seem to be important to humans, but otherwise, they are the source of many unknown viral and infectious diseases (zoonosis) in the urban settlements. Parasites or pathogens comprise >50% of the known species on the planet (Brooks and Hoberg 2006) and 60% of total species are zoonotic in nature (Taylor et al. 2001). "A zoonosis is an infectious disease that has jumped from a non-human animal to humans. Zoonotic pathogens may be bacterial, viral or parasitic, or may involve unconventional agents and can spread to humans through direct contact or food, water or the environment" (World Health Organization 2022). After the introduction of a pathogen to a novel place, it will actually spread more rapidly as its habitat gets distracted resulting in increased transmission using human beings as hosts (Kilpatrick and Randolph 2012; Wilkinson et al. 2018).

The global human urban population is expected to rise to nearly 66% by 2050 and thus the interaction between wildlife and human is likely to increase, which will ultimately lead to an increase in the emergence of zoonotic diseases such as

R. S. Dubey
Central University of Gujarat, Gandhinagar, India

S. Kalyan · B. Pathak (✉)
School of Environment and Sustainable Development, Central University of Gujarat, Gandhinagar, India
e-mail: bhawana.pathak@cug.ac.in

B. Pathak, R. S. Dubey (eds.), *Climate Change and Urban Environment Sustainability*, Disaster Resilience and Green Growth,
https://doi.org/10.1007/978-981-19-7618-6_17

Babesiosis disease in Europe and Asia (Yabsley and Shock 2013; United Nations, Department of Economic and Social Affairs, Population Division 2015). The emergence of zoonotic diseases has two stages (1) contact of humans with the infectious agents and (2) transmission of infectious agents in the cross-species and these both stages are essentially required for the successful transition of the pathogen (Childs et al. 2007). Three factors: demographic change, infrastructure and governance are the governing aspects in any region for the spread of zoonotic disease. As urban areas are extending to the sub/peri-urban areas, chances of the disease spread have increased (Connolly et al. 2021). The pathogenic species may also use humans as a host and can transfer the disease variants, as anthropogenic processes are taking place at a large scale (Moreno 2014). The zoonotic diseases which spread in livestock are mainly dependent on the agricultural practices, and the changes in these practices affect the pathogen habitats causing the pathogens to spread among the human population (Slingenbergh et al. 2004). The altitudinal range shift of the species due to climate change has also been studied and it has been observed that a shift can occur across the species in the range as small as 40 m (Shoo et al. 2006). Rat fleas are the ectoparasites feeding on warm-blooded animals, and their population index is correlated with the mean annual precipitation and temperature. Their population index and density of reservoir host are positively correlated with the climatic indicators (Chotelersak et al. 2015). The urban environment is altering the physical characteristics of the Quenda species, thus disturbing the natural ecosystem sustainability even for the species that are well adjusted to the urban ecosystem (Hillman et al. 2017).

Zoonotic diseases have shown increasing incidence as a result of urbanization and climate change and also it has been observed that these diseases spread through the imbalance in the food chain. For example, the *Campylobacteriosis* is a food-based poultry product disease in the state of Assiut Governorate, Egypt which is causing serious threat to human health (Hafez and Hauck 2015; Abushahba et al. 2018). The immunity of the potential host of the zoonotic species is the only factor that defines the impact magnitude of zoonosis (Albery and Becker 2021). It has been observed that urban mammals have a greater load of pathogens as they might have a connection with the pathogens due to greater historical contact with humans (Albery et al. 2021). This could be due to the increasing heterospecific social relationship of bats among themselves and with humans (Ancillotto et al. 2015). Biodiversity loss and forest fragmentation over the years have resulted in parasite richness in humans (Bordes et al. 2015). The potential mechanisms leading to occurrence of more zoonoses in the urban settlements with more proximity of pathogens with human beings and more ultimate contact of humans with zoonotic species, pathogen diversity have been presented in Fig. 17.1.

Land-use change has a significant role in disease dynamics as it alters the species behaviour, demographic conditions, vector–host relationships, movement, etc. (Gottdenker et al. 2014). The alteration in the peri-urban region increased the risk of HPAI H5N1 virus in Vietnam as it disturbed the natural habitat of carrier species such as duck and geese (Saksena et al. 2014). Social phenomena such as migration towards urban regions, economic conditions, inadequate housing and sanitation

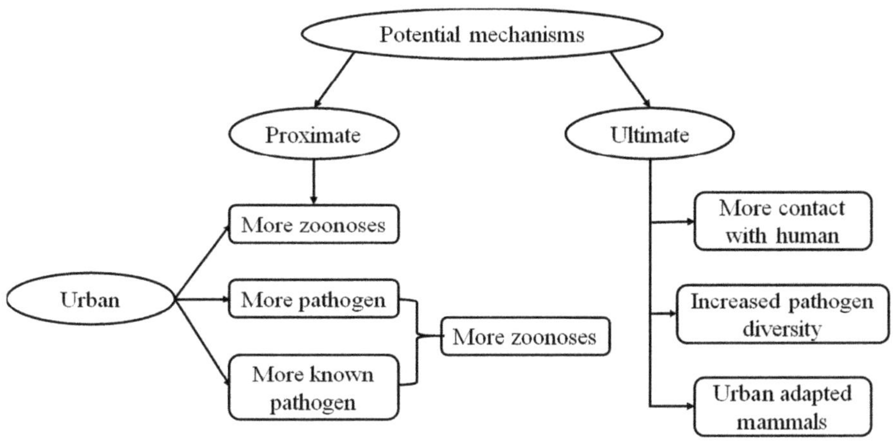

Fig. 17.1 Potential mechanisms for occurrence of more zoonoses due to greater zoonotic diversity in urban mammals (modified from Albery et al. 2021)

levels are also the determining factors for the epidemiological pattern of parasitic disease in the rural and urban areas (Mott et al. 1990). The gradient of urbanization is also affecting the zoonotic risk of *Nosopsyllus fasciatus* fleas in the selected sites of Berlin as using mice and voles as a host poses health concern to the residing population (Maaz et al. 2018).

The migration of monarchs in the southern US and Hawaii is expected to lead pathogen spread as their natural habitat has been distracted and they have started moving to the urban gardens (Majewska et al. 2019). Not only the migrating but also native species such as red foxes, etc. are also needed to be taken into account as they serve as disease carrier of *Enterocytozoon bieneusi* genotype which is liable to infect the local human population and animals (Perec-Matysiak et al. 2021). Temperature is the one factor that has affected the population of Kuhl's pipistrelle (*Pipistrellus kuhlii*) species and it has been observed that the population is on an increasing trend in the coldest quarter and its geographical range has increased by 394% as precipitation also has an impact on the species distribution in Russian range from 1980 to 2013 (Ancillotto et al. 2016).

The parameters such as relative humidity, temperature, deforestation, forest degradation, precipitation and socio-economic status of the region have been shown to alter the distribution of zoonotic species and their habitat (Lau et al. 2010; Ullmann & Langoni 2011; Li et al. 2010; Wilder-Smith et al. 2017; Asad and Carpenter 2018; Tunali et al. 2021; Ortiz et al. 2022). *Aedes aegypti* species with the characteristics of causing multiple infections appear to be more suited for their growth in urban areas due to the congenial environmental conditions, presence of water bodies, etc. (Wilke et al. 2019). Zoonotic cutaneous leishmaniasis (ZCL) is a vector-borne disease prevalent in Iran and Turkey. El Nino Southern Oscillation (ENSO) is a driver of climate change at a global level and it is related to the increased incidence of zoonotic diseases in humans and increase in the population of pathogen species (Wang et al. 1999; Glass et al. 2002; Kelly-Hope et al. 2004). Under various climatic conditions, when the habitat suitability for pathogens was assessed in Iran, it was observed that nearly 63–67% of the total area of Iran will provide a suitable condition for the growth of *Phlebotomus papatasi* which is a vector for the virus that causes sand-fly fever disease (Charrahy et al. 2021). It has also been observed that excessive rainfall conditions drive zoonotic transmission level of Rift Valley fever virus, western equine encephalomyelitis virus and Ross River virus and cause increased chances of virus spillover in urbanized regions (Lindsay et al. 1993; Linthicum et al. 1999; Wegbreit and Reisen 2000).

17.2 Urbanization and Climate Change Induced Emergence of Zoonotic Species

17.2.1 West Nile Virus (WNV)

The most prominent reasons behind the exposure of Ebola in urban and peri-urban areas of West Africa were the improper service of the clean water supply and sanitation, temporary housing and the high population density and close contact between humans and animals in the suburban area (Waldman 2015; Robertson et al. 2020). The natural habitat of the zoonotic species is also disturbed as a result of climate change, urbanization, land use/land cover (LULC) and other such natural as well anthropogenic activities. Such activities cause spreading of pathogens such as Ebola virus in bats, West Nile virus (WNV) in songbirds and nuclear polyhedrosis viruses in insects that affect the human health (White et al. 2000; Thompson 2013).

Vector-borne zoonosis has become prevalent in regions where biodiversity and ecosystem get distracted due to anthropogenic activities (Ostfeld and Keesing 2000). In the case of WNV and Lyme disease (LD), the biodiversity has played a significant role in the spread of the disease. Biodiversity loss will lead to the invasion of new exotic species, increase in pollution and direct human exploitation that would also ultimately result in zoonosis (Ostfeld 2009). WNV disease has global importance and gets influenced due to variation in climatic conditions and urbanization. MNV also infects horses and other mammals (Paz 2015). An NDVI-based analysis showed that 33% of the study regions and 54 out of 75 locations were found under high-risk of West Nile virus-positive mosquitoes. The results were cross-validated and found to be significant at $p < 0.05$ (Brownstein et al. 2002).

17.2.2 Dengue Fever Virus

During the recent years, dengue disease has spread almost in all regions of the world, however it is more prevalent in tropical and subtropical climates. It is a mosquito born viral infection and is transmitted through the female mosquito species *Aedes aegypti*. The risk of dengue in Brazil has been correlated with rainfall and drought intensity and with urbanization level, and the spread has multiplied with extreme wet and dry conditions (Lowe et al. 2021). It has been observed that climatic parameters such as a monthly average of minimum and maximum temperature, precipitation, humidity and socio-economic parameters such as population density, urbanization level and pollution, urban heat island (UHI), habitat loss/fragmentation, host density, biodiversity loss, altered interspecific interactions, life cycle flexibility, interaction of biotic and abiotic factors, parasite reproductive strategies, parasite longevity, resilience and plasticity, infectivity and virulence, spatial patterns and habitat connectivity, host switching, birth edging, evolution, adaptive management and predictive epidemiology, evolutionary implications and other environmental parameters, etc. are significantly affecting the risk of dengue fever virus and transmission of other parasites in Europe and other parts of the world (Lambin et al. 2010; Bouzid et al. 2014; Estrada-Peña et al. 2014; Simons et al. 2014; LaDeau et al. 2015; Cable et al. 2017).

17.2.3 Rabies

Dogs are the most common carriers of rabies virus and are responsible for its transmission in human beings. It has been observed that with temperature rise the ecological niche of dogs gets disturbed that leads to significant increase of dog bite cases with the climate change, i.e. with increase in temperature (Ahmed et al. 2019). Bats are also posing serious concerns as they are the carriers of several infections such as rabies virus and also bats have been shown to be sensitive to climate change in Brazil which is also a matter of concern for the people living in the urban patches of the country (Nunes et al. 2017). Vampire bats are feeding on the mammal blood that makes it an efficient host for the rabies spread. Due to the changing climatic conditions, these bats are potentially moving towards the southern US, which has been observed during the last few years (Hayes and Piaggio 2018).

A cyclic rabies occurrence in Alaska has been observed, as the peak of spread has been noticed in winter and spring and also it is related to high altitude regions. If the temperature is on a rise then a decrease in the spread of rabies is expected in the region (Kim et al. 2014). The cattle industry in Mexico loses over $2.6 million annually due to the rabies infection through vampire bats (*Desmodus rotundus*). It has been analysed that this species will lose 20% of its habitat due to climate change and will likely move to the northern and central regions by 2050–2070 (Zarza et al. 2017). Species habitat loss due to the urbanization and agricultural expansion is expected to wipe out the resisting elements of zoonotic species and this will ultimately cause migration of these species and spreading of the disease (Wilcove 2010).

17.2.4 Monkeypox

Monkeypox disease has been observed in rain forest countries in central and West Africa and is caused by *Monkeypox* virus. In monkeys, this disease was detected in 1958 and after that sporadic cases since 1978 were observed in Nigeria. Multiple numbers of cases have been noticed from September 2017 to June 2019, and the most surprising result was that the majority of cases were in the urban settlements (Akar et al. 2020). The 2003 US monkeypox outbreak was due to the import of squirrel (*Funisciurus*) and dormice (*Graphiurus*) species and it was evident that from squirrels the virus got transferred to dogs and then to humans and that led to a massive wave of monkeypox in North America (Hutson et al. 2007). The monkeypox outbreak which took place in South Sudan in 2005, the transmission of the virus was through infected animals or humans entering Sudan from the Congo Basin and person to person transmission of virus, rather than transmission of indigenous virus from infected animals to humans (Nakazawa et al. 2013). The presence of parasites and other zoonotic species were also prevalent in the raccoon dogs in Europe and 19 of 32 helminth species found in dogs appeared to be zoonotic which were able to spread many diseases leading to human health problems at large scale (Laurimaa et al. 2016).

17.2.5 Zika Fever Virus

The *Aedes* mosquitos transmit Zika virus and are tolerant to temperature between 10 and 35 °C and relative humidity of >42% (Climate Central 2020). A study by Tesla et al. (2018) concluded that the Zika virus had optimum transmission temperature of 29 °C with an average thermal transmission range of 22.7–34.7 °C. Therefore, this virus would tend to shift towards those regions of the world that favour its optimum temperature range, and the environmental conditions with temperature lesser than optimum would be no longer suitable. The impact of the Zika virus has been observed even in the offspring as the virus infection is also able to alter the normal physiology of the foetus and can cause microcephaly. The virus spread has been observed in and around Kanpur region in India which has a large number of tanneries that provide a suitable environment for the growth of mosquito species such as *Aedes aegypti* and *Aedes albopictus* (Zika virus-carrying species). It has also been predicted that by 2050 almost all Asian and African countries could face a year-round outbreak of the Zika virus due to increasing temperature (Mishra et al. 2021).

Earlier the virus was only found in tropical and subtropical regions but now it is prevalent all over the continents except Antarctica. It has been analysed that the virus spread is at its peak when extended monsoon takes place in India. Light pollution, increasing GHG emission and rising temperature are also providing suitable conditions for the virus spread. Many researchers also connected the Zika outbreak in

Brazil in 2016 with El Nino as well (Gulich 2016). An interesting fact is that the virus spread is also triggered by natural disasters such as an earthquake. In Ecuador, a drastic surge in increasing cases of Zika virus has been seen as the El Nino event took place just after an earthquake event in 2016 that created favourable climatic conditions for the virus to spread (Sorensen et al. 2017).

17.2.6 Hantavirus

The human threat to biodiversity is in the form of globalization, species loss, climate change which lead to increased tsunamis, hurricanes, increased erosion and precipitation in some places and drought in others. These events lead to ecosystem destruction and might be the reason for the spreading of the pathogens such as Hantavirus, g RNA viruses and other vector-borne diseases and also these events lead to the habitat and vertebrate species loss and extinction (Aguirrea and Taborb 2008; Aguirre 2017; Powers and Jetz 2019).

It is also very important to explore a relationship between wildlife host diversity and the risk of spread of zoonotic diseases for a sustainable urban development (Fig. 17.2). As land use and land cover change, wildlife host diversity changes, resulting in more risk of spread of zoonotic diseases. A special emphasis on de-urbanization process is also needed as it has impacted the disease spread and also contributed significantly to human health (Eskew and Olival 2018).

17.2.7 Lassa Fever Virus

Forest destruction has been seen in many regions of the world due to deforestation, overgrazing, agricultural activities, urbanization and many such anthropogenic activities that lead to the destruction of the natural habitat of many organisms and one such organism is the rodent having Lassa fever virus which has a great historical outbreak record of 69.8% (Patz et al., 2004; Mackenstedt et al. 2015; Redding et al. 2016). Due to increase in the population of rodents, they are moving towards urban areas in search of food and shelter, thus they spread the virus and cause health hazards to the human population (Adetola and Adebisi 2019). The natural habitat of zoonotic species has been disturbed by various types of human interventions, such as through deforestation, destruction of ecosystem, etc. that has ultimately led to the increased probability of zoonotic diseases among humans (Fig. 17.3). Wild animals such as raccoons and foxes are more attracted to the urban or suburban regions in search of food and work and serve as a mediator for the spread of many pathogens in the urban population (Mackenstedt et al. 2015).

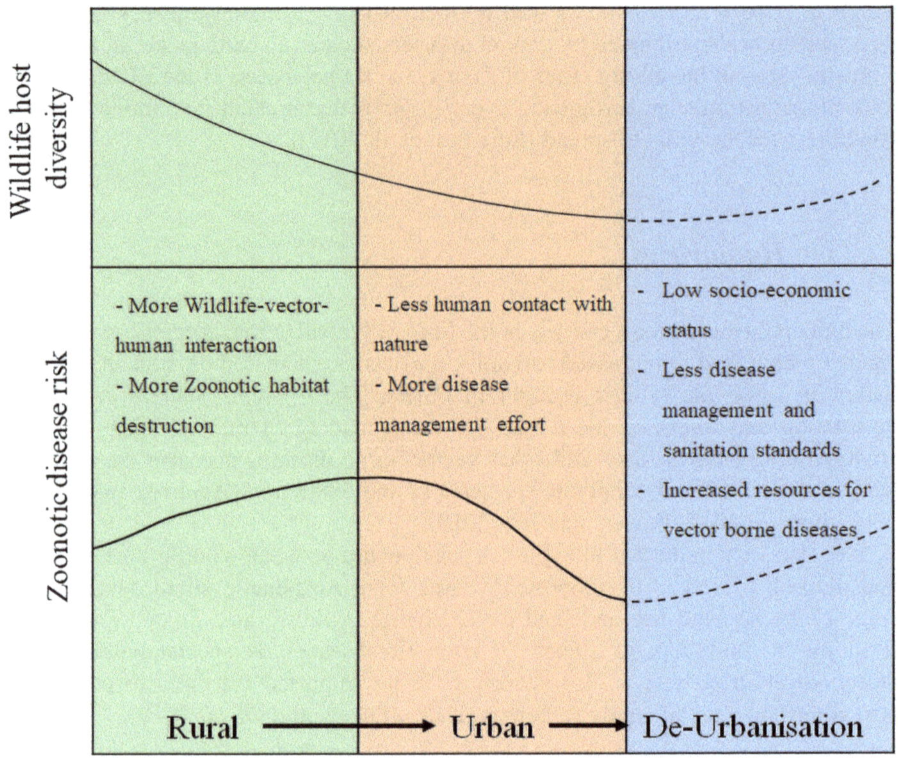

Fig. 17.2 Hypothesized wildlife host diversity and zoonotic disease risk under a consequent land-use change

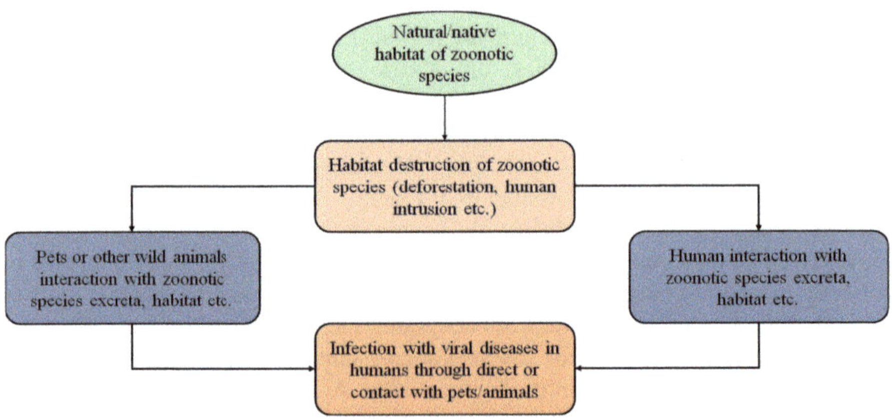

Fig. 17.3 Habitat destruction and spread of zoonosis due to anthropogenic activities

17.2.8 Covid-SARS

During the years 2020–2022, the world has witnessed several waves of the deadly coronavirus disease 2019 (COVID-19) caused due to severe acute respiratory syndrome related coronavirus (SARS-CoV-2). This pandemic has claimed millions of lives across the world. This was found to spread through Rhinolophidae Asian horseshoe bats which have high density and number in China. Though the mode of transmission of the virus is not yet so clear, based on high-resolution data it has been observed that increase in livestock concentration, forest fragmentation, increase in population, encroachment of wildlife area by human population, as well as more interaction of human population with wildlife are the reasons for the outbreaks of this viral disease (Rulli et al. 2021). Corona disease follows a linear relationship with population density and spreads more in urban fragments than in rural areas and with more wildlife–human interactions the flow of pathogen to human increases (Abdullahi et al. 2020; Chauhan 2020; Bhat et al. 2021).

It has been observed that the unsustainable environment favours a healthy microclimate for the growth of COVID-19. The climatic variables such as temperature, rainfall, pressure and wind patterns determine the local environment, and it has been observed that COVID patients show a greater recovery with better air quality or open space (Fisher et al. 2009; Cheval et al. 2020; Espejo et al. 2020; Fezi 2021; von Seidlein et al. 2021). The infection rate of COVID can be reduced if the indoor relative humidity is maintained at $>40\%$ but high density of buildings with high relative humidity could accelerate the disease spread (Noti et al. 2013; Yu et al. 2021).

17.2.9 Malaria

Malaria is a disease prevalent in tropical climates, which spreads due to the bite of mosquito vectors *Aedes aegypti* and *Aedes albopictus* which serve as malarial vectors. Due to variation in temperature, the intensity of infection varies. Transmission of the virus varies between 21.3–34.0 °C and 19.9–29.4 °C for *Ae. Aegypti* and *Ae. Albopictus,* respectively, in Europe and high-elevated tropical and subtropical regions based on general circulation models (GCMs) (Ryan et al. 2019). Malarial disease-carrying vectors *Anopheles marajoara* and *Anopheles darlingi* seem to have a strong relationship with several anthropogenic activities such as land-use change invasion of primary breeding sites, agricultural intensification, human interventions and climate change. Forest habitat destruction due to burning and deforestation and increased agricultural activities lead to the creation of marsh and sunlit pools which cause increased population of *An. marajoara* and *An. darlingi* vector and create increasing threats to human health. Such activities have resulted in increased incidence of malarial disease infection in the regions of northeastern Amazonia (Conn et al. 2002).

17.2.10 Helminthiasis

Helminthiasis is considered as less harmful to human health compared to other infectious diseases, as it has a low level of severity in young and healthy populations. But it can cause significant health impact in school-aged children and thereby may lead to cognitive, intellectual development and physical disorders (Chaisiri et al. 2015). Mean helminth species richness and abundance are found to be significantly higher in crop fragments, nematodes and cestode which are prevalent in anthropogenic linked landscapes (Froeschke and Matthee 2014). Helminth infection prevalence in Bangkok is recorded as 74.6%, the reason behind this might be a significant positive correlation between host boy mass and helminth species richness and also the abundance of the children in public parks where they get enough exposure to get the disease spread in more and more people (Paladsing et al. 2020). With deforestation, habitat fragmentation is also occurring and as result, it affects species occupancy, diversity and co-occurrence between the rodent and helminth (Schleuning et al. 2011). Rodent *Necromys lasiurus* is a helminth species from the mammalian fauna primarily found in urban fragments but due to several environmental disturbances, this species also spreads to rural areas where it can be a possible carrier of many diseases as high plantation or homogenous vegetation conditions are more favouring the parasite exchange among hosts and also provide a subsequent food supply for the rodents (Gheler-Costa et al. 2012; Costa et al. 2019). It has also been observed that helminth communities show a greater species richness in the fragmented or agricultural areas (Gillespie and Chapman 2008; Chaisiri et al. 2010).

17.2.11 Leptospirosis

Leptospirosis is a bacterial zoonotic disease that spreads due to *Leptospira* that emerges in tropical regions (Dubey et al. 2021). Flooding, extreme weather conditions, humid environments, infected urine, and climate change influence the incidence of the disease (White and Razgour 2020). Leptospirosis is one of the most widespread and severe diseases that spreads through zoonotic species with approx. 60,000 deaths and over 10 lakh cases reported annually (Costa et al. 2015). It has been observed that *Leptospira* species that cause spread of leptospirosis are commonly found at the sites where high anthropogenic activities take place, particularly in residential areas. *L. interrogans* and *L. borgpetersenii* species are found in the area with low forest cover and without natural water bodies, respectively, which indicates how these species are associated with anthropogenic activities particularly with urbanization (Blasdell et al. 2019). The incidence of this disease has also been related to the drought-driven severe 1997–1998 El Nino Southern Oscillation (ENSO) event in Colombia as the cases increased and decreased by 25% and 17% during La Nina and El Nino, respectively (Arias-Monsalve and Builes-Jaramillo 2019). Brazil has experienced the occurrence of majority of disease cases during the rainy season and the rain has influenced the cases in all regions except south Brazil from 2007 to 2019 (Costa et al. 2021). In the Philippines, the maximum cases were recorded during May, when there was a peak in temperature (Sumi et al. 2017). After a major flooding event in Malaysia, cases were noticed more in clustered patterns in the post-flood period than pre-flood with a mean distance of 1139 and 1666 m, respectively (Radi et al. 2018).

17.2.12 Yellow Fever Disease

Yellow fever disease is caused by arbovirus infection. This virus is transmitted to the humans with the bite of infected *Aedes* and *Haemagogus* mosquitoes. It has been shown that the *Aedes aegypti* originated in Africa and got migrated to the rest of the world through trade and movement of humans and got adopted to new environmental conditions and impacted multiple evolution processes (Brown et al. 2014). Before 350 years, *Ae. Aegypti* was the only species found in Africa, however now 19 loci of this species have been found in the world as it has gone under multiple evolution processes (Tabachnick and Powell 1979). Hermes element from the housefly is also equally responsible in the transmission of yellow fever virus (YFV) which causes a serious threat to the human population at a large scale (Jasinskiene et al. 1998). Mosquito species are greatly impacted by seasonal variations, and it has been observed that maximum species richness of the mosquitos were found in the peri-urban regions during the dry season in Cameroon (Mayi et al. 2020). Amazonia forest especially the Brazil region possesses great importance in the regulation of global climate change but due to deforestation, mining and extreme weather conditions, the forest ecosystem has been fragmented and therefore the chances of vector-borne diseases through cats and dogs in urban regions have increased (Ellwanger and Chies 2019). As a result, yellow fever disease, which has been basically forest associated, has been noticed nowadays in various urban areas (Ellwanger et al., 2020). Analytical data reveal that the death due to YFV in Africa may increase by more than 95% by the year 2070 (Gaythorpe et al. 2020).

17.2.13 Other Diseases

Birds are also major carriers of several pathogens and parasites in urban areas and it has been observed that in more urbanized regions the infection rate is more intense (Delgado-V and French, 2012). Many pathogens like arboviruses get spread to human beings through the animals such as birds, reptiles, rodents, etc. which are being imported or traded for pet purposes. Exposure of humans and horses to different infections more specially by arboviruses, Usutu virus (USUV), Rift Valley fever virus (RVFV) is rising and the diseases like Western equine encephalomyelitis, Venezuelan equine encephalitis, etc. are becoming more prevalent (Pfeffer and

Dobler 2010; Durand et al. 2013; Wu et al. 2017). *Leishmania infantum* is associated with an endemic disease in the Mediterranean basin, Spain which has an acute impact on human and animal health. Wild American mink is an invasive species in Spain that has been found to be an active carrier of the *L. infantum*. The life cycle of the species is around 10–12 years and with its consistent contact with the sand-fly population could be the medium of the transfer of the infection to the human settlements, as sand-fly has a transient relationship with humans and the wild American mink (Azami-Conesa et al. 2021). The migration of the wild American mink could be due to habitat destruction due to agricultural or urbanization activities, climate change or better food opportunities which result in the spread of this infectious disease.

A hookworm species *Ancylostoma ceylanicum* is dominant species in cats and dogs and causes severe diarrhoea for a long period. It has been observed that the worm gets transferred to humans through dog's excreta during gardening and creates a threat to the human health (Heo et al. 2021). *Bartonella* bacteria is also one such example which spreads through rats of the genus *Rattus* as a consequence of habitat destruction and migration. The spatial distribution of the species shows how it has spread in the urban cities of major countries around the globe such as Taiwan, Thailand, Turkey, Kenya, Madagascar, Spain, France, United States, Brazil, Australia, Singapore, China, Japan, etc. (Kosoy and Bai 2019).

Foxes in the urban areas of Estonia show a different behaviour as they mostly feed on domestic animals, cats and poultry during night-time in the patches of urban settlements. The foxes serve as carrier of the sarcoptic mange and a tapeworm *Echinococcus multilocularis* which is life-threatening to dogs. Therefore, it could also cause serious infections to humans due to dense urbanization taking place (Plumer et al. 2014).

Tomato yellow leaf curl virus (TYLCV) is a disease having *Solanum lycopersicum* as a host plant and the virus alters its growth properties and habitat suitability under different climatic conditions. Using the habitat suitability model, it has been observed that in the scenario of climate change the large regions of tomato cultivation will become prone to TYLCV infection leading to drastic decline in tomato yield by the years 2050–2070 (Ramos et al. 2019).

17.3 Methodology Used for the Impact Study

Several methodologies have been used by various researchers to study the possible impact of urbanization and climate change on zoonosis and zoonotic species. These models and methods used as well as types of measurements involved have been summarized in Table 17.1.

Table 17.1 Methods used for the assessment of urbanization and climate change impact on zoonotic species

S. no.	Methods	Type of measurements	Sources
1	Generalized additive model (GAM) and geographical information system (GIS)	Climatic and socio-economic conditions	Bouzid et al. (2014)
2	Normalized difference vegetation index (NDVI)	Vegetation, green cover, photosynthetic activity	Brownstein et al. (2002)
3	Maximum entropy modeling (MaxEnt)	Climate change	Ramos et al. (2019) and Charrahy et al. (2021)
4	Polymerase chain reaction (PCR)	DNA primers to detect incidental hosts of the disease	Azami-Conesa et al. (2021)

17.4 Conclusion

Climatic change and global warming influence the severity of zoonotic diseases in urban settlements by causing alterations in the dynamics of hosts and pathogens. Extremes of climatic conditions cause more incidence of many zoonotic diseases in urban areas due to more favourable environments for growth of pathogens and their vector hosts. The prevalence of pathogens and other zoonotic species varies depending on the habitat structure and the environment of urban settlements. Knowing the distribution pattern of pathogens in urban areas would help in urban planning to avoid disease spread as well as to design a comprehensive strategy to maintain a healthy and sustainable urban environment (Rothenburger et al. 2017). In many cases, urbanization has caused loss of biodiversity and reduction in the population of carrier hosts, leading to a direct contact of pathogens with humans and in turn more incidence of zoonotic diseases. The urban population will be around 6.3 billion by 2050. Therefore, a well thought proper town planning would be essential to reduce or prevent the spread of zoonoses (Alirol et al. 2011). The management of many zoonotic diseases like leptospirosis, *Toxoplasma gondii* infection, angiostrongyliasis, and toxocariasis in humans has been achieved by introducing mammal species such as rodents, cats and dogs for the ecological restoration (de Wit et al. 2017).

The socio-economic imbalance in the society is also responsible for zoonosis as it has been observed that respiratory and gastrointestinal infection is lower in the economically sustained regions (Li et al. 2010). In the present days using various mathematical modelling methods, it is possible to predict potential pandemics which can help in reducing spread of the disease (Morse et al. 2012). It has been estimated that 36.6% of Emerging Infectious Zoonotic Diseases (EIZD) are associated with the sources of our food which we eat every day (Otte and Pica-Ciamarra 2021). Urban education, medical and health care, better public services, public health entities and employment could help in the reduction of infectious diseases (López et al. 2019; Yu et al. 2021). The best way to combat leptospirosis is to control the infection in animals. *Leptospira interrogans* serovar Hardjo (L. Hardjo) generally spreads through dairy cows, and these cows directly come into contact with human beings leading to the spread of the disease and causing threat to human health (Rajala et al. 2017).

An assessment of impact of natural and anthropogenic activities on the spread of zoonotic diseases will help us in formulating strategies for proper urban planning so that the spread of diseases is prevented or minimized. During the process of urbanization, it is essential to consider the natural habitats of the zoonotic species, so that the natural habitat and biodiversity are not disturbed otherwise unplanned urbanization will surprisingly invite many zoonotic diseases, as the whole world has recently witnessed alarming and deadly effects of COVID-19 pandemic.

References

Abdullahi H, Sinniah GK, Siong HC (2020) Urban growth air pollution, CO, NO_2 and SO_2 emission and COVID-19 in Kano Metropolis Nigeria. Science Press, Riga

Abushahba MF, Ahmed SO, Ibrahim AA, Mosa HA (2018) Prevalence of zoonotic species of Campylobacter in broiler chicken and humans in Assiut governorate, Egypt. Approaches Poult Dairy Vet Sci 3(4):260–268

Adetola OO, Adebisi MA (2019) *Impacts of deforestation on the spread of Mastomys natalensis* in Nigeria. World Scientific News 130:286–296

Aguirre AA (2017) Changing patterns of emerging zoonotic diseases in wildlife, domestic animals, and humans linked to biodiversity loss and globalization. ILAR J 58(3):315–318

Aguirrea AA, Taborb GM (2008) Global factors driving emerging infectious diseases. Virus 1:7

Ahmed T, Asghar MW, Mushtaq MH (2019) A short report on epidemiological investigation of dog bite cases in association with temperature rise as a part of climate change. Advancements in Life Sciences 6(3):106–109

Akar S, Adesola YO, Burga J, Oluwafemi B, Akinrogbe J, Ihekweazu C (2020) Descriptive epidemiology of monkeypox in Nigeria, September 2017–June 2019. Int J Infect Dis 101: 219–220

Albery GF, Becker DJ (2021) Fast-lived hosts and zoonotic risk. Trends Parasitol 37(2):117–129

Albery, G. F., Carlson, C. J., Cohen, L. E., Eskew, E. A., Gibb, R., Ryan, S. J., . & Becker, D. J. (2021). Urban-adapted mammal species have more known pathogens. Nat Ecol Evol 6(6):794-801

Alirol E, Getaz L, Stoll B, Chappuis F, Loutan L (2011) Urbanisation and infectious diseases in a globalised world. Lancet Infect Dis 11(2):131–141

Altizer S, Bartel R, Han BA (2011) Animal migration and infectious disease risk. Science 331(6015):296–302

Ancillotto L, Allegrini C, Serangeli MT, Jones G, Russo D (2015) Sociality across species: spatial proximity of newborn bats promotes heterospecific social bonding. Behav Ecol 26(1):293–299

Ancillotto L, Santini L, Ranc N, Maiorano L, Russo D (2016) Extraordinary range expansion in a common bat: the potential roles of climate change and urbanisation. Sci Nat 103(3–4):15

Arias-Monsalve C, Builes-Jaramillo A (2019) Impact of El Niño-southern oscillation on human leptospirosis in Colombia at different spatial scales. J Infect Dev Ctries 13(12):1108–1116

Asad H, Carpenter DO (2018) Effects of climate change on the spread of zika virus: a public health threat. Rev Environ Health 33(1):31–42

Azami-Conesa I, Sansano-Maestre J, Martínez-Díaz RA, Gómez-Muñoz MT (2021) Invasive species as hosts of zoonotic infections: the case of American mink (Neovison vison) and *Leishmania infantum*. Microorganisms 9(7):1531

Bhat SA, Bashir O, Bilal M, Ishaq A, Dar MUD, Kumar R et al (2021) Impact of COVID-related lockdowns on environmental and climate change scenarios. Environ Res 195:110839

Blasdell KR, Morand S, Perera D, Firth C (2019) Association of rodent-borne Leptospira spp. with urban environments in Malaysian Borneo. PLoS Negl Trop Dis 13(2):e0007141

Bordes F, Morand S, Pilosof S, Claude J, Krasnov BR, Cosson JF et al (2015) Habitat fragmentation alters the properties of a host–parasite network: rodents and their helminths in South-East Asia. J Anim Ecol 84(5):1253–1263

Bouzid M, Colón-González FJ, Lung T, Lake IR, Hunter PR (2014) Climate change and the emergence of vector-borne diseases in Europe: case study of dengue fever. BMC Public Health 14(1):1–12

Brooks DR, Hoberg EP (2006) Systematics and emerging infectious diseases: from management to solution. J Parasitol 92(2):426–429

Brown JE, Evans BR, Zheng W, Obas V, Barrera-Martinez L, Egizi A et al (2014) Human impacts have shaped historical and recent evolution in Aedes aegypti, the dengue and yellow fever mosquito. Evolution 68(2):514–525

Brownstein JS, Rosen H, Purdy D, Miller JR, Merlino M, Mostashari F, Fish D (2002) Spatial analysis of West Nile virus: rapid risk assessment of an introduced vector-borne zoonosis. Vector Borne Zoonot Dis 2(3):157–164

Cable J, Barber I, Boag B, Ellison AR, Morgan ER, Murray K et al (2017) Global change, parasite transmission and disease control: lessons from ecology. Philos Trans R Soc B Biol Sci 372(1719):20160088

Chaisiri K, Chaeychomsri W, Siruntawineti J, Bordes F, Herbreteau V, Morand S (2010) Human-dominated habitats and helminth parasitism in southeast Asian murids. Parasitol Res 107(4): 931–937

Chaisiri K, Siribat P, Ribas A, Morand S (2015) Potentially zoonotic helminthiases of murid rodents from the Indo-Chinese peninsula: impact of habitat and the risk of human infection. Vector Borne Zoonot Dis 15(1):73–85

Charrahy Z, Yaghoobi-Ershadi MR, Shirzadi MR, Akhavan AA, Rassi Y, Hosseini SZ (2021) Climate change and its effect on the vulnerability to zoonotic cutaneous leishmaniasis in Iran. Transbound Emerg Dis 69(3):1506–1520

Chauhan S (2020) Comprehensive review of coronavirus disease 2019 (COVID-19). Biomed J 43(4):334–340

Cheval S, Mihai Adamescu C, Georgiadis T, Herrnegger M, Piticar A, Legates DR (2020) Observed and potential impacts of the COVID-19 pandemic on the environment. Int J Environ Res Public Health 17(11):4140

Childs JE, Richt JA, Mackenzie JS (2007) Introduction: conceptualizing and partitioning the emergence process of zoonotic viruses from wildlife to humans. In: Wildlife and emerging zoonotic diseases: the biology, circumstances and consequences of cross-species transmission. Springer, New York, pp 1–31

Chotelersak K, Apiwathnasorn C, Sungvornyothin S, Panasoponkul C, Samung Y, Ruangsittichai J (2015) Correlation of host specificity, environmental factors and oriental rat flea abundance. Southeast Asian J Trop Med Public Health 46(2):198

Climate Central (2020) This news bites: more mosquito days. https://medialibrary.climatecentral.org/resources/more-mosquito-days

Conn JE, Wilkerson RC, Nazare M, Segura O, De Souza RT, Schlichting CD, Povoa MM (2002) Emergence of a new neotropical malaria vector facilitated by human migration and changes in land use. University of Vermont, Burlington

Connolly C, Keil R, Ali SH (2021) Extended urbanisation and the spatialities of infectious disease: demographic change, infrastructure and governance. Urban Stud 58(2):245–263

Costa F, Hagan JE, Calcagno J, Kane M, Torgerson P, Martinez-Silveira MS, Ko AI (2015) Global morbidity and mortality of leptospirosis: a systematic review. PLoS Negl Trop Dis 9(9): e0003898

Costa NA, dos Santos Cardoso T, da Costa Neto SF, Maldonado A, Gentile R (2019) Metacommunity structure of helminths of *Necromys lasiurus* (Rodentia: Sigmodontinae) in different land use areas in the Brazilian Cerrado. J Parasitol 105(2):271–282

Costa ACTRB, Pereira CR, Sáfadi T, Heinemann MB, Dorneles EMS (2021) Climate influence the human leptospirosis cases in Brazil, 2007–2019: a time series analysis. Trans R Soc Trop Med Hyg 116(2):124–132

de Wit LA, Croll DA, Tershy B, Newton KM, Spatz DR, Holmes ND, Kilpatrick AM (2017) Estimating burdens of neglected tropical zoonotic diseases on islands with introduced mammals. Am J Trop Med Hyg 96(3):749

Delgado-V CA, French K (2012) Parasite–bird interactions in urban areas: current evidence and emerging questions. Landsc Urban Plan 105(1–2):5–14

Dubey S, Singh R, Gupta B, Patel R, Soni D, Dhakad BMS et al (2021) Leptospira: an emerging zoonotic pathogen of climate change, global warming and unplanned urbanization: a review. Architecture 10(17):54

Durand B, Lecollinet S, Beck C, Martínez-López B, Balenghien T, Chevalier V (2013) Identification of hotspots in the European Union for the introduction of four zoonotic arboviroses by live animal trade. PLoS One 8(7):e70000

Ellwanger JH, Chies JAB (2019) The triad "dogs, conservation and zoonotic diseases"—an old and still neglected problem in Brazil. Perspect Ecol Conserv 17(3):157–161

Ellwanger JH, Kulmann-Leal B, Kaminski VL, Valverde-Villegas JA, Veiga ABG, Spilki FR et al (2020) Beyond diversity loss and climate change: impacts of Amazon deforestation on infectious diseases and public health. An Acad Bras Cienc 92:e20191375

Eskew EA, Olival KJ (2018) De-urbanization and zoonotic disease risk. Ecohealth 15(4):707–712

Espejo W, Celis JE, Chiang G, Bahamonde P (2020) Environment and COVID-19: pollutants, impacts, dissemination, management and recommendations for facing future epidemic threats. Sci Total Environ 747:141314

Estrada-Peña A, Ostfeld RS, Peterson AT, Poulin R, de la Fuente J (2014) Effects of environmental change on zoonotic disease risk: an ecological primer. Trends Parasitol 30(4):205–214

Fezi BA (2021) The role of architecture and urbanism in preventing pandemics. https://www.intechopen.com/chapters/76976

Fisher B, Turner RK, Morling P (2009) Defining and classifying ecosystem services for decision making. Ecol Econ 68(3):643–653

Froeschke G, Matthee S (2014) Landscape characteristics influence helminth infestations in a peri-domestic rodent-implications for possible zoonotic disease. Parasit Vectors 7(1):1–13

Gaythorpe KA, Hamlet A, Cibrelus L, Garske T, Ferguson NM (2020) The effect of climate change on yellow fever disease burden in Africa. elife 9:e55619

Gheler-Costa C, Vettorazzi CA, Pardini R, Verdade LM (2012) The distribution and abundance of small mammals in agroecosystems of southeastern Brazil. Mammalia 76(2):185–191

Gillespie TR, Chapman CA (2008) Forest fragmentation, the decline of an endangered primate, and changes in host–parasite interactions relative to an unfragmented forest. Am J Primatol 70(3): 222–230

Glass GE, Yates TL, Fine JB, Shields TM, Kendall JB, Hope AG et al (2002) Satellite imagery characterizes local animal reservoir populations of Sin Nombre virus in the southwestern United States. Proc Natl Acad Sci 99(26):16817–16822

Gottdenker NL, Streicker DG, Faust CL, Carroll CR (2014) Anthropogenic land use change and infectious diseases: a review of the evidence. EcoHealth 11(4):619–632

Gulich GA (2016) Epidemiology, driving factors, transmission and control options of Zika virus: a review. J Infect Dis Ther 04:1000278

Hafez HM, Hauck R (2015) Zoonoses with public health relevance in poultry. In: Zoonoses-infections affecting humans and animals. Springer, Dordrecht, pp 103–123

Hayes MA, Piaggio AJ (2018) Assessing the potential impacts of a changing climate on the distribution of a rabies virus vector. PLoS One 13(2):e0192887

Heo CC, Rafiz AR, Ngui R (2021) A case of zoonotic *Ancylostoma ceylanicum* infection in a suburban area of Selangor, Malaysia. Acta Parasitol 67:564–568

Hillman AE, Lymbery AJ, Elliot AD, Thompson RA (2017) Urban environments alter parasite fauna, weight and reproductive activity in the quenda (*Isoodon obesulus*). Sci Total Environ 607:1466–1478

Hutson CL, Lee KN, Abel J, Carroll DS, Montgomery JM, Olson VA et al (2007) Monkeypox zoonotic associations: insights from laboratory evaluation of animals associated with the multi-state US outbreak. Am J Trop Med Hyg 76(4):757–768

Jasinskiene N, Coates CJ, Benedict MQ, Cornel AJ, Rafferty CS, James AA, Collins FH (1998) Stable transformation of the yellow fever mosquito, *Aedes aegypti*, with the Hermes element from the housefly. Proc Natl Acad Sci 95(7):3743–3747

Kelly-Hope LA, Purdie DM, Kay BH (2004) El Nino southern oscillation and Ross River virus outbreaks in Australia. Vector Borne Zoonot Dis 4(3):210–213

Kilpatrick AM, Randolph SE (2012) Drivers, dynamics, and control of emerging vector-borne zoonotic diseases. Lancet 380(9857):1946–1955

Kim BI, Blanton JD, Gilbert A, Castrodale L, Hueffer K, Slate D, Rupprecht CE (2014) A conceptual model for the impact of climate change on fox rabies in Alaska, 1980–2010. Zoonoses Public Health 61(1):72–80

Kosoy M, Bai Y (2019) Bartonella bacteria in urban rats: a movement from the jungles of Southeast Asia to metropoles around the globe. Front Ecol Evol 7:88

LaDeau SL, Allan BF, Leisnham PT, Levy MZ (2015) The ecological foundations of transmission potential and vector-borne disease in urban landscapes. Funct Ecol 29(7):889–901

Lambin EF, Tran A, Vanwambeke SO, Linard C, Soti V (2010) Pathogenic landscapes: interactions between land, people, disease vectors, and their animal hosts. Int J Health Geogr 9(1):1–13

Lau CL, Smythe LD, Craig SB, Weinstein P (2010) Climate change, flooding, urbanisation and leptospirosis: fuelling the fire? Trans R Soc Trop Med Hyg 104(10):631–638

Laurimaa L, Süld K, Davison J, Moks E, Valdmann H, Saarma U (2016) Alien species and their zoonotic parasites in native and introduced ranges: the raccoon dog example. Vet Parasitol 219: 24–33

Li X, Gao L, Dai L, Zhang G, Zhuang X, Wang W, Zhao Q (2010) Understanding the relationship among urbanisation, climate change and human health: a case study in Xiamen. Int J Sustain Dev World Ecol 17(4):304–310

Lindsay MD, Broom AK, Wright AT, Johansen CA, Mackenzie JS (1993) Ross River virus isolations from mosquitoes in arid regions of Western Australia: implication of vertical transmission as a means of persistence of the virus. Am J Trop Med Hyg 49(6):686–696

Linthicum KJ, Anyamba A, Tucker CJ, Kelley PW, Myers MF, Peters CJ (1999) Climate and satellite indicators to forecast Rift Valley fever epidemics in Kenya. Science 285(5426): 397–400

López MS, Müller GV, Lovino MA, Gómez AA, Sione WF, Pomares LA (2019) Spatio-temporal analysis of leptospirosis incidence and its relationship with hydroclimatic indicators in northeastern Argentina. Sci Total Environ 694:133651

Lowe R, Lee SA, O'Reilly KM, Brady OJ, Bastos L, Carrasco-Escobar G, Gasparrini A (2021) Combined effects of hydrometeorological hazards and urbanisation on dengue risk in Brazil: a spatiotemporal modelling study. Lancet Planet Health 5(4):e209–e219

Maaz D, Krücken J, Blümke J, Richter D, McKay-Demeler J, Matuschka FR, von Samson-Himmelstjerna G (2018) Factors associated with diversity, quantity and zoonotic potential of ectoparasites on urban mice and voles. PLoS One 13(6):e0199385

Mackenstedt U, Jenkins D, Romig T (2015) The role of wildlife in the transmission of parasitic zoonoses in peri-urban and urban areas. Int J Parasitol Parasites Wildl 4(1):71–79

Majewska AA, Satterfield DA, Harrison RB, Altizer S, Hepinstall-Cymerman J (2019) Urbanization predicts infection risk by a protozoan parasite in non-migratory populations of monarch butterflies from the southern coastal US and Hawaii. Landsc Ecol 34(3):649–661

Mayi MPA, Bamou R, Djiappi-Tchamen B, Fontaine A, Jeffries CL, Walker T, Tchuinkam T (2020) Habitat and seasonality affect mosquito community composition in the west region of Cameroon. Insects 11(5):312

Mishra V, Varshney V, Singh N (2021) Aedes the menace. Cover story, vector borne disease. Down to earth, Net result. pp 28–38

Moreno E (2014) Retrospective and prospective perspectives on zoonotic brucellosis. Front Microbiol 5:213

Morse SS, Mazet JA, Woolhouse M, Parrish CR, Carroll D, Karesh WB, Daszak P (2012) Prediction and prevention of the next pandemic zoonosis. Lancet 380(9857):1956–1965

Mott KE, Desjeux P, Moncayo A, Ranque P, De Raadt P (1990) Parasitic diseases and urban development. Bull World Health Organ 68(6):691

Nakazawa Y, Emerson GL, Carroll DS, Zhao H, Li Y, Reynolds MG et al (2013) Phylogenetic and ecologic perspectives of a monkeypox outbreak, southern Sudan, 2005. Emerg Infect Dis 19(2): 237

Noti JD, Blachere FM, McMillen CM, Lindsley WG, Kashon ML, Slaughter DR, Beezhold DH (2013) High humidity leads to loss of infectious influenza virus from simulated coughs. PLoS One 8(2):e57485

Nunes H, Rocha FL, Cordeiro-Estrela P (2017) Bats in urban areas of Brazil: roosts, food resources and parasites in disturbed environments. Urban Ecosyst 20(4):953–969

Ortiz DI, Piche-Ovares M, Romero-Vega LM, Wagman J, Troyo A (2022) The impact of deforestation, urbanization, and changing land use patterns on the ecology of mosquito and tick-borne diseases in Central America. Insects 13(1):20

Ostfeld RS (2009) Biodiversity loss and the rise of zoonotic pathogens. Clin Microbiol Infect 15: 40–43

Ostfeld RS, Keesing F (2000) Biodiversity series: the function of biodiversity in the ecology of vector-borne zoonotic diseases. Can J Zool 78(12):2061–2078

Otte J, Pica-Ciamarra U (2021) Emerging infectious zoonotic diseases: the neglected role of food animals. One Health 13:100323

Paladsing Y, Boonsri K, Saesim W, Changsap B, Thaenkham U, Kosoltanapiwat N et al (2020) Helminth fauna of small mammals from public parks and urban areas in Bangkok metropolitan with emphasis on community ecology of infection in synanthropic rodents. Parasitol Res 119(11):3675–3690

Patz JA, Daszak P, Tabor GM, Aguirre AA, Pearl M, Epstein J, Working Group on Land Use Change Disease Emergence (2004) Unhealthy landscapes: policy recommendations on land use change and infectious disease emergence. Environ Health Perspect 112(10):1092–1098

Paz S (2015) Climate change impacts on West Nile virus transmission in a global context. Philos Trans R Soc B Biol Sci 370(1665):20130561

Perec-Matysiak A, Leśniańska K, Buńkowska-Gawlik K, Merta D, Popiołek M, Hildebrand J (2021) Zoonotic genotypes of *Enterocytozoon bieneusi* in wild living invasive and native carnivores in Poland. Pathogens 10(11):1478

Pfeffer M, Dobler G (2010) Emergence of zoonotic arboviruses by animal trade and migration. Parasit Vectors 3(1):1–15

Plumer L, Davison J, Saarma U (2014) Rapid urbanization of red foxes in Estonia: distribution, behaviour, attacks on domestic animals, and health-risks related to zoonotic diseases. PLoS One 9(12):e115124

Powers RP, Jetz W (2019) Global habitat loss and extinction risk of terrestrial vertebrates under future land-use-change scenarios. Nat Clim Change 9(4):323–329

Radi MFM, Hashim JH, Jaafar MH, Hod R, Ahmad N, Nawi AM et al (2018) Leptospirosis outbreak after the 2014 major flooding event in Kelantan, Malaysia: a spatial-temporal analysis. Am J Trop Med Hyg 98(5):1281

Rajala EL, Sattorov N, Boqvist S, Magnusson U (2017) Bovine leptospirosis in urban and peri-urban dairy farming in low-income countries: a "one health" issue? Acta Vet Scand 59(1):1–4
Ramos RS, Kumar L, Shabani F, Picanço MC (2019) Risk of spread of tomato yellow leaf curl virus (TYLCV) in tomato crops under various climate change scenarios. Agric Syst 173:524–535
Redding DW, Moses LM, Cunningham AA, Wood J, Jones KE (2016) Environmental-mechanistic modelling of the impact of global change on human zoonotic disease emergence: a case study of Lassa fever. Methods Ecol Evol 7(6):646–655
Robertson LJ, Johansen ØH, Kifleyohannes T, Efunshile AM, Terefe G (2020) Cryptosporidium infections in Africa—how important is zoonotic transmission? A review of the evidence. Front Vet Sci 7:575881
Rothenburger JL, Himsworth CH, Nemeth NM, Pearl DL, Jardine CM (2017) Environmental factors and zoonotic pathogen ecology in urban exploiter species. EcoHealth 14(3):630–641
Rulli MC, D'Odorico P, Galli N, Hayman DT (2021) Land-use change and the livestock revolution increase the risk of zoonotic coronavirus transmission from rhinolophid bats. Nature Food 2(6): 409–416
Ryan SJ, Carlson CJ, Mordecai EA, Johnson LR (2019) Global expansion and redistribution of Aedes-borne virus transmission risk with climate change. PLoS Negl Trop Dis 13(3):e0007213
Saksena S, Fox J, Epprecht M, Tran C, Castrence M, Nong D, Wilcox B (2014) Role of urbanization, land-use diversity, and livestock intensification In zoonotic emerging infectious diseases. Academia, San Francisco
Schleuning M, Farwig N, Peters MK, Bergsdorf T, Bleher B, Brandl R, Böhning-Gaese K (2011) Forest fragmentation and selective logging have inconsistent effects on multiple animal-mediated ecosystem processes in a tropical forest. PLoS One 6(11):e27785
Shoo LP, Williams SE, HERO, J. M. (2006) Detecting climate change induced range shifts: where and how should we be looking? Austral Ecol 31(1):22–29
Simons RR, Gale P, Horigan V, Snary EL, Breed AC (2014) Potential for introduction of bat-borne zoonotic viruses into the EU: a review. Viruses 6(5):2084–2121
Slingenbergh J, Gilbert M, Balogh KD, Wint W (2004) Ecological sources of zoonotic diseases. Revue Sci Tech 23(2):467–484
Sorensen CJ, Borbor-Cordova MJ, Calvello-Hynes E, Diaz A, Lemery J, Stewart-Ibarra AM (2017) Climate variability, vulnerability, and natural disasters: a case study of Zika virus in Manabi, Ecuador following the 2016 earthquake. GeoHealth 1(8):298–304
Sumi A, Telan EFO, Chagan-Yasutan H, Piolo MB, Hattori T, Kobayashi N (2017) Effect of temperature, relative humidity and rainfall on dengue fever and leptospirosis infections in Manila, the Philippines. Epidemiol Infect 145(1):78–86
Tabachnick WJ, Powell JR (1979) A world-wide survey of genetic variation in the yellow fever mosquito, *Aedes aegypti*. Genet Res 34(3):215–229
Taylor LH, Latham SM, Woolhouse ME (2001) Risk factors for human disease emergence. Philos Trans R Soc Lond B Biol Sci 356(1411):983–989
Tesla B, Demakovsky LR, Mordecai EA, Ryan SJ, Bonds MH, Ngonghala CN, Murdock CC (2018) Temperature drives Zika virus transmission: evidence from empirical and mathematical models. Proc R Soc B Biol Sci 285:0795
Thompson RA (2013) Parasite zoonoses and wildlife: one health, spillover and human activity. Int J Parasitol 43(12–13):1079–1088
Tunali M, Radin AA, Başıbüyük S, Musah A, Borges IVG, Yenigun O, Campos LC (2021) A review exploring the overarching burden of Zika virus with emphasis on epidemiological case studies from Brazil. Environ Sci Pollut Res 28(40):55952–55966
Ullmann LS, Langoni H (2011) Interactions between environment, wild animals and human leptospirosis. J Venom Anim Toxins Includ Trop Dis 17:119–129
United Nations, Department of Economic and Social Affairs, Population Division (2015) World Urbanization Prospects: The 2014 Revision (ST/ESA/SER.A/366). https://population.un.org/wup/publications/files/wup2014-report.pdf

von Seidlein L, Alabaster G, Deen J, Knudsen J (2021) Crowding has consequences: prevention and management of COVID-19 in informal urban settlements. Build Environ 188:107472

Waldman L (2015) Urbanisation, the peri-urban growth and zoonotic disease. https://opendocs.ids.ac.uk/opendocs/bitstream/handle/20.500.12413/5855/ID563%20Online.pdf?sequence=1&isAllowed=y

Wang HJ, Zhang RH, Cole J, Chavez F (1999) El Niño and the related phenomenon Southern Oscillation (ENSO): the largest signal in interannual climate variation. Proc Natl Acad Sci 96(20):11071–11072

Wegbreit JENY, Reisen WK (2000) Relationships among weather, mosquito abundance, and encephalitis virus activity in California: Kern County 1990–1998. J Am Mosq Control Assoc 16(1):22–27

White RJ, Razgour O (2020) Emerging zoonotic diseases originating in mammals: a systematic review of effects of anthropogenic land-use change. Mammal Rev 50(4):336–352

White A, Watt AD, Hails RS, Hartley SE (2000) Patterns of spread in insect-pathogen systems: the importance of pathogen dispersal. Oikos 89(1):137–145

Wilcove DS (2010) No way home: the decline of the world's great animal migrations. Island Press, Washington, DC

Wilder-Smith A, Gubler DJ, Weaver SC, Monath TP, Heymann DL, Scott TW (2017) Epidemic arboviral diseases: priorities for research and public health. Lancet Infect Dis 17(3):e101–e106

Wilke AB, Chase C, Vasquez C, Carvajal A, Medina J, Petrie WD, Beier JC (2019) Urbanization creates diverse aquatic habitats for immature mosquitoes in urban areas. Sci Rep 9(1):1–11

Wilkinson DA, Marshall JC, French NP, Hayman DT (2018) Habitat fragmentation, biodiversity loss and the risk of novel infectious disease emergence. J R Soc Interface 15(149):20180403

World Health Organization (2022). https://www.who.int/news-room/fact-sheets/detail/zoonoses

Wu T, Perrings C, Kinzig A, Collins JP, Minteer BA, Daszak P (2017) Economic growth, urbanization, globalization, and the risks of emerging infectious diseases in China: a review. Ambio 46(1):18–29

Yabsley MJ, Shock BC (2013) Natural history of zoonotic Babesia: role of wildlife reservoirs. Int J Parasitol Parasites Wildl 2:18–31

Yu D, Li X, Yu J, Shi X, Liu P, Tian P (2021) Whether urbanization has intensified the spread of infectious diseases—renewed question by the COVID-19 pandemic. Front Public Health 9: 699710

Zarza H, Martínez-Meyer E, Suzán G, Ceballos G (2017) Geographic distribution of *Desmodus rotundus* in Mexico under current and future climate change scenarios: implications for bovine paralytic rabies infection. Vet México 4(3):1–16

Chapter 18
Roles of Community Resilience and Risk Appraisal in Climate Change Adaptation

Purabi Sarkar

18.1 Introduction

There are three major factors which are responsible for climate change in the Himalayan regions: Asian monsoon, inner Asian high-pressure systems, and winter westerlies and these create changes in the sudden climate shift in the Himalayan regions.

Uttarakhand is a Himalayan state in India which is prone to lot of climate change issues since many years.

Cloudburst related flash floods in Kedarnath in the year 2013, collapse of a hanging glacier in Chamoli in the year 2021, landslides in Pithoragarh, heavy rainfall, earthquake, heavy snowfall, water scarcity and more are common affair in the state and this has created havoc and casualties for the communities, the livestock, and the livelihoods.

The region has been a travel destination for treks and Himalayan tourism but in lieu of that lot of construction and urbanisation have happened. Communities have started on homestays and responsible tourism and that is helping in also protecting the environment, flora, fauna, and natural resources.

There is a dire need to accelerate the shift to sustainable urbanisation and focus on energy sources in the region, which can improve the efficiency of currently existing systems and protect vulnerable populations.

Revisiting the problem statement and rationale in the high-altitude regions of the Himalayas each household consumes around 1500 kg of wood every year for space heating application. The population in the Hindu Kush Himalayas (HKH) was around 237 million people as of 2017 and is expected to grow to more than

---

P. Sarkar (✉)
Tirambhapur Askote Raj Ecosystem Private Limited (TARE), Pithoragarh, Uttarakhand, India
e-mail: purabi7@alumni.iimk.ac.in

B. Pathak, R. S. Dubey (eds.), *Climate Change and Urban Environment Sustainability*, Disaster Resilience and Green Growth,
https://doi.org/10.1007/978-981-19-7618-6_18

300 million by 2030. As biomass is the primary fuel used for heating, approximately 70 million tonnes of biomass is required every year leading to deforestation. Furthermore, it has been estimated that space heating of the entire Himalayan region amounts to 16 million tonnes (MT) of CO_2 emission annually as reported in June 2020 [1]. The magnitude of the problem can be seen as this is equivalent to cumulative emissions being generated from 27,200 MW capacity thermal power plants. To add to the problem, current traditional heating solutions result in increased Indoor Air Pollution (IAP) substantially impacting the health of the household members. In addition, the collection of wood at the onset of the winter season involves high drudgery and time investment which ranges from 2 to 4 h a day for 1 month. Moreover, the community is unable to use this time for livelihood generation activities leading to low income during the period.

As per a recent TERI paper on Himachal Pradesh and Uttarakhand, these two states are major contributors of CO_2 emissions in the Himalayan (38%) due to larger share of population in IHR; greater % of population at higher altitudes compared to other states (exception J&K); Uttarakhand population is more dispersed in higher altitude with higher percentage of rural population. Estimated CO_2 emissions between the two states have been showcasing the 78%contribution from the rural residential sector and only 19% from urban residential and 3% from commercial emission (Teri report).

The community plays a big role in tackling the situational challenges that have erupted due to the sudden outbursts due to climate change. Their resilience is unparalleled, and the community has been proactively trying to protect the region as much as possible, but there is more to be done so that the calamities can be handled and better policies and contingency planning need to be in place. Clean energy is the need of the situation, and also permaculture driven practices need to be ingrained in the ecology therein, awareness programs, and also involvement of participatory organisations and relevant funding to create the necessary developmental changes based on the need of the community.

It is important to work with leaders in government, the private sector, and civil society to drive further, faster action toward achievement of Sustainable Development Goal 7 in the region and many regions are going towards the Private–Public Partnership (PPP) model so that the change and timeline of implementation are much faster in the region.

18.1.1 Climate Change and the Impact

18.1.1.1 Livelihoods

Majority of the urban and rural population is part of an agriculture-based economy. Framing is practiced by communities, even forestry is an age-old practice by locals. Carpet making, yarn, tourism, carpentry, woodwork, bell craft, bamboo basketry, herbs drying, natural dyes etc. are the skillsets of the region. Communities have started diverting from traditional livelihoods due to lack of natural resources and raw materials; also due to lack of market linkage.

18.1.1.2 Farming

Baranaja: a traditional climate resilient farming practice which is still practiced in many regions of Uttarakhand by communities but is diminishing now. Baranaja is a mixed crop farming practice from Uttarakhand based on traditional wisdom. Millets and pulses are the main crops cultivated through this system. There is a need to revive such farming practices as they also help in landslide management by the cropping patterns (Baranaja: a climate resilient farming practice (downtoearth.org.in)).

Farming of hemp is much needed as it grows like wild, it can also help in addressing climate change owing to the benefits of the hemp plant and also help farmers in generating livelihoods due to the good demand of hemp in global market.

18.1.1.3 Water Scarcity in the Hindu Kush Himalayas Region Has Led to an Uncertain Future for Water

The Himalayan region has a unique topography, climate, and hydrology, and these factors are responsible for water availability in the region. The regions have lot of traditional water recharge and systems in place like springs, naulas (water storage systems), watermill, and many more. Lot of springs have dried up due to no maintenance and lot of construction work in the terrain. But the regions are now experiencing water insecurity in majority of areas due to socio-economic changes and urbanisation. Urbanisation leads to economic growth but since these regions have an unplanned urbanisation and the attention to detail on water conservation, water recharge, check dams are not being taken care of.

Scarcity of water has also led local farmers to leave farming and selling their landholdings to others and also hoteliers.

An excerpt from an article published at countercurrents.org says ‘From drying up of springs to deforestation several other factors have led to acute shortage of water in the region. A recent report published by the National Rural Drinking Water Program (NRDWP) reveals a grave situation of drinking water shortage in Uttarakhand. Out of 39,202 villages only 21,363 villages have drinking water facility. The remaining 17,839 villages face severe water shortage either due to the dried-up water sources or the failed drinking water projects. Another UNDP report also suggests that Uttarakhand is facing acute water crisis wherein discharges of 500 water supply sources including springs, streams, ponds etc. have reportedly reduced by more than 50%’.

Clean drinking water for both communities and livestock is the need of the region. Communities walk miles to get access to drinking water and they carry potable water and many villages don't even have a dedicated water source since lot of nearby water spruces have dried up.

In Uttarakhand, out of total 16,000 villages, 8800 villages have been placed as water scarce villages. The districts like Almora, Pauri, Tehri, Pithoragarh, and Chamoli are facing drinking water crisis (Joshi 2004). 72% women and 14% children have to bear the responsibility of carrying potable water (source: Water Crisis in Uttarakhand I Hindi Water Portal) (indiawaterportal.org).

18.1.1.4 Landslide

Climate change has led to landslides every year, and this is serious hazard and posing rick to lot of communities. Many villages get cut off from road access after monsoons and there has been incidences when entire villages have been flushed off in rains and landslides, there has been immense harm to humanity and to livestock and forests.

A lot of these landslides are due to extreme deforestation, urbanisation, and lack of policies to do contingency planning and also preparedness models are the need of the hour.

There is also a dire need to do plantations of oak trees, Indian butter tree (chyura in local language in Pithoragarh) and other crops and also grass varieties (Napier grass) that hold the soil intact and also aid in natural water recharge.

18.1.1.5 Clean Cooking

Women play the role of being the major workforces in Uttarakhand. Drudgery can be easily defined looking at the roles and responsibilities they play in entirety. They work from morning till evening catering to different tasks from taking care of the household chores, look after children, cooking, cleaning, washing, collect water for drinking and bathing for human beings and cattle at house, collecting fodder grass from forests and also collecting backload of firewood for cooking at home.

They are exposed to black smoke that emanates from the non-efficient mud chulhas (stoves) and they inhale lot of carbon smoke due to the same. This constant exposure to day-in and out cooking in these stoves affects their health in long run leading to vision issues and health issues for lungs. The children and the elderly people also get affected as the households have a constricted design with respect to ventilation outlet for the smoke.

Also, the women and girls walk miles carrying approximately 30 kg of firewood on their back from the forest almost every day. This data came out as a part of the survey that was done for the clean energy implementation project for Uttarakhand by TARE (Tirambhapur Askote Raj Ecosystem Private Limited) for Himalayan rocket stoves clean energy project.

TARE (Tirambhapur Askote Raj ecosystem Private limited), organisation, based at Pithoragarh, Uttarakhand implemented 160 smokeless stove cum heating equipment in the year 2021 in the region to address the issue and implement clean energy for clean cooking aspect for the households and communities. Also, these stoves are energy efficient and consume less firewood and so that also addresses the issue of drudgery on the women in the Himalayan communities.

18.1.2 Community Kitchen and Events

The communities have lot of social gatherings, and there is a concept of community cooking together many a times. There is a need for efficient cooking mechanisms so that reduction of usage of firewood can be achieved.

18.1.3 Project Implementation—Case Study

Target audience—Urban and rural communities primarily village community.

Financial model-subsidised cost under a community implementation project.

Region Specifics-Disaster prone, heavy rainfall, harsh winters, heavy snowfall, landslides, flash flood, altitude—2400–3470 m.

Villages—Darma Valley, Johar Valley, Munsiari of Pithoragarh district.

Primary occupation/Livelihood of target Community-Small landholding agriculture, labour, local small business, Tourism (Village homestay and trek)—seasonal livelihood.

Local demonstration carried out by TARE field team

Intervention impact on community after ECOMINI stove implementation as a clean energy initiative is showing positive impact on the areas:

1. Environment.
2. Health.
3. Alternate skill learning.
4. Drudgery of carrying excessive firewood by women from the forest.
5. Less consumption of fuel.
6. Since it's a heating cum cooking equipment, it saves on grid and firewood consumption in harsh winter areas.
7. Helps reduce fuel expenditure.
8. Women are less prone to wild animal attack in forest.

Implementation applicability:
Such type of innovation and implementation can be helpful in

1. Longer winter months in Darma Valley (more than 6 months).
2. Not easy access to firewood in winter months due to heavy snowfall.
3. Even monsoons get chilling in some areas and require heating in houses at times.
4. Tourism can be enhanced in extreme remote areas by using these stoves for quick cooking and for heating purposes.

Dual benefit of cooking & heating has been a big positive in Uttarakhand

18.1.4 Groundwork Methodology and Challenges

- Local Team Leader-Versatile role and responsibility
- New product implementation challenge
- Challenges in terrain
- Challenges in Covid time implementation

- Storage of inventory
- Logistics cost challenge
- Part payment handling with villagers

18.2 Conclusion

Innovative ecologies are important, and implementation of such climate change and clean energy initiatives in rural areas will be beneficial and will be helpful to mitigate the climate change. CO_2 emission reduction stoves are low smoke biomass-based stoves that run on wood, agro-waste, cow dung cakes, etc. as fuel for cooking. The design is sleek, and it uses natural air ventilation for supplying air to fuel for cooking. Innovations and technologies could be implemented through community participation and contribute towards the mitigation of the climate change. Clean energy initiatives help to reduce pollution, better health security, less utilisation of tradition sources and contribute for sustainability.

Further Reading

Baranaja: a climate resilient farming practice (downtoearth.org.in) - This explains about the traditional multicropping models that farmers used to adopt and it was beneficial for rotational harvest of different crops , also helped in mitigating climate change

ICIMOD, Annual Report (2019) Chapter 2: knowledge generation and use – stories our solutions are in nature. http://www.icimod.org/reports/ar2019/index.php?p=knowledge_generation; www.icimod.org/reports/ar2019/index.php?p=knowledge_generation

Sherpa PY (2022) The Himalayan climate change: global warming in the globe's highest region, interview, Harvard International review (harvard.edu). https://hir.harvard.edu/himalayan-climate-change/. 28 Jan 2022

Think Hazard – Pithoragarh – Landslide. https://thinkhazard.org/en/report/17932-india-uttarakhand-pithoragarh/LS

Julia, Bhatt V, Pandey P (2016) Water crisis in Uttarakhand. https://hindi.indiawaterportal.org/articles/water-crisis-uttarakhand. 19 April 2016

Dogra B (2021) Urgent need for proper priorities in regions of water-scarcity in India. https://countercurrents.org/2021/06/urgent-need-for-proper-priorities-in-regions-of-water-scarcity/. 01 June 2021